AF621253

Springer Proceedings in Mathematics & Statistics

Volume 338

Springer Proceedings in Mathematics & Statistics

This book series features volumes composed of selected contributions from workshops and conferences in all areas of current research in mathematics and statistics, including operation research and optimization. In addition to an overall evaluation of the interest, scientific quality, and timeliness of each proposal at the hands of the publisher, individual contributions are all refereed to the high quality standards of leading journals in the field. Thus, this series provides the research community with well-edited, authoritative reports on developments in the most exciting areas of mathematical and statistical research today.

More information about this series at http://www.springer.com/series/10533

Frank Nijhoff · Yang Shi · Da-jun Zhang
Editors

Asymptotic, Algebraic and Geometric Aspects of Integrable Systems

In Honor of Nalini Joshi On Her 60th Birthday, TSIMF, Sanya, China, April 9–13, 2018

Springer

Editors
Frank Nijhoff
School of Mathematics
University of Leeds
Leeds, UK

Yang Shi
College of Science and Engineering
Flinders at Tonsley
Adelaide, SA, Australia

Da-jun Zhang
Department of Mathematics
Shanghai University
Shanghai, China

ISSN 2194-1009 ISSN 2194-1017 (electronic)
Springer Proceedings in Mathematics & Statistics
ISBN 978-3-030-56999-0 ISBN 978-3-030-57000-2 (eBook)
https://doi.org/10.1007/978-3-030-57000-2

Mathematics Subject Classification: 33xx, 39xx, 14xx, 32xx, 14H70, 37K10, 34M55

© Springer Nature Switzerland AG 2020
This work is subject to copyright. All rights are reserved by the Publisher, whether the whole or part of the material is concerned, specifically the rights of translation, reprinting, reuse of illustrations, recitation, broadcasting, reproduction on microfilms or in any other physical way, and transmission or information storage and retrieval, electronic adaptation, computer software, or by similar or dissimilar methodology now known or hereafter developed.
The use of general descriptive names, registered names, trademarks, service marks, etc. in this publication does not imply, even in the absence of a specific statement, that such names are exempt from the relevant protective laws and regulations and therefore free for general use.
The publisher, the authors and the editors are safe to assume that the advice and information in this book are believed to be true and accurate at the date of publication. Neither the publisher nor the authors or the editors give a warranty, expressed or implied, with respect to the material contained herein or for any errors or omissions that may have been made. The publisher remains neutral with regard to jurisdictional claims in published maps and institutional affiliations.

This Springer imprint is published by the registered company Springer Nature Switzerland AG
The registered company address is: Gewerbestrasse 11, 6330 Cham, Switzerland

Dedicated to Nalini Joshi on the occasion of her 60th birthday

Contents

Quadrangular Sets in Projective Line and in Moebius Space, and Geometric Interpretation of the Non-commutative Discrete Schwarzian Kadomtsev–Petviashvili Equation

Adam Doliwa and Jarosław Kosiorek

Abstract We present geometric interpretation of the discrete Schwarzian Kadomtsev–Petviashvili equation in terms of quadrangular set of points of a projective line. We give also the corresponding interpretation for the projective line considered as a Moebius chain space. In this way we incorporate the conformal geometry interpretation of the equation into the projective geometry approach via Desargues maps.

Keywords Discrete Schwarzian KP equation · Desargues maps · Projective line · Chain geometry · Moebius–Veblen configuration

2010 Mathematics Subject Classification 51B10, 51A20

1 Introduction

In the present paper we address two questions concerning geometric interpretation of the following discrete integrable system

$$(\phi_{(jk)} - \phi_{(k)})(\phi_{(jk)} - \phi_{(j)})^{-1}(\phi_{(ij)} - \phi_{(j)})(\phi_{(ij)} - \phi_{(i)})^{-1}(\phi_{(ik)} - \phi_{(i)})(\phi_{(ik)} - \phi_{(k)})^{-1} = 1, \tag{1}$$

where $\phi\colon \mathbb{Z}^N \to \mathbb{F}$ is a map from N-dimensional integer lattice to a division ring $\mathbb{F}$, and indices in brackets denote shifts in the corresponding variables, i.e. $\phi_{(i)}(n_1, \ldots, n_N) = \phi(n_1, \ldots, n_i + 1, \ldots, n_N)$. The above equation appeared first as the generalized lattice spin equation in Nijhoff and Capel (1990), and was called the

---

A. Doliwa (✉) · J. Kosiorek
Faculty of Mathematics and Computer Science, University of Warmia and Mazury, ul. Słoneczna 54, 10-710 Olsztyn, Poland
e-mail: doliwa@matman.uwm.edu.pl

J. Kosiorek
e-mail: kosiorek@matman.uwm.edu.pl

© Springer Nature Switzerland AG 2020

F. Nijhoff et al. (eds.), *Asymptotic, Algebraic and Geometric Aspects of Integrable Systems*, Springer Proceedings in Mathematics & Statistics 338,
https://doi.org/10.1007/978-3-030-57000-2_1

non-commutative discrete Schwarzian Kadomtsev–Petviashvili (SKP) equation in Bogdanov and Konopelchenko (1998); Konopelchenko and Schief (2005). As being one of various forms of the discrete Kadomtsev-Petviashvili (KP) system (Hirota 1981; Kuniba et al. 2011), Eq. (1) plays pivotal role in the theory of integrable systems and its applications.

Relation between geometry of submanifolds and integrable systems is an ongoing research subject which can be dated back to second half of XIX-th century (Darboux 1887–1896). In fact, geometric approach to discrete integrable systems initiated in Bobenko and Pinkall (1996), Doliwa and Santini (1997), Konopelchenko and Schief (1998), see also Bobenko and Suris (2009) for a review, demonstrates that the basic principles of the theory are encoded in incidence geometry statements, some of them known in antiquity.

For example, complex $\mathbb{F} = \mathbb{C}$ version of Eq. (1) was identified in Konopelchenko and Schief (2002) as a multi-ratio condition which describes generalization to conformal geometry of circles of the Menelaus theorem in the metric geometry (Coxeter and Greitzer 1967). Quaternionic version of the equation was studied in Konopelchenko and Schief (2005, 2009), see also King and Schief (2003) for other geometric interpretations of the multi-ratio condition in relation to integrable discrete systems.

The more recently introduced notion of Desargues maps (Doliwa 2010), as underlying property of discrete KP equation considers collinearity of three points. This approach works in projective geometries over division rings and leads directly to the linear problem for the equation in its non-Abelian Hirota–Miwa form (Nimmo 2006). We remark that the Desargues maps give new understanding (Doliwa 2010, 2013) of the previously studied discrete conjugate nets (Doliwa and Santini 1997). These are characterized by planarity of elementary quadrilaterals (see also Sauer 1937; Doliwa 1997). The compatibility condition for Desargues maps gives projective Menelaus theorem, but leaves open the following **Question 1: Can the conformal geometry interpretation of the discrete Schwarzian Kadomtsev–Petviashvili equation be incorporated into the Desargues map approach?** Notice that the recent generalization of the Desargues theorem to context of conformal geometry (King and Schief 2012) may suggest something opposite.

When studying reductions of the Desargues maps, as for example in Doliwa (2013), one is forced to restrict dimension of the ambient projective space up to "Desargues maps into projective line". Even if the linear problem is well defined there the geometric condition, which defines the maps, is empty. This leads to **Question 2: What should replace the Desargues map condition for the ambient space being projective line?** We remark that the analogous problem for discrete conjugate nets in the ambient space being a plane was successfully solved in Adler (2006).

Our answer for both questions is based on the notion of the quadrangular sets, which was introduced by von Staudt in his seminal work (von Staudt 1847) as a tool to provide axiomatization of the projective geometry. We remark that quadrangular sets of points appeared in integrable discrete geometry in theory of the B-quadrilateral lattice (Doliwa 2007a), but in the context of the Pappus theorem and the Moebius pair of tetrahedra, which is outside of the interest of the present paper.

In Sect. 2 we first recall basic ingredients of the geometry of the projective line, in particular the notion of cross-ratio in the general non-commutative case (Baer 1952). We also formulate the corresponding concept of the multi-ratio of six points on the projective line over a division ring, which generalizes the definition known for commutative case in terms of two cross-ratios or determinants. We show that quadrangular sets of points are fully characterized by the "multiratio equals one" condition also in the non-commutative case (as the commutative case is well known Richter-Gebert 2011). This gives our answer to Question 2, which we present in Sect. 3.

Our answer to Question 1, which we present in Sect. 4, is also implied by geometry of the projective line, but this time the line is equipped with additional structure. When the division ring $\mathbb{F}$ contains a subfield $\mathbb{K}$ in its center then $\mathbb{F}$-projective images of the canonically embedded $\mathbb{K}$-line form the so called chains. This leads to the concept of Moebius chain geometry (Benz 1973; Herzer 1995). We show that in such spaces certain quadrangular sets have particular interpretation in terms of the so called Moebius–Veblen chain configuration. In the simplest case of the classical Moebius geometry, where $\mathbb{K} = \mathbb{R} \subset \mathbb{C} = \mathbb{F}$ the chains are circles (homographic images of the real line in the complex conformal plane), and our approach gives that of Konopelchenko and Schief (2002).

2 Projective Geometry of a Line

2.1 *Cross-Ratio and Multi-ratio in Projective Geometry over Division Rings*

A *right linear space* $(\mathbb{F}, \mathbb{V})$ consists of a division ring $\mathbb{F}$ and an additive abelian group $\mathbb{V}$ such that $\mathbb{F}$ acts on $\mathbb{V}$ from the right satisfying usual axioms. The corresponding projective geometry studies linear subspaces of the $\mathbb{F}$-space $\mathbb{V}$. The points of the corresponding projective space $\mathbb{P}(\mathbb{F}, \mathbb{V})$ are one dimensional subspaces of $(\mathbb{F}, \mathbb{V})$.

Remark For simplicity we assume that $\mathbb{F}$ is of characteristic zero, but we expect that also finite characteristic may be relevant and give interesting results (Białecki and Doliwa 2005).

A collineation of the linear space $(\mathbb{F}, \mathbb{V})$ upon the linear space $(\mathbb{F}, \mathbb{W})$ is a bijective and order preserving mapping σ of the partially ordered (by inclusion) set of subspaces of $\mathbb{V}$ upon the set of subspaces of $\mathbb{W}$. When dimension of $\mathbb{V}$ is at least three, any such collineation is given by a semi-linear map, i.e. linear map $\mathbb{V} \to \mathbb{W}$ and supplemented by an automorphism of the division ring.

The case of two dimensional linear spaces (i.e. projective lines) needs a special treatment. Then any bijection of projective line can be called collineation. There arises the problem of characterizing those maps which are induced by semi-linear maps of $(\mathbb{F}, \mathbb{V})$. The first step in that direction (the full answer can be found in Baer 1952) makes use of a generalization of the classical notion of cross ratio.

Definition 1 (Baer 1952) Suppose that P, Q, R, S are four distinct points on the line L. Then the number $c \in \mathbb{F}$ belongs to the cross ratio $\begin{bmatrix} P & Q \\ S & R \end{bmatrix} \subset \mathbb{F}$ if there exist elements $\boldsymbol{p}, \boldsymbol{q} \in \mathbb{V}$ such that

$$P = \langle \boldsymbol{p} \rangle, \qquad Q = \langle \boldsymbol{q} \rangle, \qquad R = \langle \boldsymbol{p} + \boldsymbol{q} \rangle, \qquad S = \langle \boldsymbol{p} + \boldsymbol{q} c \rangle.$$

Below we present the known expression of the cross-ratio in terms of non-homogeneous coordinates.

Theorem 1 *If $\boldsymbol{k}, \boldsymbol{l}$ are two independent elements of $\in \mathbb{V}$, and p, q, r, s are four distinct elements of $\mathbb{F}$ then*

$$\begin{bmatrix} \langle \boldsymbol{k} + \boldsymbol{l} p \rangle & \langle \boldsymbol{k} + \boldsymbol{l} q \rangle \\ \langle \boldsymbol{k} + \boldsymbol{l} s \rangle & \langle \boldsymbol{k} + \boldsymbol{l} r \rangle \end{bmatrix} = \left[(s-q)^{-1}(p-s)(p-r)^{-1}(r-q) \right],$$

where by for $c \in \mathbb{F}$ by $[c] = \{aca^{-1} \mid a \in \mathbb{F} \setminus \{0\}\}$ we denote the equivalence class of conjugate elements.

Remark Given three distinct points $P, Q, R \in L$ and given $c \in \mathbb{F} \setminus \{0, 1\}$ there exists the fourth point $S \in L$, distinct from the previous ones, such that $c \in \begin{bmatrix} P & Q \\ S & R \end{bmatrix}$; $c = 0$ corresponds to $S = P$, $c = 1$ corresponds to $S = R$, while to admit $S = Q$ we need to give the value $c = \infty$.

Remark When points P, Q, R have been taken as projective basis of the line, i.e. $p = 0, q = \infty, r = 1$, then $\begin{bmatrix} P & Q \\ S & R \end{bmatrix} = [s]$.

The following result for $\dim \mathbb{V} \geq 3$ justifies the use of cross-ratio in describing geometry of the projective line.

Theorem 2 *Suppose that (P, Q, R, S) and (P', Q', R', S') are quadruples of distinct collinear points. There exists collineation π of the linear space $(\mathbb{F}, \mathbb{V})$, $\dim \mathbb{V} \geq 3$. such that $\pi(P) = P', \pi(Q) = Q', \pi(R) = R', \pi(S) = S'$ if and only if there exists an automorphism α of the division ring $\mathbb{F}$ such that*

$$\begin{bmatrix} P & Q \\ S & R \end{bmatrix}^{\alpha} = \begin{bmatrix} P' & Q' \\ S' & R' \end{bmatrix}.$$

We present below the analogous geometric notion of multi-ratio in the non-commutative case, which we adapted from known definition in the commutative case in terms of two cross-ratios (Morley and Musselman 1937; King and Schief 2003; Richter-Gebert 2011). Like for the non-commutative cross-ratio our geometric definition leads to a class of conjugate elements of the division ring.

Definition 2 Suppose that P, Q, R, S, T, U are six distinct points on the line L. Then the number $m \in \mathbb{F}$ belongs to the multi-ratio $\begin{bmatrix} P & Q & S \\ U & R & T \end{bmatrix} \subset \mathbb{F}$ if there exist elements $\boldsymbol{p}, \boldsymbol{q}, \boldsymbol{s} \in \mathbb{V}$ such that

$$P = \langle \boldsymbol{p} \rangle, \quad Q = \langle \boldsymbol{q} \rangle, \quad S = \langle \boldsymbol{s} \rangle, \quad R = \langle \boldsymbol{p} + \boldsymbol{q} \rangle, \quad T = \langle \boldsymbol{p} + \boldsymbol{s} \rangle,$$
$$U = \langle \boldsymbol{p} + \boldsymbol{q}a \rangle, \quad U = \langle \boldsymbol{p} + \boldsymbol{s}b \rangle \quad m = ab^{-1}.$$

Proposition 3 *If $\boldsymbol{k}, \boldsymbol{l}$ are two independent elements of $\in \mathbb{V}$, and p, q, r, s, t, u are six distinct elements of $\mathbb{F}$ then*

$$\begin{bmatrix} \langle \boldsymbol{k} + \boldsymbol{l}p \rangle & \langle \boldsymbol{k} + \boldsymbol{l}q \rangle & \langle \boldsymbol{k} + \boldsymbol{l}s \rangle \\ \langle \boldsymbol{k} + \boldsymbol{l}u \rangle & \langle \boldsymbol{k} + \boldsymbol{l}r \rangle & \langle \boldsymbol{k} + \boldsymbol{l}t \rangle \end{bmatrix} = \left[(r-p)^{-1}(q-r)(q-u)^{-1}(s-u)(s-t)^{-1}(t-p) \right]. \quad (2)$$

Proof Let the vectors $\boldsymbol{p}, \boldsymbol{q} \in \mathbb{V}$ be such as in Definition 2, define the factors $\lambda, \mu, \sigma, \rho \in \mathbb{F} \setminus \{0\}$ for points P, Q, R, U by

$$(\boldsymbol{k} + \boldsymbol{l}p)\lambda = \boldsymbol{p}, \quad (\boldsymbol{k} + \boldsymbol{l}q)\mu = \boldsymbol{q}, \quad (\boldsymbol{k} + \boldsymbol{l}r)\sigma = \boldsymbol{p} + \boldsymbol{q}, \quad (\boldsymbol{k} + \boldsymbol{l}u)\rho = \boldsymbol{p} + \boldsymbol{q}a.$$

Elimination of the factors μ, σ, ρ leads to the relation

$$a = \lambda^{-1}(r-p)^{-1}(q-r)(q-u)^{-1}(u-p)\lambda.$$

Similar reasoning, but for points P, S, T, U leads to similar relation

$$b = \lambda^{-1}(t-p)^{-1}(s-t)(s-u)^{-1}(u-p)\lambda,$$

which, combined with the previous one, concludes the proof. □

Remark When the division ring $\mathbb{F}$ is commutative, our definition of the multi-ratio $\begin{bmatrix} P & Q & S \\ U & R & T \end{bmatrix} \in \mathbb{F}$ reduces to the product $\begin{bmatrix} P & Q \\ U & R \end{bmatrix} \begin{bmatrix} P & S \\ U & T \end{bmatrix}^{-1}$ of two cross-ratios.

Proposition 4 *The multi-ratio is an invariant of the collineations induced by linear transformations of the space* $(\mathbb{F}, \mathbb{V})$.

Proof Fix coordinate system, like in Proposition 3, and use the fact, that such collineations are generated by affine transformations $\phi \mapsto a\phi + b$ and inversions $\phi \mapsto \phi^{-1}$

$$(a\phi + b)(c\phi + d)^{-1} = ac^{-1} + (b - ac^{-1}d)(c\phi + d)^{-1}.$$

By direct calculation, both transformations preserve the multi-ratio (2) understood as the class of conjugate elements. □

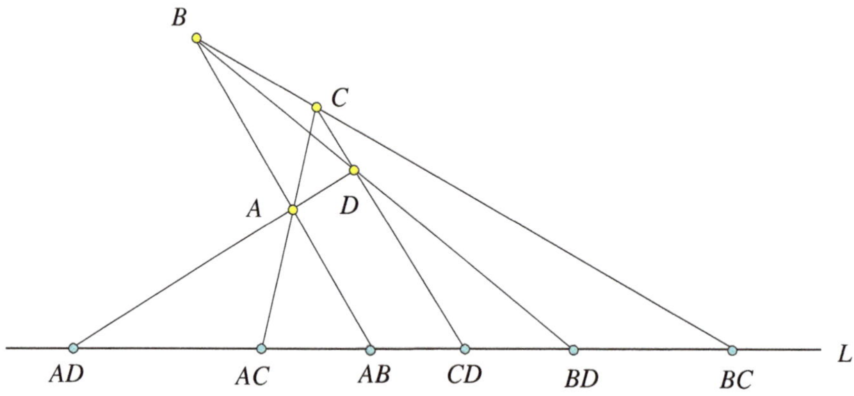

Fig. 1 Quadrangular set of points

2.2 *Quadrangular Set of Points on Projective Line*

A *complete quadrangle* is a projective figure formed by four points (vertices) A, B, C, D in the plane, no three of which are collinear, and the six distinct lines (sides) that are produced by joining them pairwise. The intersection points $AB, \ldots, CD$ of the six lines with a line not incident with vertices of the quadrangle form the *quadrangular set* (von Staudt 1847), see Fig. 1.

It is known that, by Desargues theorem, any five points of the quadrangular set (labeling fixed) determine uniquely the sixth point of the set. Moreover, collineations map quadrangular sets into quadrangular sets.

Remark The ordering of the points is important, up to permutation of the letters A, B, C, D. By combinatorial arguments one can show that given five generic points of the projective line there are 30 positions of the sixth point such that for appropriate ordering the six points form a quadrangular set.

Proposition 5 *The six distinct points $AB, \ldots, CD$ form a quadrangular set if and only if their non-homogeneous coordinates $\Phi_{AB}, \ldots, \Phi_{CD}$ satisfy the multi-ratio condition*

$$(\Phi_{AD} - \Phi_{AC})(\Phi_{AD} - \Phi_{AB})^{-1}(\Phi_{BD} - \Phi_{AB})(\Phi_{BD} - \Phi_{BC})^{-1}(\Phi_{CD} - \Phi_{BC})(\Phi_{CD} - \Phi_{AC})^{-1} = 1. \quad (3)$$

Proof Given five points of the set, we reconstruct the planar quadrilateral which allows to obtain the sixth point. It is known (Veblen and Young 1910) that the construction is independent on the freedom in choice of the quadrilateral. Because we were not able to find the multi-ratio characterization of the quadrangular sets in the non-commutative case we present its detailed derivation.

In the general case fix coordinate system on the line $L = \{(a, 0) | a \in \mathbb{F}\}$. Choose an arbitrary point $A \neq L$, which can be given then non-homogeneous coordinates

(0, 1). The last freedom in the construction is the choice of a point B on the line $\langle A, AB\rangle$, which fixes its coordinates

$$(\Phi^1_B, \Phi^2_B) = (\Phi_{AB}(1-\sigma), \sigma), \qquad \sigma \in \mathbb{F} \setminus \{0, 1\}.$$

Then the coordinates of the point $C = \langle A, AC\rangle \cap \langle B, BC\rangle$ are

$$(\Phi^1_C, \Phi^2_C) = (\Phi_{AC}(1-\mu), \mu), \qquad \mu = \left[\Phi_{AC} - \Phi_{AB} + (\Phi_{AB} - \Phi_{BC})\sigma^{-1}\right]^{-1} (\Phi_{AC} - \Phi_{BC});$$

notice identity

$$\mu^{-1} - 1 = (\Phi_{AC} - \Phi_{BC})^{-1}(\Phi_{AB} - \Phi_{BC})(\sigma^{-1} - 1). \tag{4}$$

Similarly, the coordinates of the point $D = \langle A, AD\rangle \cap \langle C, CD\rangle$ read

$$(\Phi^1_D, \Phi^2_D) = (\Phi_{AD}(1-\lambda), \lambda), \qquad \lambda = \left[\Phi_{AD} - \Phi_{AC} + (\Phi_{AC} - \Phi_{CD})\mu^{-1}\right]^{-1} (\Phi_{AD} - \Phi_{CD}),$$

and then

$$\lambda^{-1} - 1 = (\Phi_{AD} - \Phi_{CD})^{-1}(\Phi_{AC} - \Phi_{CD})(\mu^{-1} - 1), \tag{5}$$

moreover

$$(\Phi_{CD} - \Phi_{AC})^{-1}(\Phi_{AD} - \Phi_{AC}) = \mu^{-1}(\mu - \lambda)(1-\lambda)^{-1}. \tag{6}$$

Finally, the non-homogeneous coordinates $(\Phi_{BD}, 0)$ of the point $BD = L \cap \langle B, D\rangle$ are given by

$$\Phi_{BD} = \Phi_{AB}(1-\sigma) + [\Phi_{AD}(1-\lambda) - \Phi_{AB}(1 - \sigma)]\,(\sigma - \lambda)^{-1}\sigma, \tag{7}$$

what implies

$$(\Phi_{BD} - \Phi_{AB})^{-1}(\Phi_{AD} - \Phi_{AB}) = \sigma^{-1}(\sigma - \lambda)(1-\lambda)^{-1}. \tag{8}$$

Equations (4)–(8) give then

$$\begin{gathered}(\Phi_{BD} - \Phi_{BC}) - (\Phi_{CD} - \Phi_{BC})(\Phi_{CD} - \Phi_{AC})^{-1}(\Phi_{AD} - \Phi_{AC})(\Phi_{AD} - \Phi_{AB})^{-1}(\Phi_{BD} - \Phi_{AB}) = \\ \left[(\Phi_{AD} - \Phi_{CD})(\lambda^{-1} - 1) - (\Phi_{AB} - \Phi_{BC})(\sigma^{-1} - 1) + (\Phi_{CD} - \Phi_{BC})(\mu^{-1} - 1)\right]\lambda(\sigma - \lambda)^{-1}\sigma = 0,\end{gathered}$$

which concludes the first part of the proof.

Because Eq. (3) is uniquely solvable for any of its six points, once other five are given, and by the analogous property of the quadrangular set, the condition described by the equation completely characterizes quadrangular sets of the projective line. □

Fig. 2 Veblen configuration

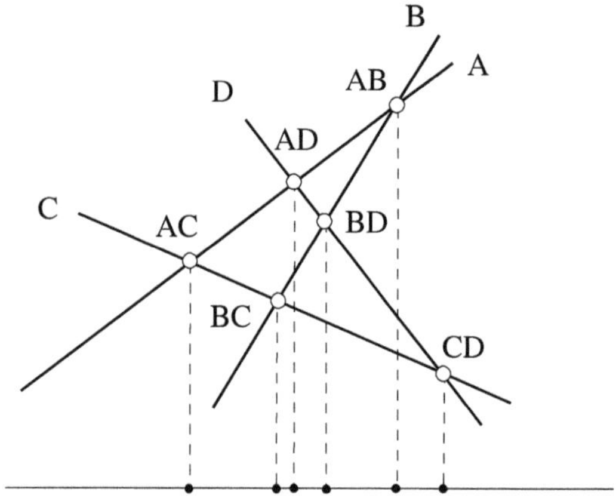

3 Desargues Maps into Projective Line

3.1 The Veblen Configuration and the Multi-ratio

Consider Veblen or Menelaus (Konopelchenko and Schief 2002) configuration $(6_2, 4_3)$ in projective space, i.e. six points and four lines with two lines incident with each point, and three points incident with each line, see Fig. 2. We label points of the configuration by two element subsets of the four element set $\{A, B, C, D\}$, and lines by single elements of the same set. A point is incident with the line if its label contains the label of the line.

Let us present an algebraic description of the Veblen configuration, which can be considered as a non-commutative version of the theorem of Menelaus (Coxeter and Greitzer 1967).

Proposition 6 *Given points $AB \in \langle BC, BD\rangle$, $AD \in \langle BD, CD\rangle$, $AC \in \langle BC, CD\rangle$ on three sides of the triangle BC, BD, CD, and distinct from its vertices. These three points are collinear if and only if the corresponding proportionality coefficients $a, b, c \in \mathbb{F} \setminus \{0, 1\}$ between their non-homogeneous coordinates, as defined by*

$$(\boldsymbol{\Phi}_{BC} - \boldsymbol{\Phi}_{AB}) = (\boldsymbol{\Phi}_{BD} - \boldsymbol{\Phi}_{AB})a, \tag{9}$$
$$(\boldsymbol{\Phi}_{CD} - \boldsymbol{\Phi}_{AD}) = (\boldsymbol{\Phi}_{BD} - \boldsymbol{\Phi}_{AD})b, \qquad (\boldsymbol{\Phi}_{BC} - \boldsymbol{\Phi}_{AC}) = (\boldsymbol{\Phi}_{CD} - \boldsymbol{\Phi}_{AC})c,$$

satisfy condition

$$a = bc. \tag{10}$$

Proof To show that the collinearity implies condition (10) assume that the vectors $\boldsymbol{\Phi}_{AB}$, $\boldsymbol{\Phi}_{AC}$, $\boldsymbol{\Phi}_{AD}$, as calculated from the above linear relations, satisfy the constraint of the form

$$\boldsymbol{\Phi}_{AB} - \boldsymbol{\Phi}_{AC} = (\boldsymbol{\Phi}_{AD} - \boldsymbol{\Phi}_{AC})\lambda.$$

The linear independence of vectors $\boldsymbol{\Phi}_{BD} - \boldsymbol{\Phi}_{CD}$ and $\boldsymbol{\Phi}_{BC} - \boldsymbol{\Phi}_{CD}$ implies then

$$\lambda = 1 - (c-1)(a-1)^{-1} = (b-1)c(a-1)^{-1},$$

which gives Eq. (10).

From the other side, insert the condition (10) into first of the above linear equations, which in conjunction with other two gives

$$(\boldsymbol{\Phi}_{AB} - \boldsymbol{\Phi}_{AD})(1 - bc) = (\boldsymbol{\Phi}_{AC} - \boldsymbol{\Phi}_{AD})(1 - c),$$

thus showing the collinearity. □

Corollary 7 *Assume that for fixed coordinate number i all components* $\Phi^i_{AB}, \ldots, \Phi^i_{CD}$ *of the points of the Veblen configuration are distinct (see Fig. 2), then the components satisfy the following multi-ratio condition*

$$(\Phi^i_{CD} - \Phi^i_{AC})(\Phi^i_{CD} - \Phi^i_{AD})^{-1}(\Phi^i_{BD} - \Phi^i_{AD})(\Phi^i_{BD} - \Phi^i_{AB})^{-1}(\Phi^i_{BC} - \Phi^i_{AB})(\Phi^i_{BC} - \Phi^i_{AC})^{-1} = 1.$$

Proof Insert expressions

$$\begin{aligned} a =& (\Phi^i_{BD} - \Phi^i_{AB})^{-1}(\Phi^i_{BC} - \Phi^i_{AB}),\\ b =& (\Phi^i_{BD} - \Phi^i_{AD})^{-1}(\Phi^i_{CD} - \Phi^i_{AD}),\\ c =& (\Phi^i_{CD} - \Phi^i_{AC})^{-1}(\Phi^i_{BC} - \Phi^i_{AC}),\end{aligned}$$

into the condition (10). □

We conclude this Section with a result, which justifies the statement that *quadrangular sets should be considered as Veblen configurations in the geometry of projective line.*

Proposition 8 *In the plane of the Veblen configuration consider point O not on lines of the configuration. The intersection points of lines joining O to vertices of the configuration with an arbitrary line not incident with O form a quadrangular set.*

Proof Take the point O as the first vertex of the quadrangle, fix a line of the Veblen configuration, and use three remaining points of the configuration as three remaining vertices of the quadrangle. On the line we have built then a quadrangular set. The lines joining the points of the Veblen configuration with point O are the lines joining O to points of the quadrangular set. Any transversal section of the lines by another line gives six points perspective with the quadrangular set. Because such transformations map quadrangular sets into quadrangular sets (Veblen and Young 1910) we obtain the statement. □

Remark The case when O is a point at infinity and the line is a coordinate line is actually visualized in Fig. 2.

3.2 Desargues Maps

From point of view of difference equations usually one considers maps of $\mathbb{Z}^N$ lattice. Recently integrable systems on other regular lattices are also of some interest, see for example (Doliwa et al. 2007). In particular, the Desargues maps, although initially defined on multidimensional integer lattice, allow for an interpretation (Doliwa 2011) as maps from multidimensional root lattice of type A. Such an approach from the very beginning takes into account the corresponding affine Weyl group symmetry of the discrete KP system.

Recall that the $N \geq 2$-dimensional root lattice $Q(A_N)$ is generated by vectors along the edges of regular N-simplex in Euclidean space (Conway and Sloane 1988; Moody and Patera 1992). Vertices of the lattice tessellate the space into N types of convex polytopes $P(k, N), k = 1, \ldots, N$, called *ambo-simplices*. It is known that the corresponding affine Weyl group $W(A_N)$ acts on the Delaunay tiling by permuting tiles within each class $P(k, N)$. The 1-skeleton of $P(k, N)$ is the so called Johnson graph $J(N + 1, k)$: its vertices are labeled by k-point subsets of $\{1, 2, \ldots, N + 1\}$, and edges are the pairs of such sets with $(k - 1)$-point intersection.

The tiles $P(1, N)$ are congruent to the initial N-simplex which generates the vertices of the root lattice. We color its faces $P(1, 2)$ in black. The faces $P(2, 2)$ of the simplex $P(N, N)$ we color white. Then $P(2, 3)$ which is regular octahedron has four black and four white triangular faces, see Fig. 3.

Definition 3 By Desargues map $\phi\colon Q(A_N) \to \mathbb{P}(\mathbb{F}, \mathbb{V})$ we mean a map for which images of vertices of simplices $P(1, 2)$ with black triangular faces are collinear.

To avoid degenerations it is implicitly assumed that images of vertices of simplices $P(2, 2)$ with white triangular faces are in generic position. Then the octahedra $P(2, 3)$

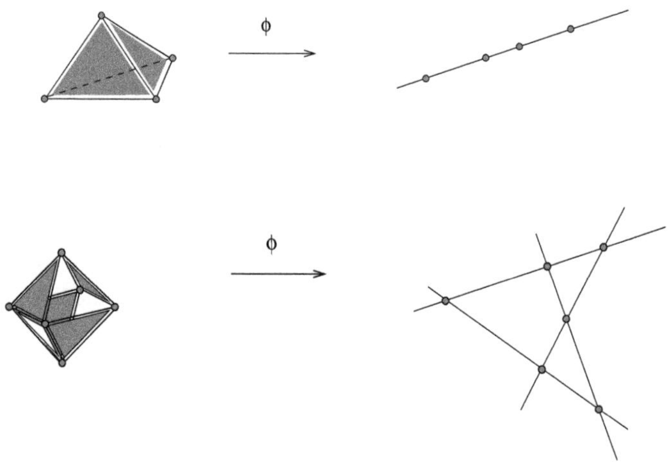

Fig. 3 Desargues map

is mapped into Veblen configurations, see Fig. 3. Moreover, after introduction of appropriate $\mathbb{Z}^N$ coordinates in the lattice the non-homogeneous coordinates of the map satisfy Eq. (1), compare with Corollary 7. Guided by Proposition 8 we give our answer to Question 2.

Definition 4 By Desargues map of root lattice $Q(A_N)$ into a projective line we mean a map for which images of vertices of ambo-simplices $P(2, 3)$ are quadrangular sets with labeling induced by that of Johnson graph $J(4, 2)$.

4 Desargues Maps in Moebius Chain Geometry

Below we describe the concept of Moebius chain geometry of a projective line, which allows for a construction of quadrangular sets which satisfy certain additional property, adjusted to the structure. This will give our answer to Question 1.

4.1 The Concept of Moebius Chain Geometry

Assume that the division ring $\mathbb{F}$ contains a proper subfield $\mathbb{K}$ in its center. The division ring $\mathbb{F}$ can be considered then as a division $\mathbb{K}$-algebra. Correspondingly, the projective line over $\mathbb{F}$ inherits additional structure, best described within the concept of chain geometry (Benz 1973; Herzer 1995; Blunck and Herzer 2005). Define the chains as images of the canonically embedded $\mathbb{K}$-line (called the standard chain) under action of the group of collineations induced by linear maps of the $\mathbb{F}$-line. Points are called cocatenal if they belong to a common chain.

Remark The simplest example for $\mathbb{K} = \mathbb{R} \subset \mathbb{C} = \mathbb{F}$ is the classical conformal Moebius geometry of circles (as chains) in the Riemann sphere (complex line or conformal plane). The notion of chain geometry applies actually to any $\mathbb{K}$-algebra, however we will be dealing exclusively with division algebras (i.e. with the so called Herzer 1995 Moebius chain geometries).

The $\mathbb{K}$-vector space $\mathbb{F}$ can be given natural affine space structure. The straight lines are the chains which pass through the infinity point of the projective line $\mathbb{F} \cup \{\infty\}$. Notice that the notion of "straight line" depends actually on the particular choice of the infinity point. Two chains are called tangent in point P when after sending P to infinity the chains are parallel in corresponding affine space. Notice following results of the Moebius chain geometry:

(1) Any three distinct points of the $\mathbb{F}$-line are contained in exactly one chain.
(2) Four distinct points are cocatenal if and only if their cross-ratio is well defined element of $\mathbb{K} \setminus \{0, 1\}$.
(3) (Miquel condition) Given four chains C_i, $i = 1, 2, 3, 4$, no three of which have a common point, but $C_i \cap C_{i+1} = \{P_i, Q_i\}$ for every i (subscripts are taken

modulo 4). Then the four points P_i are cocatenal if and only if the four points Q_i are cocatenal.

Remark One can consider (Blunck and Havlicek 2000) generalizations of the Moebius chain geometry for which $\mathbb{K}$ is a subdivision ring of $\mathbb{F}$ not necessarily contained in its center (like for example in the case $\mathbb{C} \subset \mathbb{H}$, where chains are two dimensional spheres S^2 in $S^4 \equiv \mathbb{H} \cup \{\infty\}$). However such a generalization may violate some of the properties above.

4.2 *Moebius–Veblen Configuration*

Let us present an analogue of the Veblen axiom/theorem, see also Blunck and Herzer (2005) for its general version (in a slightly different formulation) for chain geometries.

Proposition 9 *Given five distinct points AB, AC, AD, BC, BD of the Moebius chain space such that the chains $\mathcal{C}(AB, AC, AD)$ and $\mathcal{C}(AB, BC, BD)$ have in common additional point $I \neq AB$. Then also the chains $\mathcal{C}(I, AC, BC)$ and $\mathcal{C}(I, AD, BD)$ have in common additional point $CD \neq I$, or alternatively the chains are tangent in I.*

Proof By sending the intersection point I to infinity, see Fig. 4, we obtain five points of the classical Veblen configuration in the affine space $(\mathbb{K}, \mathbb{F})$ what allows to construct the sixth point of the "straightened configuration", and then eventually to go back to the original one. □

Remark Notice that in the classical Moebius geometry of the complex projective line the assumption about existence of the point I is not needed. However this assumption is essential, if we would like to perform the construction in the case of quaternionic projective line with $\mathbb{R} \subset \mathbb{H}$.

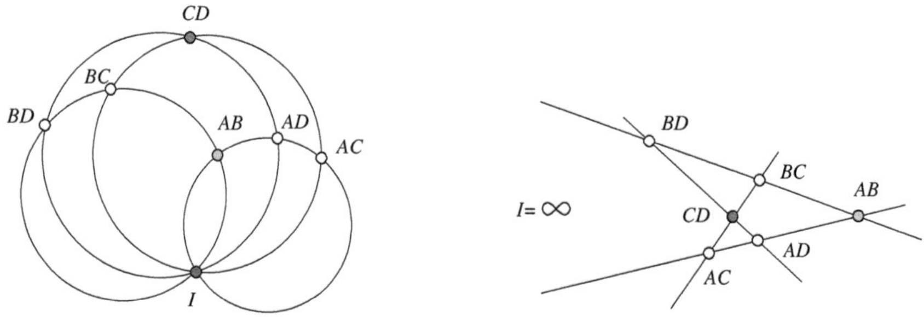

Fig. 4 Moebius–Veblen configuration in Moebius chain geometry

The points AB, AC, AD, CD, BC, BD will be called ordinary points of the Moebius–Veblen configuration, while I is called the infinity point of the configuration.

Proposition 10 *In the Moebius–Veblen configuration the constructed point CD is the sixth quadrangular point of the initial five ordinary points in the standard labeling.*

Proof In the "straightened configuration" apply Proposition 5 to get coefficients a, b, c which belong to $\mathbb{K} \setminus \{0, 1\}$ and satisfy condition (10). But now one can eliminate the coefficients directly on the level of Eq. (9) in order to get the multi-ratio condition (3) in $\mathbb{F}$. Proposition 4, which gives invariance of the condition with respect to projective collineations of the $\mathbb{F}$-line, and Proposition 5 imply the statement. □

Remark Even if we do not have the point I to our disposal the sixth point CD of the quadrangular set exists, by the general construction described in Sect. 2. Then one can consider also the corresponding initial chains $\mathcal{C}(AB, AC, AD)$, $\mathcal{C}(AB, BC, BD)$, which contain point AB, and the resulting chains $\mathcal{C}(AC, BC, CD)$ and $\mathcal{C}(AD, BD, CD)$ containing the point CD. These new chains can be given only *after* construction of the point.

Finally we give our answer to Question 1 presenting the special type of quadrangular sets for which the construction of the sixth point follows from Moebius chain geometry principles, thus incorporating the conformal geometry interpretation of the discrete Schwarzian Kadomtsev–Petviashvili equation into the Desargues map approach.

Definition 5 By Moebius quadrangular set we mean the six ordinary points of the Moebius–Veblen configuration with the labeling as in Proposition 9.

It is well known (Clifford 1871; Konopelchenko and Schief 2002) that the Moebius–Veblen configuration can be supplemented by four chains $\mathcal{C}(AB, BC, AC)$, $\mathcal{C}(AB, BD, AD)$, $\mathcal{C}(AC, CD, AD)$ and $\mathcal{C}(BC, CD, BD)$ which then all intersect in the so called Clifford point. The new circle-point configuration of 8 points and 8 circles, with each point/circle incident with 4 circles/points, is called the Clifford configuration. Actually, as it was described in Doliwa (2007b), this result in an equivalent version was known already to Miquel (Miquel 1838).

5 Conclusion

The projective structure of the line and the notion of quadrangular sets can be used to provide geometric meaning to non-commutative discrete Schwarzian KP system. When the thiner structure of the underlying division ring is considered then the theory becomes more intriguing. Up to now the case of complex and quaternionic Moebius spaces (with the subfield of real numbers) have been thoroughly examined (Konopelchenko and Schief 2002, 2009). There are known works, see for example (Bazhanov et al. 2008), where the Miquel condition has been used to study particular algebras of quantum integrable systems. We expect that the chain geometry may become useful platform to investigate other quantum algebras of mathematical physics.

Acknowledgements Both authors acknowledge numerous discussions with Andrzej Matraś on various aspects of chain geometries. The research of A.D. was supported by National Science Centre, Poland, under grant 2015/19/B/ST2/03575 *Discrete integrable systems – theory and applications*.

References

Adler, V.E.: Some incidence theorems and integrable discrete equations. Discrete Comput. Geom. **36**, 489–498 (2006)

Baer, R.: Linear Algebra and Projective Geometry. Academic Press, New York (1952)

Bazhanov, V.V., Mangazeev, V.V., Sergeev, S.M.: Quantum geometry of three-dimensional lattices. J. Stat. Mech.: Th. Exp. P07004 (2008)

Benz, W.: Vorlesungen über Geometrie der Algebren. Springer, Berlin (1973)

Białecki, M., Doliwa, A.: Algebro-geometric solution of the discrete KP equation over a finite field out of a hyperelliptic curve. Comm. Math. Phys. **253**, 157–170 (2005)

Blunck, A., Havlicek, H.: Extending the concept of chain geometry. Geometriae Dedicata **83**, 119–130 (2000)

Blunck, A., Herzer, A.: Kettengeometrien. Shaker Verlag, Aachen, Eine Einführung (2005)

Bobenko, A., Pinkall, U.: Discrete isothermic surfaces. J. Reine Angew. Math. **475**, 187–208 (1996)

Bobenko, A.I., Suris, YuB: Discrete Differential Geometry: Integrable Structure. AMS, Providence (2009)

Bogdanov, L.V., Konopelchenko, B.G.: Analytic-bilinear approach to integrable hiererchies II. Multicomponent KP and 2D Toda hiererchies. J. Math. Phys. **39**, 4701–4728 (1998)

Clifford, W.K.: A synthetic proof of Miquels's theorem. Oxford, Cambridge and Dublin Messenger Math. **5**, 124–141 (1871)

Conway, J.H., Sloane, N.J.A.: Sphere Packings, Lattices and Groups. Springer, Berlin (1988)

Coxeter, H.S.M., Greitzer, S.L.: Geometry Revisited. Random House, New York (1967)

Darboux, G.: Leçons sur la théorie générale des surfaces. I–IV. Gauthier – Villars, Paris (1887–1896)

Doliwa, A.: Geometric discretisation of the Toda system. Phys. Lett. A **234**, 187–192 (1997)

Doliwa, A.: The B-quadrilateral lattice, its transformations and the algebro-geometric construction. J. Geom. Phys. **57**, 1171–1192 (2007)

Doliwa, A.: Generalized isothermic lattices. J. Phys. A: Math. Theor. **40**, 12539–12561 (2007)

Doliwa, A.: Desargues maps and the Hirota-Miwa equation. Proc. R. Soc. A **466**, 1177–1200 (2010)

Doliwa, A.: The affine Weyl group symmetry of Desargues maps and of the non-commutative Hirota-Miwa system. Phys. Lett. A **375**, 1219–1224 (2011)

Doliwa, A.: Desargues maps and their reductions. In: Ma, W.X., Kaup, D. (eds.) Nonlinear and Modern Mathematical Physics. AIP Conference Proceedings, vol. 1562, pp. 30–42. AIP Publishing (2013)

Doliwa, A.: Non-commutative lattice modified Gel'fand-Dikii systems. J. Phys. A: Math. Theor. **46**, 205202 (14pp) (2013)

Doliwa, A., Santini, P.M.: Multidimensional quadrilateral lattices are integrable. Phys. Lett. A **233**, 365–372 (1997)

Doliwa, A., Nieszporski, M., Santini, P.M.: Integrable lattices and their sublattices. II. From the B-quadrilateral lattice to the self-adjoint schemes on the triangular and the honeycomb lattices. J. Math. Phys. **48**, 113506 (2007)

Herzer, A.: In: Buekenhout, F. (ed.) Chain Geometries. Handbook of Incidence Geometry, pp. 781–842. Elsevier (1995)
Hirota, R.: Discrete analogue of a generalized Toda equation. J. Phys. Soc. Jpn. **50**, 3785–3791 (1981)
King, A.D., Schief, W.K.: Tetrahedra, octahedra and cubo-octahedra: integrable geometry of multi-ratios. J. Phys. A: Math. Gen. **36**, 785–802 (2003)
King, A.D., Schief, W.K.: Clifford lattices and a conformal generalization of Desargues' theorem. J. Geom. Phys. **62**, 1088–1096 (2012)
Konopelchenko, B.G., Schief, W.K.: Three-dimensional integrable lattices in Euclidean spaces: conjugacy and orthogonality. Proc. Roy. Soc. Lond. A **454**, 3075–3104 (1998)
Konopelchenko, B.G., Schief, W.K.: Menelaus' theorem, Clifford configuration and inversive geometry of the Schwarzian KP hierarchy. J. Phys. A: Math. Gen. **35**, 6125–6144 (2002)
Konopelchenko, B.G., Schief, W.K.: Conformal geometry of the (discrete) Schwarzian Davey-Stewartson II hierarchy. Glasgow Math. J **47A**, 121–131 (2005)
Konopelchenko, B.G., Schief, W.K.: A novel generalization of Clifford's classical point-circle configuration, geometric interpretation of the quaternionic discrete Schwarzian Kadomtsev-Petviashvili equation. Proc. R. Soc. A **465**, 1291–1308 (2009)
Kuniba, A., Nakanishi, T., Suzuki, J.: T-systems and Y-systems in integrable systems. J. Phys. A: Math. Theor. **44**, 103001 (146pp) (2011)
Miquel, A.: Théorèmes sur les intersections des cercles et des sphères. J. Math. Pur. Appl. (Liouville J.) **3**, 517-522 (1838). http://portail.mathdoc.fr/JMPA/
Moody, R.V., Patera, J.: Voronoi and Delaunay cells of root lattices: classification of their facets by Coxeter-Dynkin diagrams. J. Phys. A: Math. Gen. **25**, 5089–5134 (1992)
Morley, F., Musselman, J.R.: On $2n$ points with a real cross-ratio. Am. J. Math. **59**, 787–792 (1937)
Nijhoff, F.W., Capel, H.W.: The direct linearization approach to hierarchies of integrable PDEs in $2+1$ dimensions: I. Lattice equations and the differential-difference hierarchies. Inverse Probl. **6**, 567–590 (1990)
Nimmo, J.J.C.: On a non-Abelian Hirota-Miwa equation. J. Phys. A: Math. Gen. **39**, 5053–5065 (2006)
Richter-Gebert, J.: Perspectives on Projective Geometry. Springer, Berlin (2011)
Sauer, R.: Projective Liniengeometrie. de Gruyter, Berlin-Leipzig (1937)
von Staudt, G.K.Ch.: Geometrie der Lage, Nürnberg (1847)
Veblen, O., Young, J.W.: Projective geometry, vol. I. Ginn and Co., Boston (1910)

Complexity and Integrability in 4D Bi-rational Maps with Two Invariants

Giorgio Gubbiotti, Nalini Joshi, Dinh Thi Tran, and Claude-Michel Viallet

Abstract In this letter we give fourth-order autonomous recurrence relations with two invariants, whose degree growth is cubic or exponential. These examples contradict the common belief that maps with sufficiently many invariants can have at most quadratic growth. Cubic growth may reflect the existence of non-elliptic fibrations of invariants, whereas we conjecture that the exponentially growing cases lack the necessary conditions for the applicability of the discrete Liouville theorem.

Keywords Integrability · Complexity · Algebraic entropy · Growth of bi-rational maps · Higher-order difference equation

1 Introduction

Bi-rational maps in two dimensions have played a crucial role in the study of integrable discrete dynamical systems since the seminal paper of Penrose and Smith (1981) and the introduction of the QRT mappings in Quispel et al. (1988, 1989). Elliptic curves and rational elliptic surfaces proved to be one of the main tools in understanding the geometry behind this kind of integrability, see Sakai (2001), Duistermaat (2011), Tsuda (2004). In this letter we give examples of higher-order maps

G. Gubbiotti (✉) · N. Joshi · D. T. Tran
School of Mathematics and Statistics F07, The University of Sydney, 2006 Sydney, NSW, Australia
e-mail: giorgio.gubbiotti@sydney.edu.au

N. Joshi
e-mail: nalini.joshi@sydney.edu.au

D. T. Tran
e-mail: dinhthi.tran@sydney.edu.au

C.-M. Viallet
LPTHE, UMR 7589 Centre National de la Recherche Scientifique and UPMC Sorbonne Université, 4 place Jussieu, 75252 Paris Cedex 05, France
e-mail: claude.viallet@upmc.fr

© Springer Nature Switzerland AG 2020
F. Nijhoff et al. (eds.), *Asymptotic, Algebraic and Geometric Aspects of Integrable Systems*, Springer Proceedings in Mathematics & Statistics 338,
https://doi.org/10.1007/978-3-030-57000-2_2

whose properties go beyond those of the two-dimensional maps, and show that the geometry of elliptic fibrations is no longer sufficient to explain their behaviour.

Up to now the QRT mappings appear to describe almost the totality of the known integrable examples in dimension two with some notable exceptions (Viallet et al. 2004; Duistermaat 2011). However, no general framework exists for higher order maps. A generalization of the QRT scheme (Quispel et al. 1988, 1989) in dimension four was given in Capel and Sahadevan (2001). Certain maps obtained in Capel and Sahadevan (2001) were shown in Hay (2007) to be autonomous reductions of members of q-Painlevé hierarchies (*multiplicative equations* in Sakai's scheme Sakai 2001). Since hierarchies are known also for the *additive* discrete Painlevé equations (Cresswell and Joshi 1999), it is clear that the cases considered in Capel and Sahadevan (2001) cannot exhaust all the possible integrable autonomous maps in four dimensions, as already shown in Joshi and Viallet (2018). It is important to mention that there are also other examples of discrete mappings of higher orders produced either by periodic or symmetry reductions of integrable partial difference equations (Papageorgiou et al. 1990; Quispel et al. 1991; van der Kamp and Quispel 2010; Levi and Winternitz 2006) or as Kahan-Hirota-Kimura discretization (Kahan 1993; Kimura and Hirota 2000) of continuous integrable systems (Petrera and Suris 2010; Celledoni et al. 2013, 2014; Petrera et al. 2009).

In this letter, we focus on the study of integrability properties of autonomous recurrence relations. Here an autonomous recurrence relation is given by a bi-rational map of the complex projective space into itself:

$$\varphi\colon [\mathbf{x}] \in \mathbb{CP}^n \to [\mathbf{x}'] \in \mathbb{CP}^n, \tag{1}$$

where $n > 1$.[1] We take $[\mathbf{x}] = \left[x_1 : x_2 : \cdots : x_{n+1}\right]$ and $[\mathbf{x}'] = \left[x'_1 : x'_2 : \cdots : x'_{n+1}\right]$ to be homogeneous coordinates on $\mathbb{CP}^n$. Moreover we recall that a bi-rational map is a rational map $\varphi\colon V \to W$ of algebraic varieties V and W such that there exists a map $\psi\colon W \to V$, which is the *inverse* of φ in the dense subset where both maps are defined (Shafarevich 1994).

Integrability for autonomous recurrence relations (discrete equations) can be characterized in different ways. In the continuous case, for finite dimensional systems, integrability is usually understood as the existence of a "sufficiently" high number of *first integrals*, i.e. of *non-trivial* functions constant along the solution of the differential system. In the Hamiltonian setting a characterization of integrability was given by Liouville (1855). In the case of map (1) the analogue of first integrals are the *invariants*. To be more precise we state the following:

Definition 1 An invariant of a bi-rational map $\varphi\colon \mathbb{CP}^n \to \mathbb{CP}^n$ is a homogeneous function $I\colon \mathbb{CP}^n \to \mathbb{C}$ such that it is left unaltered by action of the map, i.e.

$$\varphi^*(I) = I, \tag{2}$$

where $\varphi^*(I)$ means the pullback of I through the map φ, i.e. $\varphi^*(I) = I(\varphi([\mathbf{x}]))$. For $n > 1$, an invariant is said to be *non-degenerate* if:

[1] Bi-rational maps in $\mathbb{CP}^1$ are just Möbius transformations so everything is trivial.

$$\frac{\partial I}{\partial x_1}\frac{\partial I}{\partial x_n} \neq 0. \tag{3}$$

Otherwise an invariant is said to be *degenerate*.

In what follows we will concentrate on a particular class of invariants:

Definition 2 An invariant I is said to be *polynomial*, if in the affine chart $[x_1 : \cdots : x_n : 1]$ the function I is a polynomial function.

A polynomial invariant in the sense of Definition 2 written in homogeneous variables is always a homogeneous rational function of degree 0. The form of the polynomial invariant in homogeneous coordinates is then given by:

$$I([\mathbf{x}]) = \frac{I'([\mathbf{x}])}{t^d}, \quad d = \deg I'([\mathbf{x}]), \tag{4}$$

where deg is the total degree.

To better characterize the properties of these invariants we introduce the following:

Definition 3 Given a polynomial function $F\colon \mathbb{CP}^n \to V$, where V can be either $\mathbb{CP}^n$ or $\mathbb{C}$, we define the *degree pattern* of F to be:

$$\operatorname{dp} F = \left(\deg_{x_1} F, \deg_{x_2} F, \ldots, \deg_{x_n} F\right). \tag{5}$$

Remark 1 The degree pattern of a polynomial function F is not invariant under general bi-rational transformations. However, the degree pattern of a polynomial function F is invariant under scaling and translations, which are transformations of the form:

$$\chi\colon [\mathbf{x}] \to [a\mathbf{x} + b], \quad a \in \mathbb{C}\setminus\{0\}, b \in \mathbb{CP}^n. \tag{6}$$

Example 1 Consider the following map in $\mathbb{CP}^2$:

$$\varphi\colon [x : y : t] \mapsto [-y(x^2 - t^2) + 2axt^2 : x(x^2 - t^2) : t(x^2 - t^2)]. \tag{7}$$

This map is known as the *McMillan map* (McMillan 1971) and possesses the following invariant:

$$t^4 I_{\mathrm{McM}} = x^2y^2 + (x^2 + y^2 - 2axy)t^2. \tag{8}$$

We have $\operatorname{dp} I_{\mathrm{McM}} = (2, 2)$, i.e. it is a *bi-quadratic* polynomial. We also note that the invariant of a QRT map (Quispel et al. 1988, 1989), I_{QRT}, which is a generalization of the McMillan map (7), is the ratio of two *bi-quadratics* in the dynamical variables of $\mathbb{CP}^2$. Hence QRT mappings leave invariant a pencil of curves of degree pattern (2, 2).

Example 2 The invariants of the maps presented in Capel and Sahadevan (2001), I_{CS}, are ratios of *bi-quadratics* in all the four dynamical variables of $\mathbb{CP}^4$, i.e. ratios of polynomial of degree pattern (2, 2, 2, 2). In this sense the classification of Capel and Sahadevan (2001) is an extension of the one in Quispel et al. (1988, 1989).

Finally we will consider invariants that are not of the most general kind, but satisfy the following condition.

Definition 4 We say that a invariant $I\colon \mathbb{CP}^n \to \mathbb{C}$ is *symmetric* if it is left unaltered by the following involution:

$$\iota\colon \left[x_1 : x_2 : \cdots : x_n : x_{n+1}\right] \to \left[x_n : x_{n-1} : \cdots : x_1 : x_{n+1}\right], \tag{9}$$

i.e. $\iota^* (I) = I$.

We then have the following characterization of integrability for autonomous recurrence relations:

(i) **Existence of invariants** An n-dimensional map is (super) integrable if *there exist $n-1$ invariants.*

(ii) **Liouville integrability** (Veselov 1991; Maeda 1987; Bruschi et al. 1991) An n-dimensional map (in affine coordinates) is integrable if *it preserves a Poisson structure of rank $2r$ and $r+n-2r=n-r$ functionally independent invariants in involution with respect to this Poisson structure*. In affine coordinates $\mathbf{w}=(w_{n-1},\dots,w_0)=\left[w_{n-1}:\cdots:w_0:1\right]$ we say that a map $\varphi\colon \mathbf{w}\mapsto \mathbf{w}'$ is called a Poisson map of rank $2r \le n$ if there is a skew-symmetric matrix $J(\mathbf{w})$ of rank $2r$ satisfies the Jacobian identity

$$\sum_{l=1}^{n}\left(J_{li}\frac{\partial J_{jk}}{\partial w_{l-1}}+J_{lj}\frac{\partial J_{ki}}{\partial w_{l-1}}+J_{lk}\frac{\partial J_{ij}}{\partial w_{l-1}}\right)=0,\quad \forall i,j,k, \tag{10}$$

and

$$\mathrm{d}\,\varphi J(\mathbf{w})\,\mathrm{d}\,\varphi^T = J(\mathbf{w}'), \tag{11}$$

where $\mathrm{d}\,\varphi$ is the Jacobian matrix of the map φ, see Capel and Sahadevan (2001), Olver (1986). The Poisson bracket of two smooth functions f and g is defined as

$$\{f,g\} = \nabla f\ J\ (\nabla g)^T, \tag{12}$$

where ∇f is the gradient of f. We can easily see that $\{w_{i-1},w_{j-1}\}=J_{ij}$. We note that in the case where the Poisson structure has full rank, i.e. $n=2r$, we only need $n/2$ invariants which are in involution. In this case the Poisson matrix is invertible, and its inverse is called a *symplectic matrix*. A symplectic matrix give raise to a *symplectic structure*.

(iii) **Existence of a Lax pair** (Lax 1968) An n-dimensional map is integrable if it *arises as compatibility condition of an overdetermined linear system*. We emphasize the fact that the Lax pair needs to provide us some integrability aspects of the maps such as invariants or solutions of the non-linear system. It is known in the literature that not all the Lax pairs satisfy such conditions (Calogero and Nucci 1991; Hay and Butler 2012, 2015; Gubbiotti et al. 2016). Lax pairs that do not satisfy such conditions are called *fake Lax pairs* and their existence cannot be used to prove integrability of a given system.

(iv) **Low growth condition** (Veselov 1992; Falqui and Viallet 1993; Bellon and Viallet 1999) An n-dimensional *bi-rational* map is integrable if *the degree of growth of the iterated map* φ^k *is polynomial with respect to the initial conditions* $[\mathbf{x}_0]$. Integrability is then equivalent to the vanishing of the *algebraic entropy*:

$$\varepsilon = \lim_{k\to\infty} \frac{1}{k} \log \deg_{[\mathbf{x}_0]} \varphi^k. \tag{13}$$

Algebraic entropy is a measure of the *complexity* of a map, analogous to the one introduced by Arnol'd (1990) for diffeomorphisms. In this sense growth is given by computing the number of intersections of the successive images of a straight line with a generic hyperplane in complex projective space (Veselov 1992).

We emphasize the fact that the above list is not completely exhaustive of all the possible definitions of integrability. Since we are focused on autonomous recurrence relations we choose to cover only the most used definition for these ones.

Remark 2 Here, we collect some observations about algebraic entropy and how to evaluate it.

(i) Algebraic entropy is invariant under bi-rational maps (Bellon and Viallet 1999).
(ii) In principle, the definition of algebraic entropy in Eq. (13) requires us to compute all the iterates of a bi-rational map φ to obtain the sequence $\left\{d_k = \deg_{[\mathbf{x}_0]} \varphi^k\right\}_{k=0}^{\infty}$. Fortunately, for the majority of applications the form of the sequence can be inferred by using generating functions (Lando 2003):

$$g(z) = \sum_{n=0}^{\infty} d_k z^k. \tag{14}$$

(iii) In almost all cases, the generating function turns out to be a rational function, which can be inferred from a finite number of iterates of the dynamical system. It then becomes a predictive tool, which can be tested using further terms of the sequence of degrees. In this paper, we find inferred generating functions for 4 cases given in Eqs. (23), (28), (47), and (53). In each case, the type of argument required to show that the given generating function is indeed the correct one may be found in Viallet (2015).

(iv) When a generating function is available, the algebraic entropy is then given by the logarithm of the smallest pole of the generating function, see Gubbiotti (2016), Grammaticos et al. (2009).

Remark 3 The condition of Liouville integrability (Maeda 1987; Veselov 1991; Bruschi et al. 1991) is stronger than the existence of invariants. Indeed, for a map, being measure preserving and preserving a Poisson/symplectic structure are very strong conditions. However, they lead to a great drop in the number of invariants needed for integrability. The same can be said for the existence of a Lax pair, since it is well known that a well-posed Lax pair gives all the invariants of the system through the spectral relations. Finally, the low growth condition means that the complexity of the map is very low, and it is known that invariants help in reducing the complexity of a map. Indeed the growth of a map possessing invariants cannot be generic since the motion is constrained to take place on the intersection of hypersurfaces defined by the invariants. For maps in $\mathbb{CP}^2$, it was proved in Diller and Favre (2001) that the growth can be only bounded, linear, quadratic or exponential. Linear cases are trivially integrable in the sense of invariants. We note that for polynomial maps, it was already known from Veselov (1992) that the growth can be only linear or exponential. It is known that QRT mappings and other maps with invariants in $\mathbb{CP}^2$ possess quadratic growth (Duistermaat 2011), so the two notions are actually equivalent for a large class of integrable systems.

Now we discuss briefly the concept of *duality* for rational maps, which was introduced in Quispel et al. (2005). Let us assume that our map φ possesses L independent invariants, i.e. I_j for $j \in \{1, \dots, L\}$. Then we can form the linear combination:

$$H = \alpha_1 I_1 + \cdots + \alpha_L I_L. \tag{15}$$

For an unspecified autonomous recurrence relation

$$\left[x_1 : x_2 : \cdots : x_{n+1}\right] \mapsto \left[x_1' : x_1 : \cdots : x_n\right], \tag{16}$$

we can write down the invariant condition for H (15):

$$\widehat{H}(x_1', [\mathbf{x}]) = H\left([\mathbf{x}']\right) - H\left([\mathbf{x}]\right) = 0. \tag{17}$$

Since we know that $[\mathbf{x}'] = \varphi\left([\mathbf{x}]\right)$ is a solution of (17) we have the following factorization:

$$\widehat{H}(x_1', [\mathbf{x}]) = A\left(x_1', [\mathbf{x}]\right) B\left(x_1', [\mathbf{x}]\right). \tag{18}$$

We can assume without loss of generality that the map φ corresponds to the annihilation of A in (18). Now since $\deg_{x_1'} \widehat{H} = \deg_{x_1} H$ and $\deg_{x_n} \widehat{H} = \deg_{x_n} H$ we have that if $\deg_{x_1} H, \deg_{x_n} H > 1$ the factor B in (18) is non constant.[2] In general, since

[2] We remark that this assertion is possible because we are assuming that all the invariants are non-degenerate. It is easy to see that degenerate invariants can violate this property.

the map φ is bi-rational, we have the following equalities:

$$\deg B_{x_1'} = \deg_{x_1'} \widehat{H} - \deg_{x_1'} A = \deg_{x_1} H - 1, \tag{19a}$$

$$\deg B_{x_n} = \deg_{x_n} \widehat{H} - \deg_{x_n} A = \deg_{x_n} H - 1. \tag{19b}$$

Therefore we have that if $\deg_{x_1} H, \deg_{x_n} H > 2$, the annihilation of B does not define a bi-rational map in general, but may define an algebraic one. However when $\deg_{x_1} H, \deg_{x_n} H = 2$ the annihilation of B defines a bi-rational projective map. We call this map the *dual map* and we denote it by $\varphi^{\vee}$.

Remark 4 We note that in principle for $\deg_{x_1} H = \deg_{x_n} H = d > 2$, more general factorizations can be considered:

$$\widehat{H}\left(x_1', [\mathbf{x}]\right) = \prod_{i=1}^{d} A_i\left(x_1', [\mathbf{x}]\right), \tag{20}$$

but we will not consider this case here.

Now assume that the invariants (and hence the map φ) depend on some *arbitrary constants* $I_i = I_i([\mathbf{x}]; a_i)$, for $i = 1, \ldots, M$. Choosing some of the a_i in such a way that there remains M arbitrary constants and such that for a subset a_{i_k} we can write Eq. (15) in the following way:

$$H = a_{i_1} J_1 + a_{i_2} J_2 + \cdots + a_{i_K} J_{a_{i_K}}, \tag{21}$$

where $J_i = J_i([\mathbf{x}])$, $i = 1, 2, \ldots, K$ are new functions. The parameters a_{i_k} do not appear in the dual maps in the same way as the parameters α_i do not appear in the main maps. Therefore, using the factorization (18) the J_i functions are invariants for the dual maps.

Remark 5 In fact, one can consider more general combinations than linear combinations given in (15) and (21). However, we only consider those linear combinations given (15) and (21) in this paper.

It is clear from Eq. (21) that even though the dual map is naturally equipped with some invariants, it is not *necessarily* equipped with a sufficient number of invariants to claim integrability. In fact there exist examples of dual maps with any possible behaviour, integrable, superintegrable and non-integrable (Joshi and Viallet 2018; Gubbiotti et al. 2020).

In a recent paper (Joshi and Viallet 2018), the authors considered the *autonomous limit* of the second member of the $\mathrm{d}P_{\mathrm{I}}$ and $\mathrm{d}P_{\mathrm{II}}$ hierarchies (Cresswell and Joshi 1999). We will denote these equations as $\mathrm{d}P_{\mathrm{I}}^{(2)}$ and $\mathrm{d}P_{\mathrm{II}}^{(2)}$ equations. These $\mathrm{d}P_{\mathrm{I}}^{(2)}$ and $\mathrm{d}P_{\mathrm{II}}^{(2)}$ equations are given by autonomous recurrence relations of order four, and showed to be integrable according to the algebraic entropy approach. They showed that both maps possess two invariants, one of degree pattern $(1, 3, 3, 1)$ and one of degree pattern $(2, 4, 4, 2)$. Using these invariants, they showed that the dual maps of the $\mathrm{d}P_{\mathrm{I}}^{(2)}$ and $\mathrm{d}P_{\mathrm{II}}^{(2)}$ equations are integrable according to the algebraic entropy test and moreover, produced some invariants, showing that these dual maps were actually superintegrable. Finally they gave a scheme to construct autonomous recurrence relations with the assigned degree pattern $(1, 3, 3, 1)$ associated with I_{low} and $(2, 4, 4, 2)$ associated with I_{high} and they provided some new examples out of this construction.

In (Gubbiotti et al. 2020) we consider the problem of finding all fourth order birational maps $\varphi\colon [x : y : z : u : t] \mapsto [x' : y' : z' : u' : t']$ possessing a polynomial symmetric invariant I_{low} such that $\mathrm{dp}\, I_{\mathrm{low}} = (1, 3, 3, 1)$ *where the only non-zero coefficients are those appearing in the* $(1, 3, 3, 1)$ *invariant of both the* $\mathrm{d}P_{\mathrm{I}}^{(2)}$ and $\mathrm{d}P_{\mathrm{II}}^{(2)}$ *equation*, and such that φ possesses a polynomial symmetric invariant I_{high} such that $\mathrm{dp}\, I_{\mathrm{high}} = (2, 4, 4, 2)$. The two invariants I_{low} and I_{high} are assumed to be functionally independent and non-degenerate. Within this class we have found the known $\mathrm{d}P_{\mathrm{I}}^{(2)}$ and $\mathrm{d}P_{\mathrm{II}}^{(2)}$ equations as well as new examples of maps with these properties.

In this letter we will present in detail four particular examples of this class. In Sect. 2, we will discuss two pairs of main-dual maps. We will discuss the integrability property of these maps in light of their invariants and of their growth. We will present maps possessing two invariants and integrable according to the algebraic entropy test with *cubic growth.* This implies that another rational invariant cannot exist. Indeed, the orbits of superintegrable maps with rational invariant are confined to *elliptic curves* and the growth is at most *quadratic* (Bellon 1999; Gizatullin 1980). From this general statement follows that a four-dimensional map with cubic growth can possess at most two rational invariants. We note that some examples of cubic growth were already presented in Joshi and Viallet (2018). However, it was pointed out that these examples can be deflated to lower dimensional maps with quadratic growth. This also holds for our maps, i.e. we can deflate them to integrable maps in lower dimension. Furthermore, we will present a map with two invariants and *exponential growth,* that is non-integrable according to the algebraic entropy test. We discuss some possible reasons why this map is non-integrable even though it possesses two invariants. In the final Section, we will give some conclusions and an outlook on the future perspectives of this approach.

2 Notable Examples

In this section we discuss two pairs of maps, which arise as part of a systematic classification to be presented in Gubbiotti et al. (2020). The interest in these particular maps arises since the relation between their invariants and growth properties is non trivial. In both cases the main maps possess two functionally independent invariants, but they behave differently. One map has *cubic* degree growth, while the other one has *exponential* degree growth. Therefore, even though these two maps have the same number of invariants with the same degree patterns, one map is integrable and the other one is non-integrable. In addition, in both cases the degree growth property of the dual maps reflect the growth of the main map. However, we note that the degree growth of the dual map does not always reflect that of the main map (Gubbiotti et al. 2020).

2.1 (P.i) and Its Dual Map (Q.i)

Consider the map $[\mathbf{x}] \mapsto \varphi_i([\mathbf{x}]) = [\mathbf{x}']$ given as follows:

$$\begin{aligned} x' &= \{[\nu t^2(x+z) + uz^2]y + t^2\mu uz + (x+z)^2y^2\}d - at^4, \\ y' &= x^2 d(t^2\mu + xy), \quad z' = yxd(t^2\mu + xy), \\ u' &= zxd(t^2\mu + xy), \quad t' = txd(t^2\mu + xy). \end{aligned} \tag{P.i}$$

This map depends on four parameters a, d and μ, ν.

From the construction in Gubbiotti et al. (2020) we know that the map (P.i) possesses the following invariants:

$$\begin{aligned} t^6 I_{\text{low}}^{\text{P.i}} &= at^4yz + d\left[\nu y^2z^2 - yz(ux - uz - xy)\mu\right]t^2 \\ &\quad - y^2z^2d(ux - xy - yz - uz), \end{aligned} \tag{22a}$$

$$\begin{aligned} t^8 I_{\text{high}}^{\text{P.i}} &= [(uz + xy - yz)\mu - \nu yz]\,at^6 \\ &\quad + \left[yz(xy + yz + uz)a + d\mu^2(uz + xy - yz)^2\right. \\ &\quad \left.+ 2d\mu\nu yz(ux - yz) - d\nu^2y^2z^2\right]t^4 \\ &\quad + \left[2dzy(uz + xy - yz)(xy + yz + uz)\mu + 2dy^2z^2\nu ux\right]t^2 \\ &\quad + dy^2z^2(xy + yz + uz)^2. \end{aligned} \tag{22b}$$

Moreover, the map (P.i) has the following degrees of iterates:

$$\begin{aligned} \{d_n\}_{\text{P.i}} = {} & 1, 4, 12, 28, 52, 86, 130, 188, 260, 348, 452, \\ & 576, 720, 886, 1074, 1288, 1528, 1796, 2092\ldots \end{aligned} \tag{23}$$

The sequence (23) is fitted by:

$$g_{\text{P.i}}(s) = \frac{s^7 - 3s^6 + s^5 - s^4 + 3s^3 + 3s^2 + s + 1}{(s+1)(s^2+1)(s-1)^4}. \tag{24}$$

This generating function can be found by using the first 15 iterates, and its validity can be confirmed by using further iterates. See Remark 2 for the justification.

Due to the presence of $(s-1)^4$ in the denominator we have that the growth of the map (P.i) is fitted by a *cubic polynomial*. As discussed in the Introduction this means

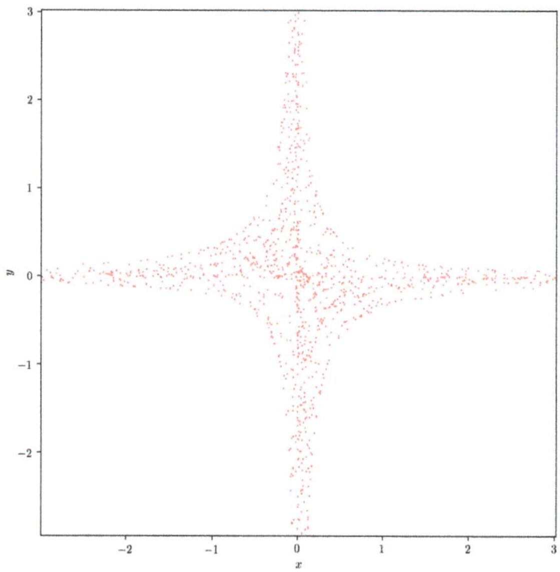

Fig. 1 Affine orbit of equation (P.i) with parameters $a = 6$, $\mu = 3$, $\nu = 4$ and $d = 6$ and initial conditions $(x, y, z, u) = (0.02, 0.05, 0.06, 0.07)$

at once that the map is integrable according to the algebraic entropy test and that another rational invariant cannot exist. This suggests that the geometry of the orbits of the map (P.i) is nontrivial, and goes beyond the existence of *elliptic fibrations*.

Explicit numerical calculations and drawings suggest that in the case of map (P.i), no additional invariant exists. Indeed, if an additional third invariant, even algebraic, existed then all the orbits of equation (P.i) would lie on a curve. On the other hand referring to Fig. 1 we see that a generic orbit of equation (P.i) does not lie on a curve. This implies that no such invariant exists.

The dual map $[\mathbf{x}] \mapsto \varphi_i^\vee([\mathbf{x}]) = [\mathbf{x}']$ of (P.i) is given by:

$$\begin{aligned} x' &= [\beta(2xy - 2yz + uz)\mu + (\beta\nu - \alpha)y(x - z)]\, t^2 \\ &\quad + \beta y(z^2 y - x^2 y + uz^2), \\ y' &= x^2\beta(t^2\mu + xy), \quad z' = yx\beta(t^2\mu + xy), \\ u' &= zx\beta(t^2\mu + xy), \quad t' = tx\beta(t^2\mu + xy). \end{aligned} \tag{Q.i}$$

This map depends on four parameters α, β, and μ, ν. The parameters μ and ν are shared with the main map (P.i).

The main map (P.i) possesses two invariants and depends on a and d whereas the dual map (Q.i) does not depend on them. Then according to (21) we can write down the invariants for the dual map (Q.i) as:

$$\alpha I_{\text{low}}^{\text{P.i}} + \beta I_{\text{high}}^{\text{P.i}} = a I_{\text{low}}^{\text{Q.i}} + d I_{\text{high}}^{\text{Q.i}}. \tag{25}$$

Therefore, we obtain the following expressions:

$$
\begin{aligned}
t^4 I_{\text{low}}^{\text{Q.i}} &= [yz\alpha + (\mu xy - yz\mu - yvz + \mu uz)\beta]t^2 \\
&\quad + \beta yz(xy + yz + uz),
\end{aligned} \tag{26a}
$$

$$
\begin{aligned}
t^8 I_{\text{high}}^{\text{Q.i}} &= \left\{\left[y^2z^2\nu - yz(ux - uz - xy)\mu\right]\alpha \right.\\
&\quad \left. + \left[(uz + xy - yz)^2\mu^2 + 2yz(ux - yz)\nu\mu - \nu^2y^2z^2\right]\beta\right\} t^4 \\
&\quad + \left\{z^2y^2(xy + yz - ux + uz)\alpha \right.\\
&\quad \left. + \left[2yz(uz + xy - yz)(xy + yz + uz)\mu + 2y^2z^2\nu ux\right]\beta\right\} t^2 \\
&\quad + z^2y^2(xy + yz + uz)^2\beta.
\end{aligned} \tag{26b}
$$

We remark that the invariant (26a) has degree pattern $(1, 2, 2, 1)$ which differs from dp $I_{\text{low}}^{\text{P.i}}$.

The map (Q.i) has the following degrees of iterates:

$$
\{d_n\}_{\text{Q.i}} = 1, 4, 12, 26, 48, 78, 118, 170, 234, 312, 406, 516, 644, 792 \ldots \tag{27}
$$

fitted by the generating function:

$$
g_{\text{Q.i}}(s) = \frac{(s^3 - 2s^2 - 1)(s^3 - s^2 - s - 1)}{(s^2 + s + 1)(s - 1)^4}. \tag{28}
$$

This means that the dual map is integrable according to the algebraic entropy test with *cubic* growth, just like the main map. See Remark 2 for the justification.

Explicit numerical calculations and drawings suggest that also in the case of map (Q.i), no additional invariant exists. Indeed, if an additional third invariant, even algebraic, existed then all the orbits of equation (Q.i) would lie on a curve. In this case we are actually able to find some orbits lying on a curve, see Fig. 2b. However,

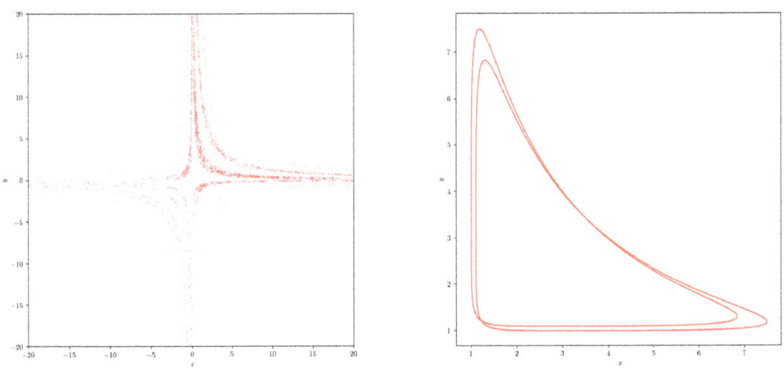

(a) Parameters $\alpha = 3$, $\mu = 3$, $\nu = 7$ and $\beta = 3$.

(b) Parameters $\alpha = 3$, $\mu = 6$, $\nu = 8$ and $\beta = 9$.

Fig. 2 Affine orbit of equation (Q.i) with different parameters but the same initial conditions $(x, y, z, u) = (3, 4, 1, 3)$

it is possible to find orbits of equation (Q.i) that do not lie on a curve. An example of such orbit is shown in Fig. 2a. Therefore, we can conclude that a *globally defined* third invariant does not exist. The existence of some closed orbits like in Fig. 2b suggests the existence of a non-analytic invariant existing only in some regions of the space.

Therefore, the pair of main-dual maps (P.i) and (Q.i) consists of two integrable equations with non-standard degree of growth. However, as remarked above the degree pattern of the invariants of the maps (P.i) and (Q.i) differ slightly.

We now consider the maps (P.i) and (Q.i) in affine coordinates, which are given by

$$\varphi : (w_3, w_2, w_1, w_0) \mapsto (w_4, w_3, w_2, w_1), \tag{29}$$

where

$$w_4 = \frac{N_1}{d w_3 (w_2 w_3 + \mu)}, \tag{AP.i}$$

$$w_4 = \frac{N_2}{\beta\, w_3 (w_2 w_3 + \mu)}, \tag{AQ.i}$$

with

$$\begin{aligned} N_1 &= -d\big(w_0 w_1^2 w_2 + w_1^2 w_2^2 + 2 w_1 w_2^2 w_3 + w_2^2 w_3^2 + \mu w_0 w_1 \\ &\quad + \nu w_1 w_2 + \nu w_2 w_3\big) - a, \end{aligned} \tag{30}$$

$$\begin{aligned} N_2 &= \beta w_0 w_1^2 w_2 + \beta \mu w_0 w_1 + \beta w_1^2 w_2^2 + (\alpha - 2\beta\mu - \beta\nu)\, w_1 w_2 \\ &\quad - \beta w_3^2 w_2^2 + (2\beta\mu + \beta\nu - \alpha)\, w_2 w_3. \end{aligned} \tag{31}$$

Invariants for these maps are obtained from I_{low} and I_{high} respectively by taking $t = 1$, $u = w_0$, $z = w_1$, $y = w_2$, and $x = w_3$.

We note that when a Poisson structure has the full rank, using Eq. (11), one gets

$$\left[\det(\mathrm{d}\,\varphi)\right]^2 = \frac{\det\big(J(\mathbf{w}')\big)}{\det\big(J(\mathbf{w})\big)}. \tag{32}$$

This implies that the map φ is either volume or anti-volume preserving.

We recall that a map φ is called (anti) volume preserving if there is a function $\Omega(\mathbf{w})$ such that the following volume form is preserved

$$\Omega(\mathbf{w})\,\mathrm{d}\,w_0 \wedge \mathrm{d}\,w_1 \wedge \cdots \wedge \mathrm{d}\,w_{n-1} = \pm \Omega(\mathbf{w}')\,\mathrm{d}\,w_0' \wedge \mathrm{d}\,w_1' \wedge \cdots \wedge \mathrm{d}\,w_{n-1}'. \tag{33}$$

Thus, we can write

$$\frac{\partial\big(w_0', w_1', \ldots, w_{n-1}'\big)}{\partial\big(w_0, w_1, \ldots, w_{n-1}\big)} = \pm \frac{\Omega(\mathbf{w})}{\Omega(\mathbf{w}')}, \tag{34}$$

where the left hand side is the determinant of the Jacobian matrix of the map φ. In Byrnes et al. (1999) it was proved that if a map in n dimension is (anti) volume preserving and possesses $n-2$ invariant, then we can construct an (anti) Poisson structure of rank 2 from these invariants. However, these invariants turn out to be Casimirs (functions that Poisson commute with all other functions) of this Poisson bracket. Therefore, in order to have Liouville integrability, we need an extra invariant in addition to the known $n-2$ invariants if we want to use Poisson structures constructed this way. In other words, the map is super integrable. Thus, to discuss Liouville integrability of the maps (AP.i) and (AQ.i), we need to find a Poisson bracket of rank 4 as we already predicted that the third invariant does not exist. We do not have that information for these maps but we can show they reduce to three dimensional Liouville integrable maps via a process called deflation (Joshi and Viallet 2018). *Mutatis mutandis*, this process will preserve the invariants, and in dimension three two invariants are sufficient to claim integrability in the general sense as discussed in the Introduction.

It is easy to check that the maps (AP.i) and (AQ.i) are volume and anti-volume preserving, respectively, with respect to the same volume form:

$$\Omega = w_1 w_2 (w_1 w_2 + \mu). \tag{35}$$

We now construct the (anti) Poisson structures for these two maps following (Byrnes et al. 1999). We consider the dual multi-vector of the volume form

$$\tau = m \frac{\partial}{w_0} \wedge \frac{\partial}{w_1} \wedge \frac{\partial}{w_2} \wedge \frac{\partial}{w_3}, \tag{36}$$

where $m = 1/\Omega$. A degenerate Poisson structure for the map (AP.i) and a degenerate anti-Poisson structure for the map (AQ.i) are given by the following contraction

$$J = \tau \rfloor \mathrm{d}\, I_{\text{low}} \rfloor \mathrm{d}\, I_{\text{high}}, \tag{37}$$

where I_{low} and I_{high} are invariants for these maps in affine coordinates. Since these (anti) Poisson structures are quite big, we do not present them here.

Remark 6 The Poisson structures which can be constructed using the method of Byrnes et al. (1999) are degenerate and cannot be used to explain the integrability of the two maps (AP.i) and (AQ.i).

We also note that the maps (AP.i) and (AQ.i) can be reduced to three dimensional maps using a deflation $v_i = w_i w_{i+1}$. The recurrences for these maps are denoted by (DP.i) and (DQ.i) and are given as follows

$$d\mu\,(v_0 + v_3) + d\nu\,(v_1 + v_2) + d\left(v_0 v_1 + v_1^2 + 2 v_1 v_2 + v_2^2 + v_2 v_3\right) + a = 0, \tag{DP.i}$$

$$\beta\mu\left(-v_0+2\beta v_1-2\beta v_2+v_3\right)+\left(\beta\nu-\alpha\right)\left(v_1-v_2\right) \\ +\beta\left(-v_0v_1-v_1^2+v_2^2+v_2v_3\right)=0. \tag{DQ.i}$$

Each of the maps (DP.i) and (DQ.i) has two functionally independent invariants which can be obtained directly from I_{low} and I_{high} even though they live in a different space. One can check that the map (DP.i) and (DQ.i) are anti-volume preserving and volume preserving with $\Omega = v_1 + \mu$. Therefore, we can construct their (anti) Poisson structure using the three dimensional version of (37). Using the following invariant from I_{low} for (DP.i)

$$I_1^{\text{P.i}} = d\mu v_0v_1 - d\mu v_0v_2 + d\mu v_1v_2 + d\nu {v_1}^2 + dv_0{v_1}^2 - dv_0v_1v_2 \\ + d{v_1}^3 + d{v_1}^2v_2 + av_1, \tag{38}$$

we have found that the map (dP.i) has an anti-Poisson structure given by

$$J_{12}^{\text{P.i}} = d(v_1 - v_0),\ J_{2,3}^{\text{P.i}} = d(v_1 - v_2),$$
$$J_{13}^{\text{P.i}} = \frac{-d\mu v_0 - d\mu v_2 - 2d\nu v_1 - 2dv_0v_1 + dv_0v_2 - 3d{v_1}^2 - 2dv_1v_2 - a}{\mu + v_1}.$$

Similarly, for the map (DQ.i) we obtain the invariant

$$I_1^{\text{DQ.i}} = \beta\mu v_0 - \beta\mu v_1 + \beta\mu v_2 - \nu\beta v_1 + \beta v_0v_1 + \beta{v_1}^2 + \beta v_1v_2 + \alpha v_1, \tag{39}$$

and the corresponding Poisson structure

$$J^{\text{Q.i}} = \begin{bmatrix} 0 & \beta & \dfrac{\beta\left(\mu+\nu-v_0-2v_1-v_2\right)-\alpha}{\mu+v_1} \\ -\beta & 0 & \beta \\ -\dfrac{\beta\left(\mu+\nu-v_0-2v_1-v_2\right)-\alpha}{\mu+v_1} & -\beta & 0 \end{bmatrix}. \tag{40}$$

For these constructions, $I_1^{\text{P.i}}$ and $I_1^{\text{Q.i}}$ are Casimirs for their associated (anti) Poisson structures. Their second (anti) Poisson structures can be obtained from the invariant I_{high} but we do not present here as they are quite big.

It is important to note that the (anti) Poisson structures of (AP.i) and (AQ.i) under inflation give us the trivial Poisson structures for (DP.i) and (DQ.i), i.e. $J = \mathbf{0}$, where $\mathbf{0}$ is the zero matrix. On the other hand, from the common factor that appears in the Poisson structure of (AP.i), we have found that there exists an anti-invariant $K^{\text{P.i}}$ for this map, i.e. $K^{\text{P.i}}(\mathbf{w}) = -K^{\text{P.i}}(\mathbf{w}')$where

$$K^{\text{P.i}} = 2d\left(w_2w_1^2w_0 + w_2^2w_1^2 + w_1w_2^2w_3 + \mu w_0w_1 - \mu w_1w_2 \\ + \mu w_2w_3 + \nu w_1w_2\right) + a. \tag{41}$$

However, $K^{\text{P.i}}$ is not independent of $I_{low}^{\text{P.i}}$ and $I_{high}^{\text{P.i}}$ since we have

$$\left(K^{\text{P.i}}\right)^2 - 4d\, I_{\text{high}}^{\text{P.i}} - 8d\, \nu I_{\text{low}}^{\text{P.i}} = a^2. \tag{42}$$

Using this anti-invariant, we obtain the following anti-invariant for the map (DP.i)

$$K^{\text{DP.i}} = 2d\mu v_0 - 2d\mu v_1 + 2d\mu v_2 + 2d\nu v_1 + 2d v_0 v_1 + 2d {v_1}^2 + 2d v_1 v_2 + a. \tag{43}$$

Therefore, using this anti-invariant, we get a Poisson structure for (DP.i) as follows (after factoring out a constant term)

$$J_2^{\text{P.i}} = \begin{bmatrix} 0 & 1 & \dfrac{\mu - \nu - v_0 - 2v_1 - v_2}{\mu + v_1} \\ -1 & 0 & 1 \\ -\dfrac{\mu - \nu - v_0 - 2v_1 - v_2}{\mu + v_1} & -1 & 0 \end{bmatrix}. \tag{44}$$

We can check directly that the invariants inherited from the affine map (AP.i) are in involution with respect to the Poisson structure (44). In the sense of the definition given in the Introduction, this means that the reduced maps (DP.i) and (DQ.i) are Liouville integrable.

Remark 7 We notice that we can always use the invariants (38) and (39) to reduce the three dimensional maps (DP.i) and (DQ.i) to two dimensional maps and relate them to QRT maps. To be more specific we have that the reduced map of (DQ.i) preserves a bi-quadratic curve so that it is of the QRT type. On the other hand, using the anti-invariant, the reduced map of (DP.i) sends a bi-quadratic to another bi-quadratic and fits in the framework of Roberts and Jogia (2015).

2.2 (P.ii) and Its Dual Map (Q.ii)

Consider the map $[\mathbf{x}] \mapsto \varphi_{\text{ii}}([\mathbf{x}]) = [\mathbf{x}']$ given as follows:

$$\begin{aligned} x' &= \left[(x^2 + z^2)y - uz^2\right]\mu - t^2(u - 2y), \\ y' &= x(t^2 + \mu x^2), \quad z' = y(t^2 + \mu x^2), \\ u' &= z(t^2 + \mu x^2), \quad t' = t(t^2 + \mu x^2). \end{aligned} \tag{P.ii}$$

This map only depends on the parameter μ.

From the construction in Gubbiotti et al. (2020) we know that the map (P.ii) has the following invariants:

$$
\begin{aligned}
t^5 I_{\text{low}}^{\text{P.ii}} &= (x-z)(u-y)\left(t^2+z^2\mu\right)\left(\mu y^2+t^2\right), && \text{(45a)}\\
t^6 I_{\text{high}}^{\text{P.ii}} &= \left[(x-z)^2 y^4 + y^2 z^4 - 2yz^4 u + u^2 z^4\right]\mu^2 \\
&\quad + 2t^2\left[\left(x^2-2xz+2z^2\right)y^2 - 2yz^2u + u^2z^2\right]\mu && \text{(45b)}\\
&\quad + t^4\left(z^2+u^2+x^2+y^2-2uy-2xz\right).
\end{aligned}
$$

Moreover, the map (P.ii) has the following degrees of iterates:

$$\{d_n\}_{\text{P.ii}} = 1, 3, 9, 21, 45, 93, 189, 381, 765, 1533\dots \tag{46}$$

fitted by the generating function:

$$g_{\text{P.ii}}(s) = \frac{1+2s^2}{(2s-1)(s-1)}. \tag{47}$$

This means that despite the existence of the two invariants (45) the map (P.ii) is non-integrable according to the algebraic entropy test: its entropy is positive and given by $\varepsilon = \log 2$. See Remark 2 for the justification.

Therefore we have that the map (P.ii) is an example of non-integrable admitting two invariants.

Again following (Byrnes et al. 1999) we can produce a Poisson structure of rank 2 for (P.ii) as the affine version of (P.ii) is volume preserving with $\Omega = (1+\mu w_1^2)(1+\mu w_2^2)$, where we have taken $t=1$, $u=w_0$, $z=w_1$, $y=w_2$, and $x=w_3$. By the construction, the two invariants (45) become Casimir functions for it, so again the existence of such Poisson structure does not imply any form of Liouville integrability. However, we notice that there are common factors appear at every non-zero entry of this structure. Thus, we have found the following anti-invariant for the map (P.ii) using these common factors

$$
\begin{aligned}
K^{\text{P.ii}} &= \left[\mu\left(w_0w_1^2 - {w_1}^2 w_2 - w_1 w_2^2 + w_2^2 w_3\right) + w_0 - w_1 - w_2 + w_3\right]\times \\
&\qquad \left[\mu\left(w_0 w_1^2 - w_1^2 w_2 + w_1 w_2^2 - w_2^2 w_3\right) + w_0 + w_1 - w_2 - w_3\right] \qquad \text{(48)}\\
&= F_1 F_2.
\end{aligned}
$$

This suggests that we should check each factor of $K^{P.ii}$ to see whether they are (anti) invariants of (P.ii). By direct calculation we can see that the first factor F_1 is an anti-invariant and F_2 is an invariant for (P.ii), but they are not functionally independent of I_{low} and I_{high}. In fact, their relations are

$$I_{\text{high}}^{\text{P.ii}} - F_1^2 + 2I_{\text{low}}^{\text{P.ii}} = 0, \text{ and } I_{\text{high}}^{\text{P.ii}} - F_2^2 - 2I_{\text{low}}^{\text{P.ii}} = 0. \tag{49}$$

Therefore, the map (P.ii) actually has two invariants of degrees $(1, 2, 2, 1)$ and $(1, 3, 3, 1)$. Nevertheless, despite the existence of such invariants the map (P.ii) is non-integrable in the sense of the algebraic entropy.

Remark 8 We can use F_1 and F_2 to construct an anti-Poisson structure for (P.ii) using the formula (37):

$$\begin{gathered} J_{1,2}=-1,\ J_{2,3}=1,\ J_{3,4}=-1,\\ J_{1,3}=\frac{2\mu w_1\left(w_2-w_0\right)}{\mu w_1^2+1},\ J_{2,4}=-\frac{2\mu w_2\left(w_3-w_1\right)}{\mu w_2^2+1},\\ J_{1,4}=-\frac{\mu^2 w_1 w_2\left[4\left(w_0 w_1-w_0 w_3+w_2 w_3\right)-3 w_1 w_2\right]+\mu\left(w_1^2+w_2^2\right)+1}{\left(\mu w_1^2+1\right)\left(\mu w_2^2+1\right)}. \end{gathered} \tag{50}$$

We have checked that F_2 and $I_{\mathrm{low}}^{P.ii}$ are in involution with respect to this anti-Poisson structure. A Poisson structure can be obtained by multiplying this anti-Poisson structure with the anti-invariant F_1.

The dual map $[\mathbf{x}] \mapsto \varphi_{ii}^{\vee}([\mathbf{x}])=[\mathbf{x}']$ of (P.ii) is given as follows:

$$\begin{aligned} x' &= \alpha\left[\left(x^2-z^2\right)y+uz^2\right]\mu+t^2\alpha u+\beta y^2\left(x-z\right)\mu \\ &\quad +t^2\beta\left(x-z\right), \\ y' &= \alpha x\left(t^2+\mu x^2\right),\quad z'=\alpha y\left(t^2+\mu x^2\right), \\ u' &= \alpha z\left(t^2+\mu x^2\right),\quad t'=\alpha t\left(t^2+\mu x^2\right). \end{aligned} \tag{Q.ii}$$

This map depends on three parameters α, β and μ. The parameter μ is shared with the main map (P.ii).

Since the main map (P.ii) possesses two invariants depending only on one parameter μ then according to (21) we can write down only a single invariant for the dual map (Q.ii):

$$I^{\mathrm{Q.ii}}=\alpha I_{\mathrm{high}}^{\mathrm{P.ii}}+\beta I_{\mathrm{low}}^{\mathrm{P.ii}}. \tag{51}$$

The invariant (51) has degree pattern (2, 4, 4, 2).

We have then that the dual map (Q.ii) has the following fast-growing degrees of iterates:

$$\{d_n\}_{\mathrm{Q.ii}}=1,3,9,21,45,93,189,381,765,1533,3069\ldots. \tag{52}$$

The growth of degrees evident in (52) is clearly exponential and is fitted by the generating function

$$g_{\mathrm{Q.ii}}(s)=\frac{1+2s^2}{(2s-1)(s-1)}. \tag{53}$$

This confirms that the algebraic entropy is positive and equal to $\varepsilon=\log 2$. See Remark 2 for the justification.

This means that the dual map is non-integrable with same rate of growth as the main map. In this case we can show that the map is anti-volume preserving with the same measure as the main map (P.ii). Moreover, we proved that the map (Q.ii) do not posses any addition invariant up to order 14. Therefore at the present stage we cannot construct a Poisson structure using the method of Byrnes et al. (1999).

3 Conclusions and Outlook

In this letter, we gave some examples of fourth order bi-rational maps with two invariants possessing interesting degree growth properties. These examples come from our classification of all the fourth-order autonomous recurrence relations possessing two invariants in a given class of degree patterns (Gubbiotti et al. 2020).

The first pair of bi-rational maps is given by the map (P.i) and its dual (Q.i) and consists of integrable maps with cubic growth. The interest in maps with cubic growth arises from geometrical considerations: maps with polynomial but higher than quadratic growth, can arise only in dimension greater than two (Diller and Favre 2001) and prove, in the case of superintegrable maps, the existence of non-elliptic fibrations of invariant varieties (Bellon and Viallet 1999). The interest in maps with this type of growth arose recently following the examples given in Joshi and Viallet (2018) and we expect them to lead to many new and interesting geometric structures.

The second pair of fourth order bi-rational maps, given by the map (P.ii) and its dual (Q.ii), consists of non-integrable maps with exponential growth. There are various possible reasons why the map (P.ii) is non-integrable despite possessing two invariants. To claim integrability with two invariants according to the discrete Liouville theorem (Maeda 1987; Bruschi et al. 1991; Veselov 1991) we need to prove that the map has a symplectic structure and that the two invariants commute with respect to this symplectic structure. Hence, either the map (P.ii) does not admit *any* symplectic structure, or the map (P.ii) admits only symplectic structures such that the two invariants (45) do not commute. Since, usually, from a set of non-commuting invariants it is possible to find a set of functionally independent commuting invariants we conjecture that Eq. (P.ii) is devoid of a non-degenerate Poisson structure.

Work is in progress to characterize the surfaces generated by the invariants in both integrable and non-integrable cases. We expect this to give new results in the geometric theory of integrable systems.

Acknowledgements The research reported in this paper was supported by an Australian Laureate Fellowship #FL120100094 and grant #DP160101728 from the Australian Research Council. CMV would like to thank the Sydney Mathematics Research Institute at the University of Sydney for its hospitality and support.

References

Arnol'd, V.I.: Dynamics of complexity of intersections. Bol. Soc. Bras. Mat. **21**, 1–10 (1990)

Bellon, M., Viallet, C.-M.: Algebraic entropy. Comm. Math. Phys. **204**, 425–437 (1999)

Bellon, M.P.: Algebraic entropy of birational maps with invariant curves. Lett. Math. Phys. **50**(1), 79–90 (1999)

Bruschi, M., Ragnisco, O., Santini, P.M., Tu, G.-Z.: Integrable symplectic maps. Physica D **49**(3), 273–294 (1991)

Byrnes, G.B., Haggar, F.A., Quispel, G.R.W.: Sufficient conditions for dynamical systems to have pre-symplectic or pre-implectic structures. Physica A **272**, 99–129 (1999)

Calogero, F., Nucci, M.C.: Lax pairs galore. J. Math. Phys. **32**(1), 72–74 (1991)

Capel, H.W., Sahadevan, R.: A new family of four-dimensional symplectic and integrable mappings. Physica A **289**, 80–106 (2001)

Celledoni, E., McLachlan, R.I., McLaren, D.I., Owren, B., Quispel, G.R.W.: Integrability properties of Kahan's method. J. Phys. A: Math. Theor. **47**(36), 365202 (2014)

Celledoni, E., McLachlan, R.I., Owren, B., Quispel, G.R.W.: Geometric properties of Kahan's method. J. Phys. A: Math. Theor. **46**(2), 025201 (2013)

Cresswell, C., Joshi, N.: The discrete first, second and thirty-fourth Painlevé hierarchies. J. Phys. A: Math. Gen. **32**, 655–669 (1999)

Diller, J., Favre, C.: Dynamics of bimeromorphic maps of surfaces. Amer. J. Math. **123**(6), 1135–1169 (2001)

Duistermaat, J.J.: Discrete Integrable Systems: QRT Maps and Elliptic Surfaces. Springer Monographs in Mathematics. Springer, New York (2011)

Falqui, G., Viallet, C.-M.: Singularity, complexity, and quasi-integrability of rational mappings. Comm. Math. Phys. **154**, 111–125 (1993)

Gizatullin, MKh: Rational g-surfaces. Izv. Akad. Nauk SSSR Ser. Mat. **44**, 110–144 (1980)

Grammaticos, B., Halburd, R.G., Ramani, A., Viallet, C.-M.: How to detect the integrability of discrete systems. J. Phys A: Math. Theor. **42**, 454002 (41 pp) (2009). Newton Institute Preprint NI09060-DIS

Gubbiotti, G.: Integrability of difference equations through algebraic entropy and generalized symmetries, chap. 3. In: Levi, D., Verge-Rebelo, R., Winternitz, P. (eds.) Symmetries and Integrability of Difference Equations: Lecture Notes of the Abecederian School of SIDE 12, Montreal 2016. CRM Series in Mathematical Physics, pp. 75–152. Springer International Publishing, Berlin (2017)

Gubbiotti, G., Joshi, N., Tran, D.T., Viallet, C.-M.: Bi-rational maps in four dimensions with two invariants. J Phys A-Math Theor. **53**(11), Art. 115201 (24 pp) (2020). https://doi.org/10.1088/1751-8121/ab72ad

Gubbiotti, G., Scimiterna, C., Levi, D.: Linearizability and fake Lax pair for a consistent around the cube nonlinear non-autonomous quad-graph equation. Theor. Math. Phys. **189**(1), 1459–1471 (2016)

Hay, M.: Hierarchies of nonlinear integrable q-difference equations from series of Lax pairs. J. Phys. A: Math. Theor. **40**, 10457–10471 (2007)

Hay, M., Butler, S.: Simple identification of fake Lax pairs (2012). arXiv:1311.2406v1

Hay, M., Butler, S.: Two definitions of fake Lax pairs. AIP Conf. Proc. **1648**, 180006 (2015)

Joshi, N., Viallet, C.-M.: Rational Maps with Invariant Surfaces. J. Int. Sys. **3**, xyy017 (14pp) (2018)

Kahan, W.: Unconventional numerical methods for trajectory calculations (1993). Unpublished lecture notes

Kimura, K., Hirota, R.: Discretization of the Lagrange top. J. Phys. Soc. Jpn. **69**, 3193–3199 (2000)

Lando, S.K.: Lectures on Generating Functions. American Mathematical Society (2003)

Lax, P.D.: Integrals of nonlinear equations of evolution and solitary waves. Comm. Pure Appl. Math. **21**(5), 467–490 (1968)

Levi, D., Winternitz, P.: Continuous symmetries of difference equations. J. Phys. A Math. Theor. **39**(2), R1–R63 (2006)

Liouville, J.: Note sur l'intégration des équations différentielles de la Dynamique, présentée au Bureau des Longitudes le 29 juin 1853. J. Math. Pures Appl. **20**, 137–138 (1855)

Maeda, S.: Completely integrable symplectic mapping. Proc. Jap. Ac. A, Math. Sci. **63**, 198–200 (1987)

McMillan, E.M.: A problem in the stability of periodic systems. In: Britton, E., Odabasi, H. (eds.) A tribute to E.U. Condon. Topics in Modern Physics, pp. 219–244. Colorado Assoc. Univ. Press, Boulder (1971)

Olver, P.J.: Applications of Lie Groups to Differential Equations. Springer, Berlin (1986)

Papageorgiou, V.G., Nijhoff, F.W., Capel, H.W.: Integrable mappings and nonlinear integrable lattice equations. Phys. Lett. A **147**(2), 106–114 (1990)

Penrose, R., Smith, C.A.B.: A quadratic mapping with invariant cubic curve. Math. Proc. Camb. Phil. Soc. **89**, 89–105 (1981)

Petrera, M., Pfadler, A., Suris, YuB: On integrability of Hirota-Kimura type discretizations: experimental study of the discrete Clebsch system. Exp. Math. **18**, 223–247 (2009)

Petrera, M., Suris, Yu.B.: On the Hamiltonian structure of HirotaâĂŘkimura discretization of the Euler top. Math. Nachr. **283**(11), 1654–1663 (2010)

Quispel, G.R.W., Capel, H.W., Papageorgiou, V.G., Nijhoff, F.W.: Integrable mappings derived from soliton equations. Physica A **173**(1), 243–266 (1991)

Quispel, G.R.W., Roberts, J.A.G., Thompson, C.J.: Integrable mappings and soliton equations. Phys. Lett. A **126**, 419 (1988)

Quispel, G.R.W., Roberts, J.A.G., Thompson, C.J.: Integrable mappings and soliton equations II. Physica D **34**(1), 183–192 (1989)

Quispel, G.W.R., Capel, H.R., Roberts, J.A.G.: Duality for discrete integrable systems. J. Phys. A: Math. Gen. **38**(18), 3965 (2005)

Roberts, J.A.G., Jogia, D.: Birational maps that send biquadratic curves to biquadratic curves. J. Phys. A Math. Theor. **48**, 08FT02 (2015)

Sakai, H.: Rational surfaces associated with affine root systems and geometry of the Painlevé Equations. Comm. Math. Phys. **220**(1), 165–229 (2001)

Shafarevich, I.R.: Basic Algebraic Geometry 1. Grundlehren der mathematischen Wissenschaften, vol. 213, 2nd edn. Springer, Berlin (1994)

Tsuda, T.: Integrable mappings via rational elliptic surfaces. J. Phys. A: Math. Gen. **37**, 2721 (2004)

van der Kamp, P., Quispel, G.W.R.: The staircase method: integrals for periodic reductions of integrable lattice equations. J. Phys. A: Math. Theor. **43**, 465207 (2010)

Veselov, A.P.: Integrable maps. Russ. Math. Surveys **46**, 1–51 (1991)

Veselov, A.P.: Growth and integrability in the dynamics of mappings. Comm. Math. Phys. **145**, 181–193 (1992)

Viallet, C.-M.: On the algebraic structure of rational discrete dynamical systems. J. Phys. A: Math. Theor. **48**, 16FT01 (2015)

Viallet, C.-M., Grammaticos, B., Ramani, A.: On the integrability of correspondences associated to integral curves. Phys. Lett. A **322**, 186–93 (2004)

A Non-linear Relation for Certain Hypergeometric Functions

Gerd Schmalz and Vladimir Ezhov

Abstract We describe a family of Gaussian hypergeometric functions that satisfy a nonlinear differential identity.

Keywords Heisenberg sphere · Essential symmetry · Gaussian hypergeometric function · Shear symmetry

The family of hypergeometric functions we are presenting in this article has its origin in the phenomenon of non-linearisable symmetries of CR manifolds, which is a special instance of essential symmetries in parabolic geometry (see e.g. Casey et al. 2013, Kruglikov and The 2017, 2018). Roughly speaking, an essential symmetry is a local diffeomorphism (or infinitesimal automorphism) that preserves the relevant geometric data and that is not determined by (a certain part of) its 1-jet. The absence of essential symmetries for non-spherical hypersurface type CR manifolds has been established by Kruzhilin and Loboda (1983) and Ezhov (1985):

Theorem 1 *Let M be a Levi non-degenerate real-analytic hypersurface in $\mathbb{C}^n$ ($n \geq 2$) and let $p \in M$. Then either there is a biholomorphic mapping in some neighbourhood U of p that maps $M \cap U$ onto a piece of the Heisenberg sphere* $\operatorname{Im} w = |z|^2$ *or any biholomorphic mapping that locally maps M to M and preserves p can be linearised in suitable coordinates.*

Notice that the Heisenberg sphere has the automorphisms

G. Schmalz
School of Science and Technology, University of New England, Armidale, NSW 2351, Australia
e-mail: schmalz@une.edu.au

V. Ezhov (✉)
College of Science and Engineering, Flinders University, 1284 South Road, Tonsley, SA 5042, Australia
e-mail: vladimir.ejov@flinders.edu.au

Department of Mathematical Analysis, Faculty of Mechanics and Mathematics, Moscow State University, Leninskie Gory, 1, 119234 Moscow, Russia

© Springer Nature Switzerland AG 2020

F. Nijhoff et al. (eds.), *Asymptotic, Algebraic and Geometric Aspects of Integrable Systems*, Springer Proceedings in Mathematics & Statistics 338,
https://doi.org/10.1007/978-3-030-57000-2_3

$$z' = \frac{z + aw}{1 - 2\mathrm{i}\bar{a}z - (r + \mathrm{i}|a|^2)w},$$
$$w' = \frac{w}{1 - 2\mathrm{i}\bar{a}z - (r + \mathrm{i}|a|^2)w},$$

which are non-linearisable when the parameters $a \in \mathbb{C}$ and $r \in \mathbb{R}$ do not both vanish.

In other words, essential symmetries are usually a privilege of the one most symmetric manifold within the particular geometry. In (Ezhov and Schmalz, 2005, 2007) the authors describe a counterexample to this principle in the class of so-called torsion-free elliptic CR manifolds of codimension 2 in $\mathbb{C}^4$. Such manifolds can be given by a complex equation

$$\frac{w_1 - \bar{w}_2}{2\mathrm{i}} = f(z_1, \bar{z}_2, w_1 + \bar{w}_2). \tag{1}$$

The most symmetric object in this class is the CR manifold given by

$$\frac{w_1 - \bar{w}_2}{2\mathrm{i}} = z_1\bar{z}_2, \tag{2}$$

with symmetry group $\mathrm{SL}(3, \mathbb{C})$. The geometry of torsion-free elliptic CR manifolds turns out to be equivalent to the geometry of complex second order ODE's with respect to holomorphic point transformations. Indeed, Eq. (1) can be interpreted as the family of solutions

$$y = a + 2\mathrm{i}f(x, b)$$

of an ODE $y'' = B(y', y, x)$ with parameters a, b, where $y = w_1$, $x = z_1$, $a = \bar{w}_2$, $b = \bar{z}_2$. The model CR manifold (2) corresponds to the ODE $y'' = 0$ with solutions $y = a + bx$.

The (linear) point symmetry

$$X = y\frac{\partial}{\partial x} \tag{3}$$

extends to the contact 1-jet-bundle with vertical coordinate $p = \frac{dy}{dx}$ as

$$\tilde{X} = y\frac{\partial}{\partial x} - p^2\frac{\partial}{\partial p}.$$

The symmetry $\tilde{X}$ is not linearisable by point transformations. The authors have proved the following theorem in Ezhov and Schmalz (2005).

Theorem 2 *A second order ODE admits the shear symmetry* (3) *if and only if it has the form*

$$y'' = f_0(y)(y - xy')^3 + f_1(y)(y - xy')^2y' + f_2(y)(y - xy')(y')^2 + f_3(y)(y')^3, \tag{4}$$

where f_0, f_1, f_2, f_3 are arbitrary functions. By suitable point transformations these ODE's can be reduced to

$$y'' = \tilde{f}_0(y)(y - xy')^3 + \tilde{f}_1(y)(y - xy')^2 y'. \tag{5}$$

The solution manifolds of the ODE's in the theorem above give rise to torsion-free elliptic CR manifolds that have an essential symmetry, without being equivalent to the model (2).

In this article we consider the more special family

$$y'' = y^k (y - xy')^3, \tag{6}$$

where $k \in \mathbb{C}$. They feature the additional symmetry

$$Y = x\frac{\partial}{\partial x} - \frac{2y}{k+2}\frac{\partial}{\partial y}.$$

Therefore the ODE's (6) are completely integrable. We will show that the solutions of these ODE's (after some point transformation) yield a class of hypergeometric functions with a remarkable non-linear symmetry.

The involutive point transformation

$$x = y^*, \qquad y = x^*, \qquad p = \frac{1}{p^*}$$

takes the ODE's (6) to

$$y'' = x^k (y - xy')^3 \tag{7}$$

with symmetries

$$X^* = x\frac{\partial}{\partial y}, \qquad Y^* = x\frac{\partial}{\partial x} - \frac{k+2}{2} y \frac{\partial}{\partial y}. \tag{8}$$

The solutions have the form

$$y(x) = -x \int \frac{\sqrt{k+2}\, dx}{x^2 \sqrt{2x^{k+2} + K}} = \frac{\sqrt{k+2}}{\sqrt{K}} F_k \left(-\frac{2x^{k+2}}{K} \right) + Cx.$$

With $s = \frac{1}{k+2}$ the functions

$$F_k(u) = u^s \int \frac{du}{u^{1+s}\sqrt{1-u}} = hypergeom \left(\frac{1}{2}, -s; 1 - s, u \right)$$

are the Gaussian hypergeometric functions with the indicated parameters. Let $K = 2$, $C = 0$ and $-x^{k+2} = u$. Then

$$w(u) = y(x)|_{x^{k+2}=u} = \sqrt{-\frac{1}{2s}}\, u^s \int \frac{du}{u^{1+s}\sqrt{1-u}}.$$

Theorem 3 *The function*

$$w(u) = \sqrt{-\frac{1}{2s}}\, u^s \int \frac{du}{u^{1+s}\sqrt{1-u}} \tag{9}$$

satisfies the hypergeometric differential equation

$$u(1-u)\frac{d^2w}{du^2} + \left(1 - s + \left(s - \frac{3}{2}\right)u\right)\frac{dw}{du} + \frac{s}{2}w = 0 \tag{10}$$

and the non-linear identity

$$u\frac{d^2w}{du^2} = s^2\left(w - \frac{u}{s}\frac{dw}{du}\right)^3 + (s-1)\frac{dw}{du}, \tag{11}$$

which is equivalent to

$$\left(w - \frac{u}{s}\frac{dw}{du}\right)' = -s\left(w - \frac{u}{s}\frac{dw}{du}\right)^3. \tag{12}$$

Proof Since the hypergeometric differential equation is linear we may drop the factor $\sqrt{-\frac{s}{2}}$ for the verification of the first claim, which readily follows by plugging

$$\begin{aligned}
w(u) &= u^s \int \frac{du}{u^{1+s}\sqrt{1-u}},\\
w'(u) &= su^{s-1}\int \frac{du}{u^{1+s}\sqrt{1-u}} + \frac{1}{u\sqrt{1-u}},\\
w''(u) &= s(s-1)u^{s-2}\int \frac{du}{u^{1+s}\sqrt{1-u}} - \frac{1-s+(s-\frac{3}{2})u}{u^2\sqrt{(1-u)^3}}
\end{aligned}$$

into (10).

We show that the differential equation (11) is equivalent to (7) by substituting $x = u^s$. Then $w(u) = y(x)|_{x=u^s}$ satisfies (11) because $y(x)$ satisfies (7). Indeed,

$$\begin{aligned}
y'' &= \frac{u^{2-2s}}{s^2}w'' + \frac{1}{s}\left(\frac{1}{s} - 1\right)u^{1-2s}w',\\
x^k(y - xy')^3 &= u^{1-2s}\left(w - \frac{u}{s}w'\right)^3,
\end{aligned}$$

which yields (11) from (7). □

Acknowledgements Research was supported by Max-Planck-Institut für Mathematik Bonn. The authors wish to thank Avgust Tsikh for inspiring discussions.

References

Casey, S., Dunajski, M., Tod, P.: Twistor geometry of a pair of second order ODEs. Comm. Math. Phys. **321**(3), 681–701 (2013). https://doi.org/10.1007/s00220-013-1729-7, MR3070033

Ezhov, V.V.: Linearization of automorphisms of a real-analytic hypersurface. Izv. Akad. Nauk SSSR Ser. Mat. **49**(4), 731–765 (1985), 894 (Russian). MR806683

Ezhov, V., Schmalz, G.: Non-linearizable CR-automorphisms, torsion-free elliptic CR-manifolds and second order ODE. J. Reine Angew. Math. **584**, 215–236 (2005). https://doi.org/10.1515/crll.2005.2005.584.215, MR2155091

Ezhov, V., Schmalz, G.: Elliptic CR-manifolds and shear invariant ordinary differential equations with additional symmetries. Ark. Mat., **45**(2), 253–268 (2007). https://doi.org/10.1007/s11512-007-0049-6, MR2342603

Kruglikov, B., The, D.: The gap phenomenon in parabolic geometries. J. Reine Angew. Math. **723**, 153–215 (2017). https://doi.org/10.1515/crelle-2014-0072, MR3604980

Kruglikov, B., The, D.: Jet-determination of symmetries of parabolic geometries. Math. Ann. **371**(3–4), 1575–1613 (2018). https://doi.org/10.1007/s00208-017-1545-z, MR3831281

Kruzhilin, N.G., Loboda, A.V.: Linearization of local automorphisms of pseudoconvex surfaces. Dokl. Akad. Nauk SSSR **271**(2), 280–282 (1983) (Russian). MR718188

An Algebraically Stable Variety for a Four-Dimensional Dynamical System Reduced from the Lattice Super-KdV Equation

Adrian Stefan Carstea and Tomoyuki Takenawa

Abstract In a prior paper the authors obtained a four-dimensional discrete integrable dynamical system by the traveling wave reduction from the lattice super-KdV equation in a case of finitely generated Grassmann algebra. The system is a coupling of a Quispel-Roberts-Thompson map and a linear map but does not satisfy the singularity confinement criterion. It was conjectured that the dynamical degree of this system grows quadratically. In this paper, constructing a rational variety where the system is lifted to an algebraically stable map and using the action of the map on the Picard lattice, we prove this conjecture. We also show that invariants can be found through the same technique.

Keywords Dynamical systems · Algebraic geometry · Integrable systems

1 Introduction

In a prior paper (Carstea and Takenawa 2019b), applying the traveling wave reduction to the lattice super-KdV equation (Carstea 2015; Xue et al. 2013) in a case of finitely generated Grassmann algebra, the authors obtained a four-dimensional discrete integrable dynamical system

A. S. Carstea
Department of Theoretical Physics, Institute of Physics and Nuclear Engineering, Reactorului 15, 077125 Magurele, Bucharest, Romania
e-mail: carstea@gmail.com

T. Takenawa (✉)
Faculty of Marine Technology, Tokyo University of Marine Science and Technology, 2-1-6 Etchu-jima, Koto-ku, Tokyo 135-8533, Japan
e-mail: takenawa@kaiyodai.ac.jp

© Springer Nature Switzerland AG 2020

F. Nijhoff et al. (eds.), *Asymptotic, Algebraic and Geometric Aspects of Integrable Systems*, Springer Proceedings in Mathematics & Statistics 338,
https://doi.org/10.1007/978-3-030-57000-2_4

$$\varphi : \begin{cases} \bar{x}_0 = x_2 \\ \bar{x}_1 = x_3 \\ \bar{x}_2 = -x_2 - x_0 + \dfrac{hx_2}{1-x_2} \\ \bar{x}_3 = -x_1 - x_3 + \dfrac{2 - x_2 + hx_3}{(1-x_2)^2} \end{cases} . \tag{1}$$

This system is a Quispel-Roberts-Thompson (QRT) map, a two dimensional map generating an automorphism of a rational elliptic surface (Quispel et al. 1989), for variables x_0, x_2 coupled with linear equations for variables x_1, x_3 with coefficients depending on x_2. This system has two invariants

$$I_1 = -hx_0^2 - hx_0x_2 + h^2x_0x_2 + hx_0^2x_2 - hx_2^2 + hx_0x_2^2, \tag{2}$$

$$\begin{aligned} I_2 =& 2hx_0 + x_0^2 - 2hx_0x_1 + 2hx_2 + x_0x_2 - hx_1x_2 + h^2x_1x_2 + 2hx_0x_1x_2 \\ &+ x_2^2 + hx_1x_2^2 - hx_0x_3 + h^2x_0x_3 + hx_0^2x_3 - 2hx_2x_3 + 2hx_0x_2x_3, \end{aligned} \tag{3}$$

but does not satisfy the *singularity confinement criterion* proposed by Grammaticos-Ramani and their collaborators (Grammaticos et al. 1991; Ramani et al. 1991). The example of this criterion is given in the next section.

In the same paper it is observed that the dynamical degree of (1) grows quadratically. This phenomena is rather unusual, since as reported in Lafortune et al. (2001), Gubbiotti (2018), the dynamical degree grows in the fourth order for generic coupled systems in the form

$$\begin{cases} \bar{x}_0 = f_0(x_0, x_1) \\ \bar{x}_1 = f_1(x_0, x_1) \\ \bar{x}_2 = f_2(x_0, x_1, x_2) \\ \bar{x}_3 = f_3(x_0, x_1, x_2, x_3) \end{cases} ,$$

where the system is a QRT map for variables x_0 and x_1, and $\bar{x}_2$ (resp. $\bar{x}_3$) depends on x_2 (resp. x_3) linearly with coefficients depending on "x_0 and x_1" (resp. "x_0, x_1 and x_2"). This type of systems is also constructed by generalising the QRT maps and referred to as "triangular" in Fordy and Kassotakis (2006).

In this paper, constructing a rational variety where System (1) is lifted to an *algebraically stable* map and using the action of the map on the Picard lattice, we prove the above conjecture. We also show that one can find invariants also using the action on the Picard group.

In the two-dimensional case, it is known that an autonomous dynamical system defined by a birational map on a projective rational variety (or more generally Kähler manifold) can be lifted to either an automorphism or an algebraically stable map on a rational variety by successive blow-ups (Diller and Fravre 2001). Here, a birational map φ from an N-dimensional rational variety $\mathcal{X}$ to itself is said to be *algebraically stable* if $(\varphi^*)^n(\mathcal{D}) = (\varphi^n)^*(\mathcal{D})$ holds for any divisor class $\mathcal{D}$ on $\mathcal{X}$ and an arbitrary positive integer n (Bedford and Kim 2008). These notions are closely related to the notion of singularity confinement criterion. While a dynamical system that can be lifted to automorphisms satisfies singularity confinement criterion (i.e. all the singularities are confined), a dynamical system that can be lifted only to algebraically stable map does not satisfy the criterion (i.e. there exists a singularity that is not confined).

In studies of higher dimensional dynamical systems, the role of automorphisms is replaced by pseudo-automorphisms, i.e. automorphisms except finite number of sub-varieties of codimension at least two (Dolgachev and Ortland 1988). In the last decade a few authors studied how to construct algebraic varieties on the level of pseudo-automorphisms (Bedford and Kim 2008; Tsuda and Takenawa 2009; Carstea and Takenawa 2019a). However, since System (1) does not satisfy the singularity confinement criterion, it is not expected that it could be lifted to a pseudo-automorphism. To authors' knowledge there are no studies (except Sect. 7 of Bedford and Kim 2008, which studies a kind of generalisation of standard Cremona transformation) on construction of an algebraic variety, in which the original system is lifted not to a pseudo-automorphism, but rather to an algebraically stable map using blow-ups along sub-varieties of positive dimensions. Since the varieties obtained by blow-ups possibly infinitely near depend on the order of blow-ups, this is not a straightforward but a challenging problem.

Since I_2 is degree $(1, 1)$ for x_1, x_3, we can restrict the phase space into 3-dimensional one as

$$\psi : \begin{cases} x_0 = x_2 \\ x_1 = \big(I_2 - (x_0^2 + x_0x_2 + x_2^2) - 2h(x_0 - x_0x_1 + x_2) \\ \qquad\quad -hx_1x_2(2x_0 + x_2 - 1 + h)\big) \\ \qquad\quad \big(h(-x_0 + hx_0 + x_0^2 - 2x_2 + 2x_0x_2)\big)^{-1} \\ x_2 = -x_2 - x_0 + \dfrac{hx_2}{1 - x_2} \end{cases} . \tag{4}$$

We also show that the degree of this 3 dimensional system grows quadratically as well.

2 Algebraically Stable Space for the 4D System

Let us consider System (1) on the projective space $(\mathbb{P}^1)^4$. In the following, we aim to obtain a four-dimensional rational variety by blowing-up procedure such that the birational map (1) is lifted to an algebraically stable map on the variety.

Let $I(\varphi)$ denote the indeterminacy set of φ. It is known that the mapping φ is algebraically stable if and only if there does not exist a positive integer k and a divisor D on $\mathcal{X}$ such that

$$\varphi(D \setminus I(\varphi)) \subset I(\varphi^k), \tag{5}$$

i.e. the image of the generic part of a divisor by φ is included in the indeterminate set of φ^k (Bedford and Kim 2008; Bayraktar 2013, Proposition 2.3 of Carstea and Takenawa 2019a). See Sect. 2 of Carstea and Takenawa (2019a) for notations and related theories used here.

The notion of singularity series of dynamics studied by Grammaticos-Ramani and their collaborators is closely related to our procedure. Let us start with a hyper-plane $x_2 = 1+\varepsilon$, where ε is a small parameter for considering Laurent series expression, and apply φ, then we have a "confined" sequence of Laurent series:

$$\begin{aligned}\cdots \to &(x_0^{(0)}, x_1^{(0)}, 1+\varepsilon, x_3^{(0)}) \to (1, x_3^{(0)}, -h\varepsilon^{-1}, (1+hx_3^{(0)})\varepsilon^{-2}) \\ \to &(-h\varepsilon^{-1}, (1+hx_3^{(0)})\varepsilon^{-2}, h\varepsilon^{-1}, -(1+hx_3^{(0)})\varepsilon^{-2}) \\ \to &(h\varepsilon^{-1}, -(1+hx_3^{(0)})\varepsilon^{-2}, 1, x_4^{(3)}) \to (1, x_1^{(4)}, x_0^{(0)}, x_3^{(4)}) \to \cdots, \end{aligned} \tag{6}$$

where $x_i^{(k)}$'s are complex constants and only the principal term is written for each entry and a hyper-surface $x_2 = 0$ is contracted to lower-dimensional varieties and returned to a hyper-surface $x_0 = 1$ after 4 steps. We can also find a cyclic sequence:

$$\begin{aligned}&(x_0^{(0)}, x_1^{(0)}, \varepsilon^{-1}, x_3^{(0)}) \to (\varepsilon^{-1}, x_3^{(0)}, -\varepsilon^{-1}, -x_1^{(0)} - x_3^{(0)}) \\ \to &(\varepsilon^{-1}, -x_1^{(0)} - x_3^{(0)}, x_0^{(0)}, x_1^{(0)}) \to (x_0^{(0)}, x_1^{(0)}, \varepsilon^{-1}, x_3^{(3)}): \text{returned},\end{aligned} \tag{7}$$

where a hyper-surface $x_2 = \infty$ is contracted to lower-dimensional varieties and returned to the original hyper-surface after 3 steps, and an "anti-confined" sequence:

$$\begin{aligned}\cdots \to &\left(\left(-1 + \frac{h}{(x_0^{(0)}-1)^2}\right)\varepsilon^{-1}, x_1^{(-1)}, x_2^{(-1)}, \varepsilon^{-1}\right) \\ \to &(x_0^{(0)}, \varepsilon^{-1}, x_2^{(0)}, x_3^{(0)}) \to (x_2^{(0)}, x_3^{(0)}, x_2^{(1)}, \varepsilon^{-1}) \\ \to &\left(x_2^{(1)}, \varepsilon^{-1}, x_2^{(2)}, \left(-1 + \frac{h}{(x_2^{(0)}-1)^2}\right)\varepsilon^{-1}\right) \to \cdots,\end{aligned} \tag{8}$$

where a lower dimensional variety is blown-up to a hyper-surface $x_1 = \infty$ and contracted to a lower dimensional variety after 3 steps.

In the following, in order to avoid anti-confined patterns, we consider $\mathbb{P}^2 \times \mathbb{P}^2$ instead of $(\mathbb{P}^1)^4$. Although there is a possibility that the anti-confined pattern can be resoluted by some blowing-down procedure, it is not easy to find the actual procedure on the level of coordinates.

The coordinate system of $\mathbb{P}^2 \times \mathbb{P}^2$ is $(x_0 : x_1 : 1,\ x_2 : x_3 : 1)$, and thus the local coordinate systems essentially consist of $3 \times 3 = 9$ charts:

$$\begin{aligned}&(x_0, x_1, x_2, x_3), (y_0, y_1, x_2, x_3), (z_0, z_1, x_2, x_3),\\ &(x_0, x_1, y_2, y_3), (y_0, y_1, y_2, y_3), (z_0, z_1, y_2, y_3),\\ &(x_0, x_1, z_2, z_3), (y_0, y_1, z_2, z_3), (z_0, z_1, z_2, z_3),\end{aligned}$$

where y_i's and z_i's are

$$(y_i, y_{i+1}) = (x_i^{-1}, x_i^{-1}x_{i+1}) \text{ and } (z_i, z_{i+1}) = (x_i x_{i+1}^{-1}, x_{i+1}^{-1})$$

for $i = 0, 2$. Then, both the cyclic Sequence (7); and the anti-confined Sequence (8) starting with $x_i^{(0)} = \varepsilon^{-1}$ do not appear, but another cyclic sequence

$$\begin{aligned} &(x_0^{(0)}, x_1^{(0)}, \varepsilon^{-1}, c^{(0)}\varepsilon^{-1}) \to (\varepsilon^{-1}, c^{(0)}\varepsilon^{-1}, -\varepsilon^{-1}, -c^{(0)}\varepsilon^{-1}) \\ \to &(-\varepsilon^{-1}, -c^{(0)}\varepsilon^{-1}, x_0^{(0)}, x_1^{(0)}) \to (x_0^{(0)}, x_1^{(0)}, \varepsilon^{-1}, c^{(0)}\varepsilon^{-1})\text{: returned} \end{aligned} \tag{9}$$

appears, where $c^{(0)}$ is also a complex constant.

In order to resolute the singularity appeared in Sequences (6) and (9), we blow up the rational variety along the sub-varieties to which some divisor is contracted to. For Sequence (6), we have three such sub-varieties whose parametric expressions are

$$\begin{aligned} V_1 &:(x_0, x_1, z_2, z_3) = (P, 1, 0, 0), \\ V_2 &:(z_0, z_1, z_2, z_3) = (0, 0, 0, 0), \\ V_3 &:(z_0, z_1, x_2, x_3) = (0, 0, P, 1), \end{aligned}$$

where P is a $\mathbb{C}$-valued parameter (independent to another sub-variety), while for Sequence (9) we have a sub-variety

$$V_4 :(z_0, z_1, z_2, z_3) = (P, 0, P, 0).$$

That is, the subvariety V_1 is the Zariski closure of $\{(x_0, x_1, x_2, x_3) = (P, 1, 0, 0) \mid P \in \mathbb{C}\}$ and V_4 is that of $\{(x_0, x_1, x_2, x_3) = (P, 0, P, 0) \mid P \in \mathbb{C}\}$ and so forth.

Since V_4 includes V_2, we have the option of blowing-up order. In the two dimensional case, resolution is unique and the order is not a matter. But in the higher dimensional case, it affects sensitively to the resulting varieties. Since we only care on the level of codimension one, the order of blow-ups does not affect the algebraical stability in some cases. However, the following results were obtained not in a straightforward manner but by trial and error.

We can resolute the singularity around V_1 by the following five blowups:

$$\begin{aligned} C_1 &:(x_0, x_1, z_2, z_3) = (1, P, 0, 0) \\ &\quad \leftarrow (s_1, t_1, u_1, v_1) := (x_0 - 1, x_1, z_2(x_0 - 1)^{-1}, z_3(x_0 - 1)^{-1}), \\ C_2 &:(s_1, t_1, u_1, v_1) = (0, P, Q, 0) \\ &\quad \leftarrow (s_2, t_2, u_2, v_2) := (s_1, t_1, u_1, v_1 s_1^{-1}), \\ C_3 &:(s_2, t_2, u_2, v_2) = (0, P, -h(1 + hP)^{-1}, Q) \\ &\quad \leftarrow (s_3, t_3, u_3, v_3) := (s_2, t_2, (u_2 + h(1 + ht_2)^{-1})s_2^{-1}, v_2), \end{aligned}$$

$$\begin{aligned}
C_4 :&(s_3, t_3, u_3, v_3) = (0, P, Q, (1+hP)^{-1}) \\
&\leftarrow (s_4, t_4, u_4, v_4) := (s_3, t_3, u_3, (v_3 - (1+ht_3)^{-1})s_3^{-1}), \\
C_5 :&(s_4, t_4, u_4, v_4) = (0, P, Q, (1+hP)^{-2}) \\
&\leftarrow (s_5, t_5, u_5, v_5) := (s_4, t_4, u_4, (v_4 - (1+ht_4)^{-2})s_4^{-1}),
\end{aligned}$$

where only one of the coordinate systems is written for each blowup. Similarly, we can resolute the singularity around V_3 by the following five blowups:

$$\begin{aligned}
C_6 :&(z_0, z_1, x_2, x_3) = (0, 0, 1, P) \\
&\leftarrow (s_6, t_6, u_6, v_6) := (x_2 - 1, x_3, z_0(x_2-1)^{-1}, z_1(x_2-1)^{-1}), \\
C_7 :&(s_6, t_6, u_6, v_6) = (0, P, Q, 0) \\
&\leftarrow (s_7, t_7, u_7, v_7) := (s_6, t_6, u_6, v_6 s_6^{-1}), \\
C_8 :&(s_7, t_7, u_7, v_7) = (0, P, -h(1+hP)^{-1}, Q) \\
&\leftarrow (s_8, t_8, u_8, v_8) := (s_7, t_7, (u_7 + h(1+ht_7)^{-1})s_7^{-1}, v_7), \\
C_9 :&(s_8, t_8, u_8, v_8) = (0, P, Q, (1+hP)^{-1}) \\
&\leftarrow (s_9, t_9, u_9, v_9) := (s_8, t_8, u_8, (v_8 - (1+ht_8)^{-1})s_8^{-1}), \\
C_{10} :&(s_9, t_9, u_9, v_9) = (0, P, Q, (1+hP)^{-2}) \\
&\leftarrow (s_{10}, t_{10}, u_{10}, v_{10}) := (s_9, t_9, u_9, (v_9 - (1+ht_9)^{-2})s_9^{-1}).
\end{aligned}$$

We need three blowups for V_4:

$$\begin{aligned}
C_{11} :&(z_0, z_1, z_2, z_3) = (0, 0, 0, 0) \\
&\leftarrow (s_{11}, t_{11}, u_{11}, v_{11}) := (z_0, z_1 z_0^{-1}, z_2 z_0^{-1}, z_3 z_0^{-1}), \\
C_{12} :&(s_{11}, t_{11}, u_{11}, v_{11}) = (P, 0, 1, 0) \\
&\leftarrow (s_{12}, t_{12}, u_{12}, v_{12}) := (s_{11}, t_{11}, (u_{11} - 1)t_{11}^{-1}, v_{11} t_{11}^{-1}), \\
C_{13} :&(s_{12}, t_{12}, u_{12}, v_{12}) = (P, 0, Q, -1) \\
&\leftarrow (s_{13}, t_{13}, u_{13}, v_{13}) := (s_{12}, t_{12}, u_{12}, (v_{12} + 1)t_{12}^{-1}),
\end{aligned}$$

where C_{11} correspond to V_2, while C_{12} and C_{13} correspond to V_4. We need additional four blowups for V_2:

$$
\begin{aligned}
C_{14}:&(s_{13}, t_{13}, u_{13}, v_{13}) = (0, 0, 1+h, 0)\\
&\leftarrow (s_{14}, t_{14}, u_{14}, v_{14}) := (s_{13}t_{13}^{-1}, t_{13}, (u_{13}-1-h)t_{13}^{-1}, v_{13}t_{13}^{-1}),\\
C_{15}:&(s_{14}, t_{14}, u_{14}, v_{14}) = (P, 0, -2Q - Ph^{-1}, Q)\\
&\leftarrow (s_{15}, t_{15}, u_{15}, v_{15}) := (s_{14}, t_{14}, v_{14}, (u_{14}+2v_{14}+s_{14}h^{-1})t_{14}^{-1}),\\
C_{16}:&(s_{15}, t_{15}, u_{15}, v_{15}) = (P, 0, -Ph^{-1}, Q)\\
&\leftarrow (s_{16}, t_{16}, u_{16}, v_{16}) := (s_{15}, t_{15}, (u_{15}+s_{15}h^{-1})t_{15}^{-1}, v_{15}),\\
C_{17}:&(s_{16}, t_{16}, u_{16}, v_{16}) = (P, 0, Q, 2^{-1}Q + (1+h)h^{-1}P)\\
&\leftarrow (s_{17}, t_{17}, u_{17}, v_{17}) := (s_{16}, t_{16}, u_{16}, (v_{16} - 2^{-1}u_{16} - (1+h)h^{-1}s_{16})t_{16}^{-1}).
\end{aligned}
$$

The (total transform of) exceptional divisor E_i of i-th blowup is described in the local chart as

$$
\begin{aligned}
E_i &: s_i = 0, \quad (i = 1, 2, 3, 4, 5, 6, 7, 8, 9, 10, 11, 14),\\
E_i &: t_i = 0, \quad (i = 12, 13, 15, 16, 17).
\end{aligned}
$$

Let us denote the total transform (with respect to blowups) of the divisors (hypersurfaces) $c_0x_0 + c_1x_1 + a = 0$ and $c_2x_2 + c_3x_3 + b = 0$ by H_a and H_b respectively, where $(c_0 : c_1 : a)$ and $(c_2 : c_3 : b)$ are constant $\mathbb{P}^2$ vectors. Let us write the classes of H_a, H_b and E_i modulo linear equivalence as $\mathcal{H}_a$, $\mathcal{H}_b$ and $\mathcal{E}_i$. Then, the Picard group of this variety $\mathcal{X}$ becomes a $\mathbb{Z}$-module:

$$
\mathrm{Pic}(\mathcal{X}) = \mathbb{Z}\mathcal{H}_a \oplus \mathbb{Z}\mathcal{H}_b \oplus \bigoplus_{i=1}^{17} \mathbb{Z}\mathcal{E}_i. \tag{10}
$$

Theorem 1 *The map* (1) *is lifted to an algebraically stable map on the rational variety obtained by blow-ups along* C_i, $i = 1, 2, \ldots, 17$, *from* $\mathbb{P}^2 \times \mathbb{P}^2$.

Proof The algebraic stability can be checked as follows. In the present case, the indeterminate set $I(\varphi)$ is given by

$$
I(\varphi) = \varphi^{-1}(E_6 - E_7) \subset E_{11},
$$

while the condition that the dimension of $\varphi(D \setminus I(\varphi))$ is at most two implies $D = E_1 - E_2$ and $\varphi(D \setminus I(\varphi)) = \varphi(E_1 - E_2) \subset E_{11}$. It can be checked that $\varphi(E_1 - E_2)$ and $I(\varphi^k)$, $k = 1, 2, 3, \ldots$, are different two-dimensional subvarieties in E_{11}, and hence (5) can not occur.

The class of proper transform of E_i is

$$
\begin{aligned}
&\mathcal{E}_i - \mathcal{E}_{i+1} \quad (i = 1, 2, 3, 4, 6, 7, 8, 9, 12, 13, 14, 15, 16),\\
&\mathcal{E}_i \quad (i = 5, 10, 17), \quad \mathcal{E}_{11} - \mathcal{E}_{15}.
\end{aligned}
$$

Since the defining function of the hyper-surface $z_1 = 0$ takes zero with multiplicities $0, 0, 0, 0, 0, 1, 2, 2, 2, 2, 1, 1, 1, 2, 2, 2, 2$ on E_i $(i = 1, \ldots, 17)$, it is decomposed as

$$\begin{aligned}\mathcal{H}_a =&\text{Proper transform}\\ &+ (\mathcal{E}_6 - \mathcal{E}_7) + 2(\mathcal{E}_7 - \mathcal{E}_8) + 2(\mathcal{E}_8 - \mathcal{E}_9) + 2(\mathcal{E}_9 - \mathcal{E}_{10}) + 2\mathcal{E}_{10}\\ &+ (\mathcal{E}_{11} - \mathcal{E}_{14}) + (\mathcal{E}_{12} - \mathcal{E}_{13}) + (\mathcal{E}_{13} - \mathcal{E}_{14}) + 2(\mathcal{E}_{14} - \mathcal{E}_{15})\\ &+ 2(\mathcal{E}_{15} - \mathcal{E}_{16}) + 2(\mathcal{E}_{16} - \mathcal{E}_{17}) + 2\mathcal{E}_{17},\end{aligned}$$

where each class enclosed in parentheses determines a prime divisor uniquely (we called such a class deterministic Carstea et al. 2017). Hence the class of its proper transform is $\mathcal{H}_a - \mathcal{E}_6 - \mathcal{E}_7 - \mathcal{E}_{11} - \mathcal{E}_{12}$. Similarly, the defining function of the hyper-surface $x_2 - 1 = 0$ takes zero with multiplicities $1, 1, 1, 1, 1, 1, 1, 1, 1, 1, 1, 0, 0, 1, 1, 1, 1$ on E_i, and therefore the class of its proper transform is $\mathcal{H}_b - \mathcal{E}_1 - \mathcal{E}_6 - \mathcal{E}_{11}$. Along the same line, the proper transform of $z_3 = 0$ can be computed as $\mathcal{H}_b - \mathcal{E}_1 - \mathcal{E}_2 - \mathcal{E}_{11} - \mathcal{E}_{12}$.

Using these data, we can compute the pull-back action of Mapping φ (1) on the Picard group. For example, the pull-back of E_1 is $(\bar{x}_1, \bar{z}_2, \bar{z}_3) = (0, 0, 0)$, whose "common factor" on each local coordinate system is $x_2 - 1$, s_6, s_7, s_8 or s_9. Thus, we have

$$\begin{aligned}\varphi(\mathcal{E}_1) =&(\mathcal{H}_2 - \mathcal{E}_1 - \mathcal{E}_6 - \mathcal{E}_{11}) + \sum_{i=6}^{9}(\mathcal{E}_i - \mathcal{E}_{i+1})\\ =&\mathcal{H}_2 - \mathcal{E}_1 - \mathcal{E}_{10} - \mathcal{E}_{11}.\end{aligned}$$

Along the same line, we have the following proposition.

Proposition 1 *The pull-back φ^* of Mapping* (1) *is a linear action on the Picard group given by*

$$\begin{aligned}&\mathcal{H}_a \to \mathcal{H}_b,\\ &\mathcal{H}_b \to \mathcal{H}_a + 3\mathcal{H}_b - 2\mathcal{E}_1 - 3\mathcal{E}_{11} - \mathcal{E}_{6,7,9,10,12,13,14},\\ &\mathcal{E}_1 \to \mathcal{H}_b - \mathcal{E}_{1,10,11}, \quad \mathcal{E}_2 \to \mathcal{H}_b - \mathcal{E}_{1,9,11}, \quad \mathcal{E}_3 \to \mathcal{H}_b - \mathcal{E}_{1,7,9,11} + \mathcal{E}_8,\\ &\mathcal{E}_4 \to \mathcal{H}_b - \mathcal{E}_{1,7,11}, \quad \mathcal{E}_5 \to \mathcal{H}_b - \mathcal{E}_{1,6,11},\\ &\mathcal{E}_6 \to \mathcal{E}_{14}, \quad \mathcal{E}_7 \to \mathcal{E}_{14}, \quad \mathcal{E}_8 \to \mathcal{E}_{15}, \quad \mathcal{E}_9 \to \mathcal{E}_{16}, \quad \mathcal{E}_{10} \to \mathcal{E}_{17},\\ &\mathcal{E}_{11} \to \mathcal{E}_{1,11} - \mathcal{E}_{14}, \quad \mathcal{E}_{12} \to \mathcal{H}_b - \mathcal{E}_{1,11,13}, \quad \mathcal{E}_{13} \to \mathcal{H}_b - \mathcal{E}_{1,11,12},\\ &\mathcal{E}_{14} \to \mathcal{E}_2, \quad \mathcal{E}_{15} \to \mathcal{E}_3, \quad \mathcal{E}_{16} \to \mathcal{E}_4, \quad \mathcal{E}_{17} \to \mathcal{E}_5,\end{aligned}$$

where $\mathcal{E}_{i_1,\ldots,i_k}$ denotes $\mathcal{E}_{i_1} + \cdots + \mathcal{E}_{i_k}$. The Jordan blocks of the corresponding matrix are

$$1,\ -1,\ 1^{\frac{1}{3}}\ (3\times 3\text{ blocks}),\ \begin{bmatrix}1&1&0\\0&1&1\\0&0&1\end{bmatrix},\ \begin{bmatrix}0&1&0&0&0\\0&0&1&0&0\\0&0&0&1&0\\0&0&0&0&1\\0&0&0&0&0\end{bmatrix}.$$

In particular, the degree of the mapping φ^n grows quadratically with respect to n.

Corollary 1 *The degree of ψ^n for the 3-dimensional map ψ (4) also grows quadratically with respect to n.*

Proof [1]Let us denote the initial values as $(x_0, x_1, x_2, x_3) = (x_0^{(0)}, x_1^{(0)}, x_2^{(0)}, x_3^{(0)})$. Map ψ^n is obtained by substituting $x_3 = h(x_0, x_1, x_2)$ to $\varphi^n : x_i^{(n)} = f_i^{(n)}(x_0, x_1, x_2, x_3)$, $i = 0, 1, 2$, where h and f_i's are some rational functions. Hence the degrees of $x_i^{(n)}$'s with respect to x_0, x_1, x_2 are bounded from the above by (degree of h) $\times$ (degree of $f_i^{(n)}$). Since the degrees of $f_i^{(n)}$'s are quadratic with respect to n, the degrees of $x_i^{(n)}$'s are at most quadratic. On the other hand, since ψ is a QRT map with respect to x_0 and x_2, its degree with regarding x_1 as a constant grows quadratically (Takenawa 2001), hence the degrees of $x_i^{(n)}$'s are at least quadratic.

The proper transforms of the conserved quantities I_1 and I_2 are

$$I_1 : 2\mathcal{H}_a + 2\mathcal{H}_b - 2\mathcal{E}_1 - 2\mathcal{E}_6 - 4\mathcal{E}_{11} - \mathcal{E}_{2,4,7,9,12,13,14,16},$$
$$I_2 : 2\mathcal{H}_a + 2\mathcal{H}_b - 3\mathcal{E}_{11} - \mathcal{E}_{1,2,4,5,6,7,9,10,12,13,14,16,17},$$

which are preserved by φ^*.

We can consider the inverse problem.

Proposition 2 *Hyper-surfaces whose class is $2\mathcal{H}_a + 2\mathcal{H}_b - 2\mathcal{E}_1 - 2\mathcal{E}_6 - 4\mathcal{E}_{11} - \mathcal{E}_{2,4,7,9,12,13,14,16}$ are given by $C_0 + C_1 I_1 = 0$ with $(C_0 : C_1) \in \mathbb{P}^1$ and $C_1 \neq 0$. Hyper-surfaces whose class is $2\mathcal{H}_a + 2\mathcal{H}_b - 3\mathcal{E}_{11} - \mathcal{E}_{1,2,4,5,6,7,9,10,12,13,14,16,17}$ are given by $C_0 + C_1 I_1 + C_2 I_2 = 0$ with $(C_0 : C_1 : C_2) \in \mathbb{P}^2$ and $C_2 \neq 0$.*

Thus, we can compute invariants by using the action of the system φ on the Picard group.

Proof The proof is straightforward but tedious. For example, the defining polynomials of a curve of the class $2\mathcal{H}_a + 2\mathcal{H}_b - 2\mathcal{E}_1 - 2\mathcal{E}_6 - 4\mathcal{E}_{11} - \mathcal{E}_{2,4,7,9,12,13,14,16}$ can be written as

[1]This kind of argument is not original. More general results can be found in Mase (2016), where it is shown that all the reduced systems from classical KP or BKP equation have the quadratic degree growth.

$$f(x_0, x_1, x_2, x_3) := \sum_{\substack{i_0, i_1, i_2, i_3 \geq 0 \\ i_0+i_1+i_2+i_3 \leq 2}} a_{i_0 i_1 i_2 i_3} x_0^{i_0} x_1^{i_1} x_2^{i_2} x_3^{i_3},$$

$$z_2^2 f(x_0, x_1, z_2 z_3^{-1}, z_3^{-1}) \quad \text{around } E_1,$$

$$z_0^2 f(z_0 z_1^{-1}, z_1^{-1}, x_2, x_3) \quad \text{around } E_5,$$

$$z_0^2 z_2^2 f(z_0 z_1^{-1}, z_1^{-1}, z_2 z_3^{-1}, z_3^{-1}) \quad \text{around } E_{11}.$$

The coefficients are determined so that defining polynomial takes zero with multiplicity 2, 3, 3, 4, 4, 2, 3, 3, 4, 4, 4, 1, 2, 7, 7, 8, 8 on E_i's; which verifies the claim.

Acknowledgements ASC was supported by Program Nuccleu, PN/2019, Romanian Ministry of Education and T. T. was supported by the Japan Society for the Promotion of Science, Grand-in-Aid (C) (17K05271).

References

Bayraktar, T.: Green currents for meromorphic maps of compact Kähler manifolds. J. Geom. Anal. **23**, 970–998 (2013)

Bedford, E., Kim, K.: Degree growth of matrix inversion: birational maps of symmetric, cyclic matrices. Discrete Contin. Dyn. Syst. **21**, 977–1013 (2008)

Carstea, A.S.: Constructing soliton solution and super-bilinear form of lattice supersymmetric KdV equation. J. Phys. A: Math. Theor. **48**, 285201 (2015)

Carstea, A.S., Dzhamay, A., Takenawa, T.: Fiber-dependent deautonomization of integrable 2D mappings and discrete Painlevé equations. J. Phys. A **50**, 405202 (2017)

Carstea, A.S., Takenawa, T.: Space of initial conditions and geometry of two 4-dimensional discrete Painlevé equations. J. Phys. A **52**, 275201 (2019)

Carstea, A.S., Takenawa, T.: Super-QRT and 4D-mappings reduced from the lattice super-KdV equation. J. Math. Phys. **60**, 093503 (2019)

Diller, J., Fravre, Ch.: Dynamics of bimeromorphic maps of surfaces. Amer. J. Math. **123**, 1135–1169 (2001)

Dolgachev, I., Ortland, D.: Point sets in projective spaces and theta functions, Astérisque **165** (Paris, 1988 i.e. 1989)

Fordy, A.P., Kassotakis, P.G.: Multidimensional maps of QRT type. J. Phys. A **39**, 10773–10786 (2006)

Grammaticos, B., Ramani, A., Papageorgiou, V.: Do integrable mappings have the Painlevé property? Phys. Rev. Lett. **67**, 1829–1832 (1991)

Gubbiotti, G., Joshi, N., Tran, D.T., Viallet, C.-M.: Complexity and integrability in 4D bi-rational maps with two invariants. arXiv:1808.04942

Lafortune, S., Carstea, A.S., Ramani, A., Grammaticos, B., Ohta, Y.: Integrable third-order mappings and their growth properties. Regul. Chaotic Dyn. **6**, 443–448 (2001)

Mase, T.: Investigation into the role of the Laurent property in integrability. J. Math. Phys. **57**, 022703 (2016)

Quispel, G.R.W., Roberts, J.A.G., Thomson, C.J.: Integrable mappings and soliton equations 2. Physica D **34**, 183–192 (1989)

Ramani, A., Grammaticos, B., Hietarinta, J.: Discrete versions of the Painlevé equations. Phys. Rev. Lett. **67**, 1829–1832 (1991)

Takenawa, T.: Algebraic entropy and the space of initial values for discrete dynamical systems. J. Phys. A: Math. Gen. **34**, 10533–10545 (2001)

Tsuda, T., Takenawa, T.: Tropical representation of Weyl groups associated with certain rational varieties. Adv. Math. **221**, 936–954 (2009)

Xue, L.L., Levi, D., Liu, Q.P.: Supersymmetric KdV equation: Darboux transformation and discrete systems. J. Phys. A: Math. Theor. **46**, 502001 (2013)

Opers for Higher States of the Quantum Boussinesq Model

Davide Masoero and Andrea Raimondo

Abstract We study the ODE/IM correspondence for all the states of the quantum Boussinesq model. We consider a particular class of third order linear ordinary differential operators and show that the generalised monodromy data of such operators provide solutions to the Bethe Ansatz equations of the Quantum Boussinesq model.

Keywords ODE/IM correspondence · Quantum Boussinesq · Opers · Bethe Ansatz

1 Introduction

The quantum Boussinesq model (Bazhanov et al. 2002) is a 2 dimensional conformal field theory with a $\mathcal{W}_3$ symmetry, and it can be exactly solved via the Bethe Ansatz equations. This model can be realised as the quantisation of a $\mathfrak{sl}_3$ Drinfeld-Sokolov hierarchy, or as the continuum limit of a $\mathfrak{sl}_3$ XXZ chain. It belongs to a large family of theories which are known as $\mathfrak{g}$-quantum KdV models; they exist for any Kac Moody algebra $\mathfrak{g}$ (Feigin and Frenkel 1996) (in the present case $\mathfrak{g} = \widehat{\mathfrak{sl}}_3$), and in the simplest case, namely $\mathfrak{g} = \widehat{\mathfrak{sl}}_2$, the Hamiltonian structure of such a theory is the quantisation of the second Poisson structure of the classical KdV equation (Bazhanov et al. 1996).

According to the celebrated ODE/IM correspondence (Dorey and Tateo 1999, 2000; Bazhanov et al. 2001, 2004; Dorey et al. 2007; Feigin and Frenkel 2011;

D. Masoero (✉)
Grupo de Física Matemática da Universidade de Lisboa, Campo Grande, Edifício C6, 1749-016 Lisboa, Portugal
e-mail: dmasoero@gmail.com

Faculdade de Ciências da Universidade de Lisboa, Campo Grande, 1749-016 Lisbon, Portugal

A. Raimondo
Dipartimento di Matematica e Applicazioni, Università di Milano-Bicocca, Via Roberto Cozzi 55, 20125 Milano, Italy
e-mail: andrea.raimondo@unimib.it

© Springer Nature Switzerland AG 2020

F. Nijhoff et al. (eds.), *Asymptotic, Algebraic and Geometric Aspects of Integrable Systems*, Springer Proceedings in Mathematics & Statistics 338,
https://doi.org/10.1007/978-3-030-57000-2_5

Masoero et al. 2016, 2017; Masoero and Raimondo 2020; Kotousov and Lukyanov 2019) to every state of the $\mathfrak{g}$ quantum KdV model there corresponds a unique $\mathfrak{g}^L$ oper (here $\mathfrak{g}^L$ is the Langlands dual of $\mathfrak{g}$) whose generalised monodromy data provide the solution of the Bethe Ansatz equations of that state.

In our previous paper (Masoero and Raimondo 2020) we constructed the opers corresponding to higher states of the $\mathfrak{g}$ quantum KdV model, for any $\mathfrak{g}$ untwisted affinization of a simply laced Lie algebra. This was done by following the definition given in Feigin and Frenkel (2011); solutions to the Bethe Ansatz were obtained based on our previous works (Masoero et al. 2016, 2017).

In this note we provide explicit and simpler formulas for opers corresponding to higher states of the quantum Boussinesq model, by specialising the results of Masoero and Raimondo (2020) to the case $\mathfrak{g} = \widehat{\mathfrak{sl}}_3$. This serves two purposes: we illustrate the general theory and its somehow heavy machinery in terms of familiar and simple objects, and we find formulas which are much closer to the original work on higher states opers of the $\widehat{\mathfrak{sl}}_2$-quantum KdV model (Bazhanov et al. 2004), where higher states are conjectured to correspond to Schrödinger operators with a *monster potential.*

As the result of the present paper, we conjecture that the level N states of the quantum Boussinesq model correspond to the following third order differential operators:

$$L = \partial_z^3 - \left(\sum_{j=1}^{N} \left(\frac{3}{(z-w_j)^2} + \frac{k}{z(z-w_j)} \right) + \frac{\bar{r}^1}{z^2} \right) \partial_z$$
$$+ \sum_{j=1}^{N} \left(\frac{3}{(z-w_j)^3} + \frac{a_j}{z(z-w_j)^2} + \frac{2(k+3)a_j - k^2}{3z^2(z-w_j)} \right) + \frac{\bar{r}^2}{z^3} + \frac{1}{z^2} + \lambda z^k, \quad (1)$$

where $-3 < k < -2$, and $\bar{r}^1, \bar{r}^2 \in \mathbb{C}$, and where the $2N$ complex variables $\{a_\ell, w_\ell\}_{\ell=1,\dots,N}$, satisfy the following system of $2N$ algebraic equations

$$a_\ell^2 - k a_\ell + k^2 + 3k - 3\bar{r}^1 = \sum_{\substack{j=1,\dots,N \\ j\neq \ell}} \left(\frac{9 w_\ell^2}{(w_\ell - w_j)^2} + \frac{3k w_\ell}{w_\ell - w_j} \right), \quad (2a)$$

$$A a_\ell + B - 9(k+2) w_\ell = \sum_{\substack{j=1 \\ j\neq \ell}}^{N} \left(\frac{18(k - a_\ell - a_j) w_\ell^3}{(w_\ell - w_j)^3} + \frac{(12k + 9k^2 - (63+6k)a_j - 9k a_\ell) w_\ell^2}{(w_\ell - w_j)^2} \right.$$
$$\left. + \frac{(9k + 16k^2 + 6(k^2 + 10k + 6)a_j - 5k a_\ell) w_\ell}{w_\ell - w_j} \right). \quad (2b)$$

The parameters A, B are given by

$$A = 14k^2 + 50k - 8\bar{r}^1 + 45,$$
$$B = 27(\bar{r}^1 - \bar{r}^2) - k(7k^2 + 7k + 9\bar{r}^2 - 13\bar{r}^1 + 9),$$

and the additional singularities w_j, $j = 1 \ldots, N$ are assumed to be pairwise distinct and nonzero. The system of algebraic equation (2) is equivalent to the requirement that the monodromy around the singular point $z = w_j$ is trivial for all $j = 1 \ldots N$, independently on the parameter λ.

The correspondence among the free parameters λ, $\bar{r}^1$, $\bar{r}^2$, k of the above equations and the free parameters c, (Δ_2, Δ_3), μ (respectively the central charge, the highest weight, the spectral parameter) of the Quantum Boussinesq model, as constructed in Bazhanov et al. (2002) (more about this below), goes as follows:

$$c = -\frac{3(4k+9)(3k+5)}{k+3}, \tag{3a}$$
$$\Delta_2 = \frac{(\bar{r}^1 - 8)k^2 + 6(\bar{r}^1 - 5)k + 9\bar{r}^1 - 27}{9(k+3)}, \tag{3b}$$
$$\Delta_3 = \frac{(k+3)^{3/2}}{27}(\bar{r}^1 - \bar{r}^2), \tag{3c}$$
$$\lambda = -i\,\Gamma(-k-2)^3\mu^3, \tag{3d}$$

where $\Gamma(s)$ denotes the Γ function with argument s. Moreover the integer N, which is the number of additional regular singularities in (1), coincides with the level of the state. Hence, system (2) is expected to possess $p_2(N)$ solutions, where $p_2(N)$ is the number of bi-coloured partitions of N.

The paper is organised as follows. In Sect. 2 we introduce the quantum KdV opers, following (Masoero and Raimondo 2020) (which in turns builds on Feigin and Frenkel 2011), and derive from the general theory of the formulas (1) and (2). In Sect. 3 we review the construction of solutions of the Bethe Ansatz equations as generalised monodromy data, following Masoero et al. (2016), Masoero and Raimondo (2020). Finally, in Sect. 4 we briefly summarise the construction of the quantum Boussinesq model provided in Bazhanov et al. (2002).

This work deals with differential equations and representation theory. We omit many proofs of the analytic results, which can be found in greater generality in Masoero and Raimondo (2020). However, we do provide all details of the algebraic calculations.

2 Quantum KdV Opers

In this section we introduce the Quantum KdV opers, as defined in Feigin and Frenkel (2011), in the special case $\mathfrak{g} = \widehat{\mathfrak{sl}}_3$, and derive the third order scalar differential operator (1). The reader should refer to Masoero and Raimondo (2020), and references therein for more details.

We begin by introducing some theory of the algebra $\mathfrak{sl}_3(\mathbb{C})$,[1] which we realise as the Lie algebra of traceless 3 by 3 matrices (in such a way that it coincides with its first fundamental representation, also known as standard representation). The algebra has the decomposition $\mathfrak{n}_- \oplus \mathfrak{h} \oplus \mathfrak{n}_+$, where $\mathfrak{n}_-$ are lower diagonal matrices, $\mathfrak{h}$ is the Cartan subalgebra of traceless diagonal matrices, and $\mathfrak{n}_+$ are upper diagonal matrices. The subalgebra $\mathfrak{b}_+ := \mathfrak{h} \oplus \mathfrak{n}_+$ is called the Borel subalgebra. We provide an explicit basis of $\mathfrak{b}_+$ as follows

$$h_1 = \begin{pmatrix} 1 & 0 & 0 \\ 0 & -1 & 0 \\ 0 & 0 & 0 \end{pmatrix}, \quad h_2 = \begin{pmatrix} 0 & 0 & 0 \\ 0 & 1 & 0 \\ 0 & 0 & -1 \end{pmatrix}, \tag{4}$$

$$e_1 = \begin{pmatrix} 0 & 1 & 0 \\ 0 & 0 & 0 \\ 0 & 0 & 0 \end{pmatrix}, \quad e_2 = \begin{pmatrix} 0 & 0 & 0 \\ 0 & 0 & 1 \\ 0 & 0 & 0 \end{pmatrix}, \quad e_\theta = \begin{pmatrix} 0 & 0 & 1 \\ 0 & 0 & 0 \\ 0 & 0 & 0 \end{pmatrix}. \tag{5}$$

We introduce three further elements, the sum of the negative Chevalley generators of the Lie algebra $f \in \mathfrak{n}_-$ (principal nilpotent element), the dual of the Weyl vector $\rho^\vee \in \mathfrak{h}$, and the dual of the highest root $\theta^\vee \in \mathfrak{h}$. We have:

$$f = \begin{pmatrix} 0 & 0 & 0 \\ 1 & 0 & 0 \\ 0 & 1 & 0 \end{pmatrix}, \quad \rho^\vee = \theta^\vee = \begin{pmatrix} 1 & 0 & 0 \\ 0 & 0 & 0 \\ 0 & 0 & -1 \end{pmatrix}. \tag{6}$$

The unipotent group $\mathcal{N} = \{\exp y,\, y \in \mathfrak{n}_+\}$ acts on $\mathfrak{sl}_3$ via the formula

$$\exp y.g = g + \sum_{k \geq 1} \frac{(\mathrm{ad}_y)^k.g}{k!}, \qquad \mathrm{ad}_y.g := [y, g],$$

and the affine subspace $f + \mathfrak{b}_+$ is preserved by the action. Following Kostant (1978), and given a vector subspace $\mathfrak{s} \subset \mathfrak{n}_+$, we say that the affine subspace $f + \mathfrak{s}$ is a *transversal space* if

1. The orbit of $f + \mathfrak{s}$ under the action of $\mathcal{N}$ coincides with $f + \mathfrak{b}_+$.
2. For each $s \in \mathfrak{s}$, then $\exp y.(f + s) \notin f + \mathfrak{s}$ unless $y = 0$.

The subspace $\mathfrak{s} = \mathbb{C}e_1 \oplus \mathbb{C}e_\theta$ satisfies the above hypotheses[2] and the transversal space $f + \mathfrak{s}$ is the space of companion matrices:

[1]For sake of simplicity we prefer to work with $\mathfrak{sl}_3$-opers, instead of $\widehat{\mathfrak{sl}_3}$-opers. We do that by considering the loop algebra variable λ as a free complex parameter. More about this in Masoero and Raimondo (2020), Sect. 4.

[2]As an example, the Cartan subalgebra $\mathfrak{h}$ satisfies the first but not the second hypothesis above.

$$f + \mathfrak{s} = \left\{ \begin{pmatrix} 0 & a & b \\ 1 & 0 & 0 \\ 0 & 1 & 0 \end{pmatrix} \mid a, b \in \mathbb{C} \right\}.$$

We fix this choice for the rest of the paper.

2.1 Opers

We denote by K the field of rational functions in the variable z, and we define

1. $\mathfrak{g}(K), \mathfrak{b}_+(K), \mathfrak{n}_+(K)$ the Lie algebras of rational functions with values in $\mathfrak{g}, \mathfrak{b}_+, \mathfrak{n}_+$ respectively.
2. The space of (global meromorphic) $\mathfrak{g}$-valued connections $\mathrm{conn}(K) = \{\partial_z + g, g \in \mathfrak{g}(K)\}$.
3. The subset $\mathrm{op}(K) = \{\mathcal{L} = \partial_z + f + b, b \in \mathfrak{b}_+(K)\} \subset \mathrm{conn}(K)$.
4. The group of unipotent Gauge transformations $\mathcal{N}(K) = \{\exp y, y \in \mathfrak{n}_+(K)\}$, acting on $\mathrm{conn}(K)$ via the formula

$$\exp y.(\partial_z + g) = \partial_z - \sum_{k \geq 0} \frac{1}{(k+1)!} (\mathrm{ad}_y)^k \frac{dy}{dz} + \exp y.g. \tag{7}$$

   Note that the above action preserves the subset $\mathrm{op}(K)$.
5. The space of $\mathfrak{sl}_3$ opers as $\mathrm{Op}(K) = \mathrm{op}(K)/\mathcal{N}(K)$.

The space of opers $\mathrm{Op}(K)$ admits a very explicit description once a transversal space $f + \mathfrak{s}$ is fixed: any element in $\mathrm{op}(K)$ is Gauge equivalent to a unique connection of the form $\partial_z + f + s, s \in \mathfrak{s}(K)$. Hence we have a bijection

$$\mathrm{Op}(K) \cong \{\partial_z + f + s, s \in \mathfrak{s}(K)\}.$$

We call $\partial_z + f + s$ the *canonical form* of any oper Gauge equivalent to it.

2.2 Opers and Scalar ODEs

It is a standard and elementary result that the space of $\mathfrak{sl}_3$ opers coincides with the space of third order linear scalar differential operators (with principal symbol equal to 1 and vanishing sub-principal symbol). Indeed, for what we have said so far, any oper has a unique representative of the form

$$\mathcal{L} = \partial_z + f + v_1(z)e_1 + v_2(z)e_\theta,$$

where v_1, v_2 are a pair of (arbitrary) rational functions. In the first fundamental representation, this oper takes the form

$$\mathcal{L} = \partial_z + \begin{pmatrix} 0 & v_1(z) & v_2(z) \\ 1 & 0 & 0 \\ 0 & 1 & 0 \end{pmatrix}. \tag{8}$$

If $\{\epsilon_1, \epsilon_2, \epsilon_3\}$ is the standard basis of $\mathbb{C}^3$, and given $\psi = \mathbb{C} \to \mathbb{C}^3$, with $\psi(z) = \psi_1(z)\epsilon_1 + \psi_2(z)\epsilon_2 + \psi_3(z)\epsilon_3$, then the matrix first order equation

$$\mathcal{L}\psi(z) = 0,$$

is easily seen to be equivalent to the following scalar ODE for the third coefficient $\Psi := \psi_3$

$$(\partial_z^3 - v_1\partial_z + v_2)\Psi(z) = 0. \tag{9}$$

We will use this scalar representation in the rest of the paper.

2.3 (Ir)Regular Singularities

Let $\mathcal{L}$ be an oper in the canonical form (8), and $w \in \mathbb{C}$ a pole of v_1 or v_2, so that

$$v_1 = \bar{s}_1(z-w)^{-\delta_1} + o((z-w)^{-\delta_1}),$$
$$v_2 = \bar{s}_2(z-w)^{-\delta_2} + o((z-w)^{-\delta_2})$$

for some $\bar{s}^1, \bar{s}^2 \neq 0$ and some $\delta_1, \delta_2 \in \mathbb{Z}$. We define (Masoero and Raimondo 2020)

- The *slope* of the singular point $w \in \mathbb{C}$ as

$$\mu = \max\left\{1, \max\left\{\frac{\delta_1}{2}, \frac{\delta_2}{3}\right\}\right\} \in \mathbb{Q}.$$

- The *principal coefficient* of the singular point w as

$$f - \rho^\vee + \bar{s}^1 e_1 + \bar{s}^2 e_2 = \begin{pmatrix} -1 & \bar{s}^1 & \bar{s}^2 \\ 1 & 0 & 0 \\ 0 & 1 & 1 \end{pmatrix} \quad \text{if} \quad \mu = 1,$$

and

$$f + \bar{s}^1 e_1 + \bar{s}^2 e_2 = \begin{pmatrix} 0 & \bar{s}^1 & \bar{s}^2 \\ 1 & 0 & 0 \\ 0 & 1 & 0 \end{pmatrix} \quad \text{if} \quad \mu > 1.$$

As proved in Masoero and Raimondo (2020), the singularity is *regular* (in the sense of linear connections) if $\mu = 1$ and *irregular* if $\mu > 1$.

Remark 1 In the case when $w = \infty$, we write $v_1 = z^{\delta_1} + o(z^{\delta_1})$, and $v_2 = \bar{s}_2 z^{\delta_2} + o(z^{\delta_2})$ for some $\bar{s}^1, \bar{s}^2 \neq 0$, and $\delta_1, \delta_2 \in \mathbb{Z}$, and define the slope of $w = \infty$ as $\mu = \max\{1, \max\{\frac{\delta_1}{2}, \frac{\delta_2}{3}\} + 2\}$. The principal coefficient is defined as above.

2.4 $\mathfrak{sl}_3$-*Quantum KdV Opers*

We define $\mathfrak{sl}_3$-quantum KdV opers following Feigin and Frenkel (2011). To this aim we fix $-3 < k < -2$ and $\bar{r}^1, \bar{r}^2 \in \mathbb{C}$ and write

$$\mathcal{L}(z, \lambda) = \mathcal{L}_{G,\mathfrak{s}}(z, \lambda) + s(z), \quad s \in K(\mathfrak{s}) . \tag{10}$$

Here $\mathcal{L}_{G,\mathfrak{s}}$ is the ground state oper

$$\mathcal{L}_{G,\mathfrak{s}}(z, \lambda) = \partial_z + \begin{pmatrix} 0 & \bar{r}^1/z^2 & \bar{r}^2/z^3 + z^{-2} + \lambda z^k \\ 1 & 0 & 0 \\ 0 & 1 & 0 \end{pmatrix} . \tag{11}$$

We notice that $\mathcal{L}_{G,\mathfrak{s}}(z, \lambda)$ has two singular points: $z = 0$ is a regular singularity with principal coefficient

$$\begin{pmatrix} -1 & \bar{r}^1 & \bar{r}^2 \\ 1 & 0 & 0 \\ 0 & 1 & 1 \end{pmatrix} ,$$

while $z = \infty$ is an irregular singularity, with slope $\mu = \frac{4}{3}$ and principal coefficient

$$\begin{pmatrix} 0 & 0 & 1 \\ 1 & 0 & 0 \\ 0 & 1 & 0 \end{pmatrix} .$$

As it will be reviewed in the next section, one can obtain solutions of the Bethe Ansatz equations by considering the differential equation $\mathcal{L}_{G,\mathfrak{s}}\psi = 0$: more precisely these are obtained as coefficients of the expansion of the subdominant solution at $+\infty$ in terms of a distinguished basis of solutions defined at $z = 0$.

In Bazhanov et al. (2004), Bazhanov, Lukyanov and Zamolodchikov proved that in the case $\mathfrak{g} = \widehat{\mathfrak{sl}}_2$, the ground state oper could be modified without altering the above global structure, so that the modified equations yield (different) solutions of the same Bethe Ansatz equations (as coefficients of the same expansion). Feigin and Frenkel (2011) extended these idea to the case of a general Kac-Moody algebra, and conjectured that the higher level opers could be uniquely specified by imposing on the $\mathfrak{s}$-valued function s the 4 conditions below. These conditions were shown to sufficient (Masoero and Raimondo 2020), and are expected to be necessary for generic values of the parameters $\hat{k}, \bar{r}^1, \bar{r}^2$ (Feigin and Frenkel 2011). We say that the oper $\mathcal{L}(z,\lambda)$ of the form (10) is a $\mathfrak{sl}_3$-quantum KdV oper if it satisfies the following 4 assumptions:

Assumption 1 The slope and principal coefficient at 0 do not depend on s.

Assumption 2 The slope and principal coefficient at ∞ do not depend on s.

Assumption 3 All additional singular points are regular and the corresponding principal coefficients are conjugated to the element $f - \rho^\vee - \theta^\vee \in f + \mathfrak{h}$.

Assumption 4 All additional singular points have trivial monodromy for every $\lambda \in \mathbb{C}$.

The following proposition, which is Proposition 4.7 in Masoero and Raimondo (2020) specialised to the case of $\mathfrak{g} = \mathfrak{sl}_3$, is a first characterisation of the Quantum KdV opers; it shows that they have the form (1).

Proposition 1 *An operator $\mathcal{L}(z,\lambda)$ of the form* (10) *satisfies the first three Assumptions if and only if there exists a (possibly empty) arbitrary finite collection of non-zero mutually distinct complex numbers $\{w_j\}_{j\in J} \subset \mathbb{C}^\times$ and a collection of numbers $\{a_{11}^{(j)}, a_{21}^{(j)}, a_{22}^{(j)}\}_{j\in J} \subset \mathbb{C}$, such that $\mathcal{L}(z,\lambda)$ has the form*

$$\mathcal{L}(z,\lambda) = \partial_z + \begin{pmatrix} 0 & W_1 & W_2 \\ 1 & 0 & 0 \\ 0 & 1 & 0 \end{pmatrix}, \tag{12}$$

where

$$W_1(z) = \frac{\bar{r}^1}{z^2} + \sum_{j\in J}\left(\frac{3}{(z-w_j)^2} + \frac{a_{11}^{(j)}}{z(z-w_j)}\right), \tag{13a}$$

$$W_2(z,\lambda) = \frac{\bar{r}^2}{z^3} + \frac{1}{z^2} + \lambda z^k + \sum_{j\in J}\left(\frac{3}{(z-w_j)^3} + \frac{a_{21}^{(j)}}{z(z-w_j)^2} + \frac{a_{22}^{(j)}}{z^2(z-w_j)}\right). \tag{13b}$$

Note that when J is empty then (12) reduces to the ground state oper (11). If J is not empty, then we set $J = \{1,\ldots,N\}$, for some $N \in \mathbb{Z}_+$. In order to fully characterise the $\mathfrak{sl}_3$-quantum KdV opers, we must impose the fourth and last Assumption on the opers of the form (12), namely the triviality of the monodromy about all the additional singularities w_j, $j = 1\ldots N$. We notice that the opers of the form (12) depend on the $4N$ complex parameters $\{a_{11}^{(j)}, a_{21}^{(j)}, a_{22}^{(j)}, w_j\}_{j\in 1\ldots N}$. We will show in the following subsection that the trivial monodromy conditions are equivalent to a complete system of $4N$ algebraic equations, which in turn are equivalent to (1) and (2).

2.5 Trivial Monodromy Conditions

We fix $\ell \in 1 \dots N$ and study under which conditions the oper $\mathcal{L}(z, \lambda)$ of the form (12) has trivial monodromy about w_ℓ. As we showed in Masoero and Raimondo (2020), Assumption 3 (more precisely, the fact that $\theta^\vee$ is a co-root) implies that the monodromy about w_ℓ is trivial if and only if it is trivial in at least one irreducible (nontrivial) representation. In other words, it is necessary and sufficient that the monodromy at $z = w_\ell$ is trivial for the solutions of the equation $\mathcal{L}(z, \lambda)\Psi = 0$ in the standard representation.

To this aim we write the above equation in the scalar form

$$(\partial_z^3 - W_1 \partial_z + W_2)\Phi(z) = 0\,, \tag{14}$$

and use the method of the Frobenius expansion, that is we look for solutions of the form

$$\Phi^{(\beta)}(z) = \sum_{m\geq 0} \Phi_m (z - w_\ell)^{\beta+m}. \tag{15}$$

Writing the Laurent expansion of (14) at w_ℓ as

$$W_1(z) = \sum_{m=0}^{+\infty} q_{1m}^{(\ell)} (z - w_\ell)^{m-2}, \qquad q_{10}^{(\ell)} = 3, \tag{16a}$$

$$W_2(z) = \sum_{m=0}^{+\infty} q_{2m}^{(\ell)} (z - w_\ell)^{m-3}, \qquad q_{20}^{(\ell)} = 3, \tag{16b}$$

expanding the Eq. (15) in powers of $z - w_\ell$, and equating to zero term-by-term we obtain

$$\begin{aligned} &\Phi_0^{(\beta)} P(\beta) = 0, \\ &P(\beta + r)\Phi_r^{(\beta)} = \sum_{m=1}^{r} \left((\beta + r - m) q_{1m}^{(\ell)} - q_{2m}^{(\ell)} \right) \Phi_{r-m}^{(\beta)}, \end{aligned} \tag{17}$$

where the indicial polynomial $P(\beta) = (\beta - 3)(\beta - 1)(\beta + 1)$. The roots of the indicial polynomial, $\beta = -1, 1, 3$, are known as *indices*. Since the indices are integers, the monodromy matrix has a unique eigenvalue, 1, with algebraic multiplicity 3, and the monodromy is trivial if and only if the recursion (17) has a solution for all the indices. Indeed, in such a case, $\Phi^\beta(e^{2\pi i} z) = \Phi^\beta(z)$ for $\beta = -1, 1, 3$; otherwise logarithmic terms must be added to the series (15) and the monodromy is not diagonalizable (Wasow 2018).

We analyse the recursion (17) separately for the three indices.

The recursion (17) for the index $\beta = 3$ admits always a unique solution, since $P(3 + r) \neq 0, \forall r \geq 1$.

In the case $\beta = 1$, we have that $P(\beta + r) = 0, r \geq 1$ if and only if $r = 2$. Hence the recursion is over-determined. Computing the first two terms we obtain

$$
\begin{aligned}
-3\Phi_1^{(1)} &= \left(q_{11}^{(\ell)} - q_{21}^{(\ell)}\right)\Phi_0^{(1)},\\
0 \times \Phi_2^{(1)} &= \left(2q_{11}^{(\ell)} - q_{21}^{(\ell)}\right)\Phi_1^{(1)} + \left(q_{12}^{(\ell)} - q_{22}^{(\ell)}\right)\Phi_0^{(1)}.
\end{aligned}
$$

It follows that the recursion for the index $\beta = 1$ has at least one solution if and only if

$$
q_{12}^{(\ell)} - q_{22}^{(\ell)} = \frac{2}{3}\left(q_{11}^{(\ell)}\right)^2 - q_{11}^{(\ell)}q_{21}^{(\ell)} + \frac{1}{3}\left(q_{21}^{(\ell)}\right)^2. \tag{18}
$$

Finally, the Frobenius method for the index $\beta_2 = -1$ gives

$$
\begin{aligned}
3\Phi_1^{(-1)} &= -\left(q_{11}^{(\ell)} + q_{21}^{(\ell)}\right)\Phi_0^{(-1)},\\
0 \times \Phi_2^{(-1)} &= -q_{21}^{(\ell)}\Phi_1^{(-1)} - \left(q_{12}^{(\ell)} + q_{22}^{(\ell)}\right)\Phi_0^{(-1)},\\
-3\Phi_3^{(-1)} &= \left(q_{11}^{(\ell)} - q_{21}^{(\ell)}\right)\Phi_2^{(-1)} - q_{22}^{(\ell)}\Phi_1^{(-1)} - \left(q_{13}^{(\ell)} + q_{23}^{(\ell)}\right)\Phi_0^{(-1)},\\
0 \times \Phi_4^{(-1)} &= \left(2q_{11}^{(\ell)} - q_{21}^{(\ell)}\right)\Phi_3^{(-1)} + \left(q_{12}^{(\ell)} - q_{22}^{(\ell)}\right)\Phi_2^{(-1)} - q_{23}^{(\ell)}\Phi_1^{(-1)} - \left(q_{14}^{(\ell)} + q_{24}^{(\ell)}\right)\Phi_0^{(-1)},
\end{aligned}
$$

and we obtain the following constraints

$$
\begin{aligned}
q_{12}^{(\ell)} + q_{22}^{(\ell)} &= \frac{1}{3}q_{21}^{(\ell)}q_{11}^{(\ell)} + \frac{1}{3}\left(q_{21}^{(\ell)}\right)^2,\\
q_{14}^{(\ell)} + q_{24}^{(\ell)} &= (2q_{11}^{(\ell)} - q_{21}^{(\ell)})\left(-\frac{1}{9}q_{22}^{(\ell)}(q_{11}^{(\ell)} + q_{21}^{(\ell)}) + \frac{1}{3}(q_{13}^{(\ell)} + q_{23}^{(\ell)})\right)\\
&\quad + \frac{1}{3}q_{23}^{(\ell)}(q_{11}^{(\ell)} + q_{21}^{(\ell)}).
\end{aligned}
$$

Combining these with (18) we obtain the following characterisation: the monodromy about w_ℓ is trivial if and only if the following system of 3 equations

$$
q_{12}^{(\ell)} = \frac{1}{3}\left(\left(q_{11}^{(\ell)}\right)^2 - q_{11}^{(\ell)}q_{21}^{(\ell)} + \left(q_{21}^{(\ell)}\right)^2\right), \tag{19}
$$

$$
q_{22}^{(\ell)} = \frac{1}{3}q_{11}^{(\ell)}\left(2q_{21}^{(\ell)} - q_{11}^{(\ell)}\right), \tag{20}
$$

$$
\begin{aligned}
q_{14}^{(\ell)} + q_{24}^{(\ell)} &= \frac{1}{3}q_{13}^{(\ell)}\left(2q_{11}^{(\ell)} - q_{21}^{(\ell)}\right) + q_{11}^{(\ell)}q_{23}^{(\ell)}\\
&\quad + \frac{1}{27}q_{11}^{(\ell)}\left(2q_{11}^{(\ell)} - q_{21}^{(\ell)}\right)\left(q_{11}^{(\ell)} - 2q_{21}^{(\ell)}\right)\left(q_{11}^{(\ell)} + q_{21}^{(\ell)}\right).
\end{aligned} \tag{21}
$$

In order to proceed further we write explicitly the coefficients q's, which appear in the above equations, in terms of the parameters of the opers (12)

$$q_{10}^{\ell}=3,\quad q_{20}^{\ell}=3,\quad q_{11}^{(\ell)}=\frac{a_{11}^{(\ell)}}{w_\ell},\quad q_{21}^{(\ell)}=\frac{a_{21}^{(\ell)}}{w_\ell},$$

$$q_{12}^{(\ell)}=\frac{\bar r^1-a_{11}^{(\ell)}}{w_\ell^2}+\sum_{\substack{j=1\\ j\neq\ell}}^N\left(\frac{3}{(w_\ell-w_j)^2}+\frac{a_{11}^{(j)}}{w_\ell(w_\ell-w_j)}\right),\quad q_{22}^{(\ell)}=\frac{a_{22}^{(\ell)}-a_{21}^{(\ell)}}{w_\ell^2},$$

$$q_{13}^{(\ell)}=\frac{a_{11}^{(\ell)}-2\bar r^1}{w_\ell^3}-\sum_{\substack{j=1\\ j\neq\ell}}^N\left(\frac{6}{(w_\ell-w_j)^3}+\frac{a_{11}^{(j)}}{w_\ell(w_\ell-w_j)^2}+\frac{a_{11}^{(j)}}{w_\ell^2(w_\ell-w_j)}\right),$$

$$q_{23}^{(\ell)}=\frac{\bar r^2+a_{21}^{(\ell)}-2a_{22}^{(\ell)}+w_\ell}{w_\ell^3}+\lambda w_\ell^k+\sum_{\substack{j=1\\ j\neq\ell}}^N\left(\frac{3}{(w_\ell-w_j)^3}+\frac{a_{21}^{(j)}}{w_\ell(w_\ell-w_j)^2}+\frac{a_{22}^{(j)}}{w_\ell^2(w_\ell-w_j)}\right),$$

$$q_{14}^{(\ell)}=\frac{3\bar r^1-a_{11}^{(\ell)}}{w_\ell^4}+\sum_{\substack{j=1\\ j\neq\ell}}^N\left(\frac{9}{(w_\ell-w_j)^4}+\frac{a_{11}^j}{w_\ell(w_\ell-w_j)^3}+\frac{a_{11}^j}{w_\ell^2(w_\ell-w_j)^2}+\frac{a_{11}^j}{w_\ell^3(w_\ell-w_j)}\right),$$

$$q_{24}^{(\ell)}=\frac{3a_{22}^{(\ell)}-a_{21}^{(\ell)}-3\bar r^2-2w_\ell}{w_\ell^4}+\lambda k w_\ell^{k-1}$$
$$-\sum_{\substack{j=1\\ j\neq\ell}}^N\left(\frac{9}{(w_\ell-w_j)^4}+\frac{2a_{21}^{(\ell)}}{w_\ell(w_\ell-w_j)^3}+\frac{a_{21}^{(\ell)}+a_{22}^{(\ell)}}{w_\ell^2(w_\ell-w_j)^2}+\frac{2a_{22}^{(\ell)}}{w_\ell^3(w_\ell-w_j)}\right).$$

We notice that while Eqs. (19) and (20) do not depend on λ, Eq. (21) is a first-order polynomial in λ. Since the trivial monodromy conditions must hold for any λ, Eq. (21) consists of a pair of independent constraints: both the constant part in λ and the linear part in λ are required to vanish independently. The vanishing of the part of (21) which is linear in λ reads:

$$q_{11}^{(\ell)}w_\ell^k-kw_\ell^{k-1}=0,\qquad \text{or}\qquad q_{11}^{(\ell)}=\frac{k}{w_\ell},$$

from which we obtain

$$a_{11}^{(\ell)}=k,\qquad \ell=1,\dots,N. \tag{22}$$

Making use of the explicit expression of the $q's$ in terms of the $a's$, as given above, and denoting

$$a_\ell=a_{21}^{(\ell)},\qquad \ell=1,\dots,N, \tag{23}$$

from (20) we obtain

$$a_{22}^{(\ell)}=\frac{2}{3}(k+3)a_\ell-\frac{k^2}{3},\qquad \ell=1,\dots,N. \tag{24}$$

Substituting (22) and (24) into the expression for the q's found above, then from (19) we obtain (2a), while the vanishing of the constant (in λ) coefficient of (21) is equivalent to (2b).

We have thus arrived to the following result: an $\mathfrak{sl}_3$ Quantum KdV oper is equivalent to a scalar third order differential operator of the form (1) such that its coefficients satisfy the system of algebraic equations (2).

2.6 *The Dual Representation. Formal Adjoint Operator*

Before we proceed further with our analysis, and we construct solutions to the Bethe Ansatz equations, we introduce a second representation of the algebra $\mathfrak{sl}_3$. This is called the second fundamental representation or dual representation, and we denote it by $\mathbb{C}^{3*}$. If $\{\epsilon_1, \epsilon_2, \epsilon_3\}$ is the standard basis of $\mathbb{C}^3$ as above, we denote by $\{\epsilon_1^*, \epsilon_2^*, \epsilon_3^*\}$, the corresponding dual basis in $\mathbb{C}^{3*}$ such that $\langle \epsilon_i^*, \epsilon_j \rangle = \delta_{ij}$. In these basis, the matrices representing the elements $h_1, h_2, \theta^\vee, \rho^\vee, e_1, e_2, e_\theta, f$ read

$$h_1 = \begin{pmatrix} 0 & 0 & 0 \\ 0 & 1 & 0 \\ 0 & 0 & -1 \end{pmatrix}, \quad h_2 = \begin{pmatrix} 1 & 0 & 0 \\ 0 & -1 & 0 \\ 0 & 0 & 0 \end{pmatrix}, \quad \rho^\vee = \theta^\vee = \begin{pmatrix} 1 & 0 & 0 \\ 0 & 0 & 0 \\ 0 & 0 & -1 \end{pmatrix},$$

$$e_1 = \begin{pmatrix} 0 & 0 & 0 \\ 0 & 0 & 1 \\ 0 & 0 & 0 \end{pmatrix}, \quad e_2 = \begin{pmatrix} 0 & 1 & 0 \\ 0 & 0 & 0 \\ 0 & 0 & 0 \end{pmatrix}, \quad e_\theta = \begin{pmatrix} 0 & 0 & -1 \\ 0 & 0 & 0 \\ 0 & 0 & 0 \end{pmatrix}, \quad f = \begin{pmatrix} 0 & 0 & 0 \\ 1 & 0 & 0 \\ 0 & 1 & 0 \end{pmatrix}.$$

As we have discussed before, the general $\mathfrak{sl}_3$ oper can be written the canonical form as the connection $\mathcal{L} = \partial_z + f + v_1 e_1 + v_2 e_\theta$, for an arbitrary pair of rational functions $v_1, v_2 \in K$. In the dual representation, we thus have

$$\mathcal{L} = \partial_z + \begin{pmatrix} 0 & 0 & -v_2 \\ 1 & 0 & v_1 \\ 0 & 1 & 0 \end{pmatrix}.$$

We showed that in the standard representation $\mathbb{C}^3$ the connection $\mathcal{L}$ is equivalent to the scalar third order operator (9). In the dual representation the same oper is equivalent to a different scalar operator, namely to its formal adjoint. Let $\psi^* =: \mathbb{C} \to \mathbb{C}^{3*}$, with $\psi^*(z) = \psi_1^*(z)\epsilon_1^* + \psi_2^*(z)\epsilon_2^* + \psi_3^*(z)\epsilon_3^*$, satisfy $\mathcal{L}\psi^* = 0$ in the dual representation. Then $\Psi^*(z) := \psi_3^*(z)$ satisfies the scalar ODE

$$(-\partial_z^3 + v_1\partial_z + (v_2 + v_1'))\Psi^*(z) = 0\,, \tag{25}$$

which is the formal adjoint of the Eq. (9).

The following standard isomorphisms (of $\mathfrak{sl}_3$-modules) will be needed later to derive the Bethe Ansatz equations: $\bigwedge^2 \mathbb{C}^3 \cong \mathbb{C}^{3*}$ and $\bigwedge^2 \mathbb{C}^{3*} \cong \mathbb{C}^3$. Explicitly,

$$\imath(\epsilon_1\wedge\epsilon_2)=\epsilon_1^*,\quad \imath(\epsilon_1\wedge\epsilon_3)=\epsilon_2^*,\quad \imath(\epsilon_2\wedge\epsilon_3)=\epsilon_3^*, \tag{26}$$

$$\imath(\epsilon_1^*\wedge\epsilon_2^*)=\epsilon_1,\quad \imath(\epsilon_1^*\wedge\epsilon_3^*)=\epsilon_2,\quad \imath(\epsilon_2^*\wedge\epsilon_3^*)=\epsilon_3. \tag{27}$$

The above isomorphisms imply that if $\psi(z),\phi(z)$ are solutions of $\mathcal{L}\psi(z)=0$, for $\psi:\mathbb{C}\to\mathbb{C}^3$ in the standard representation then $\imath\big(\psi(z)\wedge\phi(z)\big)$ is a solution of the dual equation $\mathcal{L}\psi^*(z)=0$, with $\psi^*:\mathbb{C}\to\mathbb{C}^{3*}$; and conversely.

In the present paper we prefer to work with solutions of the equations in the scalar form (9) and (25). Recall that the solution of the equations in the scalar form is just the third component of the solution of the vector equation. If $\psi(z)=\psi_1(z)\epsilon_1+\psi_2(z)\epsilon_2+\psi_3(z)\epsilon_3$ and $\phi(z)=\phi_1(z)\epsilon_1+\phi_2(z)\epsilon_2+\phi_3(z)\epsilon_3$, then a simple calculation shows that

$$\langle\imath(\psi\wedge\phi),\epsilon_3\rangle=Wr[\psi_3,\phi_3]$$

where $Wr[\cdot,\cdot]$ denotes the usual Wronskian $Wr[f(z),g(z)]=f(z)g'(z)-f'(z)g(z)$. Similarly, for $\psi^*(z)=\psi_1^*(z)\epsilon_1^*+\psi_2^*(z)\epsilon_2^*+\psi_3^*(z)\epsilon_3^*$ and $\phi^*(z)=\phi_1^*(z)\epsilon_1^*+\phi_2^*(z)\epsilon_2^*+\phi_3^*(z)\epsilon_3^*$ we have

$$\langle\epsilon_3^*,\imath(\psi^*\wedge\phi^*)\rangle=Wr[\psi_3^*,\phi_3^*].$$

To prove the above relations, it is sufficient to note that from the matrix first order equations $\mathcal{L}\psi(z)=0$, $\mathcal{L}\psi^*(z)=0$ we obtain the identities $\psi_2(z)=-\psi_3'(z)$ and $\psi_2^*=-\psi_3^{*\prime}(z)$. We have thus shown that the Wronskian of two solutions of (9) satisfies (25), and conversely the Wronskian of two solutions of (25) satisfies (9).

2.7 Relation with Previous Works

The ground state $\mathfrak{sl}_3$-quantum KdV oper, given by Eq. (11), was also considered – in the scalar form – by Dorey and Tateo (2000), and by Bazhanov et al. (2002), who wrote the following third order scalar operator

$$\widetilde{L}(x,E)=\partial_x^3+\frac{\widetilde{w}_1}{x^2}\partial_x+\frac{\widetilde{w}_2}{x^3}+x^{3M}-E, \tag{28}$$

with $\widetilde{w}_1=\tilde{\ell}_1\tilde{\ell}_2+\tilde{\ell}_1\tilde{\ell}_3+\tilde{\ell}_2\tilde{\ell}_3-2$, $\widetilde{w}_2=-\tilde{\ell}_1\tilde{\ell}_2\tilde{\ell}_3$ and where the $\tilde{\ell}_i$'s are constrained by the equation $\tilde{\ell}_1+\tilde{\ell}_2+\tilde{\ell}_3=3$. In addition, in our previous paper (Masoero et al. 2016) we considered the ground state oper in the following form

$$\mathcal{L}(x,E)=\partial_x+\begin{pmatrix}\ell_1/x & 0 & x^{3M}-E\\ 1 & (\ell_2-\ell_1)/x & 0\\ 0 & 1 & -\ell_2/x\end{pmatrix} \tag{29}$$

for arbitrary $\ell_1, \ell_2 \in \mathbb{C}$ and $M > 0$. We now show that the differential operators (11), (28), and (29) are equivalent under appropriate change of coordinates and Gauge transformations, once the parameters are correctly identified. To show that the differential operators (28) and (29) are equivalent, we write the operator (28) in the oper form

$$\partial_x + \begin{pmatrix} 0 & -\widetilde{w}_1/x^2 & \widetilde{w}_2/x^3 + x^{3M} - E \\ 1 & 0 & 0 \\ 0 & 1 & 0 \end{pmatrix}. \tag{30}$$

It is then a simple computation to show that (30) and (29) are Gauge equivalent if we set $\tilde{\ell}_1 = -\ell_1 + 2$, $\tilde{\ell}_2 = \ell_1 - \ell_2 + 1$ and $\tilde{\ell}_3 = \ell_2$.

Next we show the equivalence between (29) and (11). As observed in Feigin and Frenkel (2011), after the change of variable

$$z = \varphi(x) = \left(\frac{k+3}{3}\right)^3 x^{\frac{3}{k+3}}, \quad k = -\frac{3M+2}{1+M}, \tag{31}$$

the operator (29) reads

$$\mathcal{L}_G(z, \lambda) = \partial_z + \begin{pmatrix} r_1/z & 0 & z^{-2} + \lambda z^k \\ 1 & (r_2 - r_1)/z & 0 \\ 0 & 1 & -r_2/z \end{pmatrix}, \tag{32}$$

where $\lambda \in \mathbb{C}$ and $r_1, r_2 \in \mathbb{C}$ are defined by the relations

$$E = -\left(\frac{k+3}{3}\right)^{3(k+2)} \lambda, \quad \ell_i = \frac{3}{k+3}(r_i - 1) + 1, \; i = 1, 2. \tag{33}$$

It is again a simple computation to show that the opers (32) and (11) are Gauge equivalent provided the coefficients $r^1, r^2, \bar{r}^1, \bar{r}^2$ satisfy the following relations

$$\begin{cases} \bar{r}^1 = (r^1)^2 - r^1 r^2 + (r^2)^2 - r^1 - r^2, \\ \bar{r}^2 = r^1 r^2 (r^1 - r^2) + r^2 (2r^2 - r^1 - 2). \end{cases} \tag{34}$$

2.8 *Weyl Group Symmetry*

The parametrisation (34) of $\bar{r}^1, \bar{r}^2$ in terms of r^1 and r^2 will be very convenient when discussing the behaviour of solutions of $\mathcal{L}(z, \lambda)\psi = 0$ in a neighbourhood of $z = 0$. The Weyl group of $\mathfrak{sl}_3$ – which is isomorphic to the group of permutations of three elements, S_3 – is a symmetry of the map (34), once its action on the parameters r^1, r^2, which is called the *dot action*, is properly defined:

$$\sigma \cdot \begin{pmatrix} r^1 \\ r^2 \end{pmatrix} = \begin{pmatrix} -r^2+2 \\ -r^1+2 \end{pmatrix}, \quad \tau \cdot \begin{pmatrix} r^1 \\ r^2 \end{pmatrix} = \begin{pmatrix} -r^2+2 \\ r^1-r^2+1 \end{pmatrix}. \tag{35}$$

We let the reader verify that σ, τ generate the group S_3 (in particular $\sigma^2 = 1, \tau^3 = 1$) and that the above action is a symmetry of (34). This phenomenon is studied in great detail and generality in Masoero and Raimondo (2020), Sect. 5.

3 The Bethe Ansatz Equations

In this section we construct solutions of the Bethe Ansatz equations as generalised monodromy data of Quantum KdV opers, $\mathcal{L}(z, \lambda)$. As proved in Sect. 2, these are opers of the form

$$\mathcal{L}(z, \lambda) = \partial_z + \begin{pmatrix} 0 & W_1 & W_2 \\ 1 & 0 & 0 \\ 0 & 1 & 0 \end{pmatrix}, \tag{36}$$

where

$$W_1(z) = \frac{\bar{r}^1}{z^2} + \sum_{j \in J} \left(\frac{3}{(z-w_j)^2} + \frac{k}{z(z-w_j)} \right), \tag{37a}$$

$$W_2(z, \lambda) = \frac{\bar{r}^2}{z^3} + \frac{1}{z^2} + \lambda z^k + \sum_{j \in J} \left(\frac{3}{(z-w_j)^3} + \frac{a_j}{z(z-w_j)^2} + \frac{2(k+3)a_j - k^2}{3z^2(z-w_j)} \right), \tag{37b}$$

and where $\{a_j, w_j\}_{j=1,\dots,N}$ satisfy the system of equations (2). We follow Masoero et al. (2016), Masoero and Raimondo (2020) closely and the reader should refer to these papers for all missing proofs. Any finite dimensional representation V of $\mathfrak{sl}_3$ defines the ODE

$$\mathcal{L}(z, \lambda)\psi = 0, \qquad \Psi : \mathbb{C} \to V.$$

Since the monodromy of $\mathcal{L}(z, \lambda)$ about w_j is trivial for any j, then the solutions of the above equation are, for fixed λ, analytic functions on the universal cover of $\mathbb{C}^*$, minus the lift of the points w_j, $j \in 1 \dots N$. We denote such a domain by $\widehat{\mathbb{C}}$. As it was originally observed by Dorey and Tateo, the appearance of the Bethe Ansatz equations is due to a discrete symmetry which acts on both the variable z and the parameter λ. It is therefore necessary to consider solutions $\psi(z, \lambda)$ as analytic functions of both variables z and λ. More precisely for our purpose $\psi(z, \cdot)$ is assumed to be an entire function of λ. We thus define a solution to be an analytic map $\psi : \widehat{\mathbb{C}} \times \mathbb{C} \to V$ which satisfies the equation $\mathcal{L}(z, \lambda)\psi(z, \lambda) = 0$ for every (z, λ).

The space of solutions, which we denote by $V(\lambda)$, is an infinite dimensional vector space which, as we showed in Masoero and Raimondo (2020), is simply isomorphic to $V \otimes O_\lambda$, where O_λ is the ring of entire functions of the variable λ. This means that an O_λ-basis of the space of solutions has cardinality dim V.

3.1 Twisted Opers

Let $\hat{k} = -k - 2$, so that $0 < \hat{k} < 1$. For any $t \in \mathbb{R}$ we define the twisted operator and twisted solution:

$$\mathcal{L}^t(z, \lambda) := \mathcal{L}(e^{2i\pi t} z, e^{2i\pi t\hat{k}} \lambda), \tag{38}$$

$$\psi_t(z, \lambda) = e^{2i\pi t\rho^\vee} \psi(e^{2\pi it} z, e^{2\pi it\hat{k}} \lambda). \tag{39}$$

Taking into account the oper change of variables (Masoero and Raimondo 2020), then from (36) we explicitly have

$$\mathcal{L}^t(z, \lambda) = \partial_z + f + e^{4\pi it} W_1(e^{2\pi it} z) e_1 + e^{6\pi it} W_2(e^{2\pi it} z, e^{2\pi it\hat{k}} \lambda) e_\theta,$$

and one easily see that the function $\psi_t(z, \lambda)$ satisfies $\mathcal{L}^t(z, \lambda)\psi_t(z, \lambda) = 0$. A crucial property of the oper (36) is the following Dorey-Tateo discrete symmetry:

$$\mathcal{L}^{t=1}(z, \lambda) = \mathcal{L}(z, \lambda) \ , \tag{40}$$

which leads us to consider the following (O_λ-linear) monodromy operator

$$M : V(\lambda) \to V(\lambda), \qquad M(\psi(z, \lambda)) = e^{2i\pi\rho^\vee} \psi(e^{2\pi i} z, e^{2\pi i\hat{k}} \lambda) \ . \tag{41}$$

In the case $\mathfrak{sl}_3$, we just need to consider the equations $\mathcal{L}^t(z, \lambda)\psi = 0$ for the standard representation and its dual. More precisely, the standard representation at 0 twist, and the dual representation at twist $t = \frac{1}{2}$

$$\mathcal{L}(z, \lambda)\psi(z, \lambda) = 0, \qquad \psi : \widehat{\mathbb{C}} \times \mathbb{C} \to \mathbb{C}^3, \tag{42}$$

$$\mathcal{L}^{\frac{1}{2}}(z, \lambda)\psi^*(z, \lambda) = 0, \qquad \psi^* : \widehat{\mathbb{C}} \times \mathbb{C} \to \mathbb{C}^{3*}. \tag{43}$$

By a slight abuse of notation we denote $\mathbb{C}^3(\lambda)$ the space of solutions of the first equation, and by $\mathbb{C}^{3*}(\lambda)$ the space of solutions of the latter equations, as well as the solutions of the same equations in the equivalent scalar form

$$\big(\partial_z^3 - W_1(z)\partial_z + W_2(z, \lambda)\big)\Psi(z, \lambda) = 0, \tag{44}$$

$$\big(\partial_z^3 - W_1(-z)\partial_z + W_2(-z, e^{\pi i\hat{k}}\lambda) - W_1'(-z)\big)\Psi^*(z, \lambda) = 0. \tag{45}$$

Since the solution of the equations in the scalar form is the third component of a solution of the equation in the matrix form, and since $\rho^\vee \epsilon_3 = -\epsilon_3, \rho^\vee \epsilon_3^* = -\epsilon_3^*$, the twist for solutions of the above scalar ODEs is defined as follows

$$\Psi_t(z,\lambda) = e^{-2i\pi t}\Psi(e^{2\pi i t}z, e^{2\pi i t\hat{k}}\lambda), \quad \Psi_t^*(z,\lambda) = e^{-2i\pi t}\Psi^*(e^{2\pi i t}z, e^{2\pi i t\hat{k}}\lambda).$$

Equation (45) is the adjoint equation to (44) twisted by $t = \frac{1}{2}$; and conversely, Eq. (44) is the adjoint equation to (45) twisted by $t = \frac{1}{2}$. As we recalled in Sect. 2.6, the Wronskian of two solutions of a scalar ODE solves the adjoint equation. It follows that

1. If $\Psi(z,\lambda), \Phi(z,\lambda) \in \mathbb{C}^3(\lambda)$ then

$$Wr[\Psi_{-\frac{1}{2}}(z,\lambda), \Phi_{\frac{1}{2}}(z,\lambda)] \in \mathbb{C}^{3^*}(\lambda),$$

2. If $\Psi^*(z,\lambda), \Phi^*(z,\lambda) \in \mathbb{C}^{3^*}(\lambda)$, then

$$Wr[\Psi^*_{-\frac{1}{2}}(z,\lambda), \Phi^*_{\frac{1}{2}}(z,\lambda)] \in \mathbb{C}^3(\lambda).$$

3.2 The Eigenbasis of the Monodromy Operator. Expansion at $z = 0$

The point $z = 0$ is a regular singularity for the Eqs. (44) and (45), but it is also a branch point of the potential W_2, because of the term λz^k. It follows that the standard Frobenius series cannot provide solution of the above equations at $z = 0$. A generalised Frobenius series, introduced in Masoero and Raimondo (2020), does however the job. The latter is defined as

$$\Phi^{(\beta)}(z,\lambda) = z^\beta \sum_{m\geq n\geq 0} c_{m,n} z^m \zeta^n, \; c_{0,0} = 1, \quad \zeta = \lambda z^{-\hat{k}}, \tag{46}$$

where the indices β are computed as in the standard Frobenius method: if the equation reads

$$\left(\partial_z^3 + \frac{a+o(1)}{z^2}\partial_z + \frac{b+o(1)}{z^3}\right)\Psi(z) = 0,$$

the indices are the roots of the indicial polynomial $P(\beta) = \beta^3 - 3\beta^2 + (2+a)\beta + b$. The following facts are proved in Masoero and Raimondo (2020), Proposition 5.1. For every finite dimensional representation V of $\mathfrak{sl}_3$, and under some genericity assumptions [3] on the triple $(\hat{k}, \bar{r}^1, \bar{r}^2)$, we have:

[3] The genericity assumptions imply that the monodromy operator M is diagonal and no logarithmic terms appear in the generalised Frobenius series.

1. The series (46) converges to a solution $\Phi^{(\beta)}(z,\lambda) \in V(\lambda)$.
2. $M\Phi^{(\beta)}(z,\lambda) = e^{2\pi i\beta}\Phi^{(\beta)}(z,\lambda)$, where M is the monodromy operator defined in (41).
3. The collection of the solutions $\Phi^{(\beta)}(z,\lambda)$ for all indices β forms an O_λ-basis of $V(\lambda)$.

In the cases under our study, namely Eqs. (44) and (45), the indicial polynomials are, respectively, given by

$$P(\beta) = \beta^3 - 3\beta^2 + (2-\bar{r}^1)\beta + \bar{r}^2,$$
$$P^*(\beta) = \beta^3 - 3\beta^2 + (2-\bar{r}^1) + 2\bar{r}^1 - \bar{r}^2.$$

Using (34), then we obtain the factorizations

$$P(\beta) = (\beta - r^2)(\beta - 1 + r^2 - r^1)(\beta - 2 + r^1),$$
$$P^*(\beta) = (\beta - r^1)(\beta - 1 + r^1 - r^2)(\beta - 2 + r^2),$$

so that the indices are given by

$$\beta_1 = r^2, \quad \beta_2 = r^1 - r^2 + 1, \quad \beta_3 = -r^1 + 2, \tag{47a}$$
$$\beta_1^* = -r^2 + 2, \quad \beta_2^* = r^2 - r^1 + 1, \quad \beta_3^* = r^1. \tag{47b}$$

We denote by

$$\{\Phi^{(\beta_1)}(z,\lambda), \Phi^{(\beta_2)}(z,\lambda), \Phi^{(\beta_3)}(z,\lambda)\}, \tag{48a}$$
$$\{\Phi^{(\beta_1^*)}(z,\lambda), \Phi^{(\beta_2^*)}(z,\lambda), \Phi^{(\beta_3^*)}(z,\lambda)\}, \tag{48b}$$

the corresponding solutions of (44) and (45) respectively. Recall that the Weyl group acting by the dot action (35) on r^1, r^2, provides a group of symmetries of $\bar{r}^1, \bar{r}^2$, hence it leaves the indicial polynomial invariant, permuting its roots.[4] The (induced) action of the generators σ, τ of the Weyl group, see (35), on the indices (47) is provided by the following permutations:

$$\sigma(\beta_i) = \beta_{\sigma(i)}, \quad \tau(\beta_i) = \beta_{\tau(i)}, \qquad i = 1,2,3, \tag{49a}$$
$$\sigma(\beta_i^*) = \beta^*_{\sigma(i)}, \quad \tau(\beta_i^*) = \beta_{\tau(i)}, \qquad i = 1,2,3, \tag{49b}$$

where

$$\sigma(1,2,3) := (3,2,1), \qquad \tau(1,2,3) := (2,3,1). \tag{50}$$

Comparing the asymptotic behaviour at $z = 0$, we deduce the following 6 quadratic identities among the (properly normalised) $\Phi^{(\beta)}$'s and $\Phi^{(\beta^*)}$'s. Let $s \in S_3$, then (we

[4]Many authors fix r^1, r^2 by imposing the conditions $\Re\beta_1 > \Re\beta_2 > \Re\beta_3$, or equivalently $\Re\beta_3^* > \Re\beta_2^* > \Re\beta_1^*$ (Bazhanov et al. 2002; Dorey et al. 2007; Masoero et al. 2016).

can find a normalisation of the solutions $\Phi^{(\beta_i)}$, $\Phi^{(\beta_i^*)}$ such that):

$$Wr[\Phi^{(\beta_{s(1)})}_{\mp\frac{1}{2}}, \Phi^{(\beta_{s(2)})}_{\pm\frac{1}{2}}] = (-1)^{p(s)} e^{\pm i\pi s(\gamma)} \Phi^{(\beta^*_{s(3)})}, \tag{51a}$$

$$Wr[\Phi^{(\beta^*_{s(3)})}_{\mp\frac{1}{2}}, \Phi^{(\beta^*_{s(2)})}_{\pm\frac{1}{2}}] = (-1)^{p(s)} e^{\pm i\pi s(\gamma^*)} \Phi^{(\beta_{s(1)})}, \tag{51b}$$

where $p(s)$ is the parity of $s \in S_3$, and where

$$s(\gamma) = \beta_{s(2)} - \beta_{s(1)}, \qquad s(\gamma^*) = \beta^*_{s(2)} - \beta^*_{s(3)}, \tag{52}$$

with the $\beta_{s(i)}$ and $\beta^*_{s(i)}$ defined by the relations (47) and (49).

3.3 Sibuya Solutions. Expansion at $z = \infty$

We let $q(z, \lambda)$ be the Puiseaux series of $\left(z^{-2}(1 + \lambda z^{-\hat{k}})\right)^{\frac{1}{3}}$ truncated after terms of z^{-1}, and $S(z, \lambda)$ be its primitive

$$q(z, \lambda) = z^{\frac{-2}{3}}\left(1 + \sum_{l=0}^{\lfloor \frac{1}{3k} \rfloor} c_l \lambda^l z^{-l\hat{k}}\right), \qquad S(z, \lambda) = \int^z q(y, \lambda) dy, \tag{53}$$

where c_l are the coefficients of Taylor series expansion at $y = 0$ of $(1 - y)^{\frac{1}{3}}$, and $\int^z y^l dl = \frac{z^{l+1}}{l+1}, l \neq -1, \int^z \frac{1}{y} = \log z$.

The Sibuya, or subdominant, solution of the Eqs. (44) and (45) is uniquely defined by the following asymptotics

$$\Psi(z, \lambda) = z^{\frac{2}{3}} e^{-S(z,\lambda)}\left(1 + o(1)\right), \quad \text{as } z \to +\infty, \tag{54a}$$

$$\Psi^*(z, \lambda) = z^{\frac{2}{3}} e^{-S(z,\lambda)}\left(1 + o(1)\right), \quad \text{as } z \to +\infty. \tag{54b}$$

Moreover we have that

$$\Psi'(z, \lambda) = -e^{-S(z,\lambda)}\left(1 + o(1)\right), \quad \text{as } z \to +\infty, \tag{55a}$$

$$\Psi^{*\prime}(z, \lambda) = -e^{-S(z,\lambda)}\left(1 + o(1)\right), \quad \text{as } z \to +\infty. \tag{55b}$$

The Sibuya solutions Ψ, Ψ^* satisfy the following properties

- It is the solution (unique up to a multiplicative constant) with the fastest decrease as $z \to +\infty$.
- The asymptotic formulas (54) hold true on the sector $|\arg z| \leq \pi + \varepsilon$, for some $\varepsilon > 0$ (Masoero et al. 2016). In other words, if we continue analytically $\Psi(z, \lambda)$, Ψ^*

(z, λ) as well as the functions $q(z, \lambda)$ and $S(z, \lambda)$ past the negative real semi-axis, the asymptotic formulas still hold.

- The solutions $\Psi(z, \lambda)$, $\Psi^*(z, \lambda)$ are entire functions of λ, i.e. $\Psi(z, \lambda) \in \mathbb{C}^3(\lambda)$ and $\Psi^*(z, \lambda) \in \mathbb{C}^{3*}(\lambda)$.
- Finally, and most importantly, the solutions $\Psi(z, \lambda)$, $\Psi^*(z, \lambda)$ satisfy the so-called Ψ-system

$$Wr[\Psi_{-\frac{1}{2}}(z, \lambda), \Psi_{\frac{1}{2}}(z, \lambda)]) = \Psi^*(z, \lambda), \tag{56a}$$

$$Wr[\Psi^*_{-\frac{1}{2}}(z, \lambda), \Psi^*_{\frac{1}{2}}(z, \lambda)]) = \Psi(z, \lambda). \tag{56b}$$

The latter identities can be checked by comparing the asymptotic expansion of the left and right hand side as $z \to +\infty$.

The Ψ-system is the last necessary ingredient to construct solutions of the Bethe Ansatz equations.

3.4 $Q\widetilde{Q}$ System and the Bethe Ansatz

As we have shown, the solutions $\{\Phi^{(\beta_1)}(z, \lambda), \Phi^{(\beta_2)}(z, \lambda), \Phi^{(\beta_3)}(z, \lambda)\}$ obtained in (48a) provide an O_λ basis of $\mathbb{C}^3(\lambda)$, and the Sibuya solution $\Psi(z, \lambda)$ belongs to the same space. It follows that there exists a unique triplet of entire functions $Q_i(\lambda) \in O_\lambda$, for $i = 1, 2, 3$, such that

$$\Psi(z, \lambda) = Q_1(\lambda)\Phi^{(\beta_1)}(z, \lambda) + Q_2(\lambda)\Phi^{(\beta_2)}(z, \lambda) + Q_3(\lambda)\Phi^{(\beta_3)}(z, \lambda)\ . \tag{57a}$$

Similarly, we have that

$$\Psi^*(z, \lambda) = Q_1^*(\lambda)\Phi^{(\beta_1^*)}(z, \lambda) + Q_2^*(\lambda)\Phi^{(\beta_2^*)}(z, \lambda) + Q_3^*(\lambda)\Phi^{(\beta_3^*)}(z, \lambda)\ , \tag{57b}$$

for a unique triplet of entire functions $Q_i^*(\lambda) \in O_\lambda$, with $i = 1, 2, 3$. Substituting the expansions (57) in the Ψ-system (56) and making use of the relations (51) we obtain the following quadratic relations among the coefficients Q's and Q^*'s, which is known as $Q\widetilde{Q}$-system. For each $s \in S_3$ we have

$$\begin{aligned}(-1)^{p(s)} Q^*_{s(3)}(\lambda) = {} & e^{i\pi s(\gamma)} Q_{s(1)}(e^{-i\pi\hat{k}}\lambda) Q_{s(2)}(e^{i\pi\hat{k}}\lambda) \\ & - e^{-i\pi s(\gamma)} Q_{s(1)}(e^{i\pi\hat{k}}\lambda) Q_{s(2)}(e^{-i\pi\hat{k}}\lambda),\end{aligned} \tag{58a}$$

$$\begin{aligned}(-1)^{p(s)} Q_{s(3)}(\lambda) = {} & e^{i\pi s(\gamma^*)} Q^*_{s(1)}(e^{-i\pi\hat{k}}\lambda) Q_{s(2)}(e^{i\pi\hat{k}}\lambda) \\ & - e^{-i\pi s(\gamma^*)} Q^*_{s(1)}(e^{i\pi\hat{k}}\lambda) Q^*_{s(2)}(e^{-i\pi\hat{k}}\lambda),\end{aligned} \tag{58b}$$

where $p(s)$ is the parity of s, and the phases $s(\gamma)$, $s(\gamma^*)$ are defined in (52).

Remark 2 System (58) was shown by Frenkel and Hernandez (2018) to be a universal system of relations in the commutative Grothendieck ring $K_0(\mathcal{O})$ of the category $\mathcal{O}$ of representations of the Borel subalgebra of the quantum affine algebra $U_q(\widehat{\mathfrak{sl}}_3)$.

Finally, the Bethe Ansatz equations is a pair of functional relations for each one of the six pairs of Q functions of the form $\{Q_{s(1)}Q^*_{s(3)}\}$, $s \in S_3$. Let λ_s denote an arbitrary zero of the function $Q_{s(1)}$,and λ^*_s an arbitrary zero of $Q^*_{s(3)}$. Evaluating the above relations at $e^{\pm i\pi\hat{k}}\lambda_s$ we obtain the Bethe Ansatz equations

$$
\begin{aligned}
-e^{2i\pi s(\gamma)}\frac{Q_{s(1)}(e^{2i\pi\hat{k}}\lambda_s)}{Q_{s(1)}(e^{-2i\pi\hat{k}}\lambda_s)} &= \frac{Q^*_{s(3)}(e^{i\pi\hat{k}}\lambda_s)}{Q^*_{s(3)}(e^{i\pi\hat{k}}\lambda_s)},\\
-e^{2i\pi s(\gamma^*)}\frac{Q^*_{s(3)}(e^{2i\pi\hat{k}}\lambda^*_s)}{Q^*_{s(3)}(e^{-2i\pi\hat{k}}\lambda^*_s)} &= \frac{Q_{s(1)}(e^{i\pi\hat{k}}\lambda^*_s)}{Q_{s(1)}(e^{-i\pi\hat{k}}\lambda^*_s)}.
\end{aligned}
$$

It is believed that each one of the 6 Bethe Ansatz equations is strong enough to characterise all of the Q's and Q^*'s, by means of the so-called Destri-De Vega equations.

4 Quantum Boussinesq Model

The quantum Boussinesq model has been described in great detail by Bazhanov et al. (2002), from which the notation of the present section is taken and to which we refer for further details. The model is defined by considering a highest weight representation $\mathcal{V}_{\Delta_2,\Delta_3}$ of the Zamolodchikov's $\mathcal{W}_3$-algebra (Zamolodchikov 1985), and it is characterized by 4 parameters: the central charge $-\infty < c < 2$, the highest weight $(\Delta_2, \Delta_3) \in \mathbb{C}^2$, and the spectral parameter $\mu \in \mathbb{C}$. For generic values of c, Δ_2, Δ_3, the representation $\mathcal{V}_{\Delta_2,\Delta_3}$ is irreducible, a condition we assume from now on. Let $\mathbf{L}_n$, $\mathbf{W}_n$, $n \in \mathbb{Z}$, denote the generators of the $\mathcal{W}_3$ algebra as in Bazhanov et al. (2002), Sect. 2. The highest weight fixes a ground state $|\Delta_2, \Delta_3\rangle \in \mathcal{V}_{\Delta_2,\Delta_3}$, satisfying $\mathbf{L}_n|\Delta_2, \Delta_3\rangle = \mathbf{W}_n|\Delta_2, \Delta_3\rangle = 0$ for $n > 0$, and

$$
\mathbf{L}_0|\Delta_2, \Delta_3\rangle = \Delta_2|\Delta_2, \Delta_3\rangle \qquad \mathbf{W}_0|\Delta_2, \Delta_3\rangle = \Delta_3|\Delta_2, \Delta_3\rangle.
$$

The $\mathcal{W}_3$-module $\mathcal{V}_{\Delta_2,\Delta_3}$ admits the level decomposition

$$
\mathcal{V}_{\Delta_2,\Delta_3} = \bigoplus_{N=0}^{\infty} \mathcal{V}^{(N)}_{\Delta_2,\Delta_3}, \qquad \mathbf{L}_0\mathcal{V}^{(N)}_{\Delta_2,\Delta_3} = (\Delta_2 + N)\mathcal{V}^{(N)}_{\Delta_2,\Delta_3}.
$$

The ground state $|\Delta_2, \Delta_3\rangle$ has level zero, the higher states are obtained by the action of products of the lowering operators $\mathbf{L}_n$, $\mathbf{W}_n$, $n < 0$. More precisely, let $\{\nu_1 \dots, \nu_l, \bar{\nu}_1 \dots, \bar{\nu}_k\}$, with $\nu_j, \bar{\nu}_j \in \mathbb{N}$, be a bicoloured integer partition of the integer

N, namely $\nu_j \leq \nu_{j+1}, \bar{\nu}_j \leq \bar{\nu}_{j+1}$ and $\sum_j \nu_j + \sum_j \bar{\nu}_j = N$; to any such a partition one associates a state of level N by the formula $\prod_j \mathbf{L}_{-\nu_j} \prod_j \mathbf{W}_{-\bar{\nu}_j} |0\rangle$.

The integrable structure of the quantum Boussinesq model can be conveniently encoded in the so-called **Q**-operators (Bazhanov et al. 2002, Sect. 2), from which the quantum integrals of motion of the model can be obtained. The **Q**-operators are more precisely operator-valued functions $\mathbf{Q}_i(t), \overline{\mathbf{Q}}_i(t), i = 1, 2, 3$, depending on the parameter $t = \mu^3$, where μ is the spectral parameter of the quantum model.[5] The level subspaces $\mathcal{V}^{(N)}_{\Delta_2,\Delta_3}$ are invariant with respect to the action of the **Q**-operators,

$$\mathbf{Q}_i(t) : \mathcal{V}^{(N)}_{\Delta_2,\Delta_3} \to \mathcal{V}^{(N)}_{\Delta_2,\Delta_3}, \qquad \overline{\mathbf{Q}}_i(t) : \mathcal{V}^{(N)}_{\Delta_2,\Delta_3} \to \mathcal{V}^{(N)}_{\Delta_2,\Delta_3},$$

and in particular (for $N = 0$), the ground state $|\Delta_2, \Delta_3\rangle$ is an eigenvector for the **Q**-operators:

$$\begin{aligned} \mathbf{Q}_i(t)|\Delta_2, \Delta_3\rangle &= P_i^{(vac)}(t)|\Delta_2, \Delta_3\rangle, \\ \overline{\mathbf{Q}}_i(t)|\Delta_2, \Delta_3\rangle &= \overline{P}_i^{(vac)}(t)|\Delta_2, \Delta_3\rangle. \end{aligned}$$

As proved in Bazhanov et al. (2002), Sect. 5, the **Q**-operators (and therefore their eigenvalues) satisfy the system of quadratic relations

$$c_1\overline{\mathbf{Q}}_1(t) = \mathbf{Q}_2(qt)\mathbf{Q}_3(q^{-1}t) - \mathbf{Q}_3(qt)\mathbf{Q}_2(q^{-1}t), \tag{59a}$$
$$c_1\mathbf{Q}_1(t) = \overline{\mathbf{Q}}_3(qt)\overline{\mathbf{Q}}_2(q^{-1}t) - \overline{\mathbf{Q}}_2(qt)\overline{\mathbf{Q}}_3(q^{-1}t), \tag{59b}$$
$$c_2\overline{\mathbf{Q}}_2(t) = \mathbf{Q}_3(qt)\mathbf{Q}_1(q^{-1}t) - \mathbf{Q}_1(qt)\mathbf{Q}_3(q^{-1}t), \tag{59c}$$
$$c_2\mathbf{Q}_2(t) = \overline{\mathbf{Q}}_1(qt)\overline{\mathbf{Q}}_3(q^{-1}t) - \overline{\mathbf{Q}}_3(qt)\overline{\mathbf{Q}}_1(q^{-1}t), \tag{59d}$$
$$c_3\overline{\mathbf{Q}}_3(t) = \mathbf{Q}_1(qt)\mathbf{Q}_2(q^{-1}t) - \mathbf{Q}_2(qt)\mathbf{Q}_1(q^{-1}t), \tag{59e}$$
$$c_3\mathbf{Q}_3(t) = \overline{\mathbf{Q}}_2(qt)\overline{\mathbf{Q}}_1(q^{-1}t) - \overline{\mathbf{Q}}_1(qt)\overline{\mathbf{Q}}_2(q^{-1}t), \tag{59f}$$

where

$$q = e^{i\pi g}, \tag{60a}$$
$$c_1 = e^{i\pi(p_1-\sqrt{3}p_2)} - e^{-i\pi(p_1-\sqrt{3}p_2)}, \tag{60b}$$
$$c_2 = e^{-2i\pi p_1} - e^{2i\pi p_1}, \tag{60c}$$
$$c_3 = e^{i\pi(p_1+\sqrt{3}p_2)} - e^{-i\pi(p_1+\sqrt{3}p_2)}, \tag{60d}$$

and the parameter g, p_1, p_2 are related to c, Δ_2, Δ_3 by the identities (Bazhanov et al. 2002, Sect. 3)

$$c = 50 - 24(g + g^{-1}), \quad \Delta_2 = \frac{p_1^2 + p_2^2}{g} + \frac{c-2}{24}, \quad \Delta_3 = \frac{2p_2(p_2^2 - 3p_1^2)}{(3g)^{3/2}}. \tag{61}$$

[5]The spectral parameter μ is denoted λ in Bazhanov et al. (2002).

4.1 From (58) to (59)

We now prove that the $Q\widetilde{Q}$ system (58) and the system (59) are equivalent. As a by-product we deduce the explicit relations (3) among the parameters of the opers, $\bar{r}^1, \bar{r}^2, k, \lambda$, and the parameters of the quantum theory, c, Δ_2, Δ_3, μ. More precisely, we derive (3a), (3b) and (3c) while (3d) can be found in Bazhanov et al. (2002).

Let Q_i, Q_i^*. $i = 1, 2, 3$ be the functions defined by the expansions (57), satisfying the $Q\widetilde{Q}$-system (58). Assume that $Q_i(0) \neq 0$ and $Q_i^*(0) \neq 0$, $i = 1, 2, 3$. Recall the definition of the indices β_i, β_i^*, $i = 1, 2, 3$ as given in (47). Then, a direct calculation shows that the functions

$$P_i(t) = t^{\beta_i} \frac{Q_i(t)}{Q_i(0)}, \qquad P_i^*(t) = t^{\beta_i^*} \frac{Q_i^*(t)}{Q_i^*(0)}, \qquad i = 1, 2, 3$$

satisfy (59), with the parameters g, p_1, p_2 appearing in (60) related to the parameters $\hat{k}, r^1, r^2$ by the relations

$$g = 1 - \hat{k} = k + 3, \quad p_1 = \frac{r^1}{2} + \frac{r^2}{2} - 1, \quad p_2 = \frac{\sqrt{3}}{2}(r^1 - r_2). \tag{62}$$

Substituting the above equation into (61) and using (34) we obtain (3a)–(3c).

Acknowledgements The authors are partially supported by the FCT Project PTDC/MAT-PUR/30234/2017 'Irregular connections on algebraic curves and Quantum Field Theory'. D. M. is supported by the FCT Investigator grant IF/00069/2015 'A mathematical framework for the ODE/IM correspondence'. The authors thank Dipartimento di Matematica dell'Università degli studi di Genova for the kind hospitality.

References

Bazhanov, V.V., Hibberd, A., Khoroshkin, S.: Integrable structure of W3 conformal field theory, quantum Boussinesq theory and boundary affine Toda theory. Nucl. Phys. B **622**(3), 475–547 (2002)

Bazhanov, V.V., Lukyanov, S.L., Zamolodchikov, A.B.: Integrable structure of conformal field theory, quantum KdV theory and thermodynamic Bethe ansatz. Commun. Math. Phys. **177**(2) (1996)

Bazhanov, V.V., Lukyanov, S.L., Zamolodchikov, A.B.: Spectral determinants for Schrodinger equation and Q operators of conformal field theory. J. Statist. Phys. **102**, 567–576 (2001)

Bazhanov, V.V., Lukyanov, S.L., Zamolodchikov, A.B.: Higher-level eigenvalues of Q-operators and Schroedinger equation. Adv. Theor. Math. Phys. **7**, 711 (2004)

Dorey, P., Dunning, C., Masoero, D., Suzuki, J., Tateo, R.: Pseudo-differential equations, and the Bethe ansatz for the classical Lie algebras. Nucl. Phys. B **772**(3), 249–289 (2007)

Dorey, P., Dunning, C., Tateo, R.: The ODE/IM correspondence. J. Phys. A. Math. Theor. **40**(32) (2007)

Dorey, P., Tateo, R.: Anharmonic oscillators, the thermodynamic Bethe ansatz, and nonlinear integral equations. J. Phys. A **32**, L419–L425 (1999)

Dorey, Patrick, Tateo, Roberto: Differential equations and integrable models: the SU(3) case. Nucl. Phys. B **571**(3), 583–606 (2000)

Feigin, B., Frenkel, E.: Integrals of motion and quantum groups. In: Integrable systems and quantum groups, pp. 349–418. Springer (1996)

Feigin, B., Frenkel, E.: Quantization of soliton systems and Langlands duality. In: Exploring new structures and natural constructions in mathematical physics. Adv. Stud. Pure Math., vol. 61, pp. 185–274. Math. Soc. Japan, Tokyo (2011)

Frenkel, E., Hernandez, D.: Spectra of quantum Kdv hamiltonians, Langlands duality, and affine opers. Commun. Math. Phys. **362**(2), 361–414 (2018)

Kostant, Bertram: On whittaker vectors and representation theory. Inventiones mathematicae **48**, 101–184 (1978)

Kotousov, G.A., Lukyanov, S.L.: Bethe state norms for the heisenberg spin chain in the scaling limit (2019). arXiv:1906.07081

Masoero, D., Raimondo, A.: Opers for higher states of quantum KdV models. *Commun. Math. Phys.* **378**, 1–74 (2020)

Masoero, D., Raimondo, A., Valeri, D.: Bethe ansatz and the spectral theory of affine lie algebra-valued connections I. The simply-laced case. Comm. Math. Phys. **344**(3), 719–750 (2016)

Masoero, D., Raimondo, A., Valeri, D.: Bethe ansatz and the spectral theory of affine lie algebra-valued connections II: the non simply-laced case. Comm. Math. Phys. **349**(3), 1063–1105 (2017)

Wasow, W.: Asymptotic Expansions for Ordinary Differential Equations. Courier Dover Publications (2018)

Zamolodchikov, A.B.: Infinite extra symmetries in two-dimensional conformal quantum field theory. Teoret. Mat. Fiz. **65**(3), 347–359 (1985)

Nonsingular Rational Solutions to Integrable Models

Gegenhasi, Xing-Biao Hu, Shi-Hao Li, and Bao Wang

Abstract In the literature, there have been considerable interests in the study of nonsingular rational solutions for nonlinear integrable models. These nonsingular rational solutions have appeared with different names in a variety of nonlinear systems, say, algebraic solitons, algebraic solitrary waves and lump solutions etc. More importantly, these nonsingular rational solutions have played a key role in the study of rogue waves. In the paper, we will develop a new procedure to generate lump solutions via Bäcklund transformations and nonlinear superposition formulae for some integrable models. It is shown that our procedure can be utilized to some well-studied equations such as KPI equation, elliptic Toda equation and BKP equation, but also to comparatively less-studied DJKM equation, Novikov-Veselov equation and negative flow of the BKP equation.

Keywords Lump solution · Bäcklund transformation · Nonlinear superposition formulae

---

This work was done when the last two authors were being PhD students. Now Shi-Hao Li is at the School of Mathematics and Statistics, ARC Centre of Excellence for Mathematical and Statistical Frontiers, The University of Melbourne, Australia and Bao Wang is at the department of Mathematics and Statistics, Brock University, Canada.

Gegenhasi
School of Mathematical Science, Inner Mongolia University, No.235 West College Road, Hohhot 010021, Inner Mongolia, China

X.-B. Hu (✉) · S.-H. Li · B. Wang
LSEC, ICMSEC, Academy of Mathematics and Systems Science, Chinese Academy of Sciences, P.O.Box 2719, Beijing 100190, China
e-mail: hxb@lsec.cc.ac.cn

School of Mathematical Sciences, University of Chinese Academy of Sciences, Beijing 100049, China

© Springer Nature Switzerland AG 2020

F. Nijhoff et al. (eds.), *Asymptotic, Algebraic and Geometric Aspects of Integrable Systems*, Springer Proceedings in Mathematics & Statistics 338,
https://doi.org/10.1007/978-3-030-57000-2_6

1 Introduction

The theory of modern integrable systems originated from the work on the celebrated Korteweg-de Vries (KdV) equation. It is a prototype water wave model involving a broad variety of mathematical methods. This theory allows one to study a wide range of phenomena and problems arising from physics, biology, and pure and applied mathematics. The special significance of integrable systems is that they combine tractability with nonlinearity. Hence, these systems enable one to explore nonlinear phenomena while working with explicit solutions. One of the interesting explicit solutions in nonlinear dynamics is that of solitons. Kruskal and Zabusky first discovered solitons in the mid-1960s when they worked on the KdV equation. A soliton is essentially a localized object that may be found in diverse areas of physics, such as gravitation and field theory, plasma and solid state physics, and hydrodynamics. The importance of solitons stems from the exhibition of particle-type interactions and the characterization of the long time asymptotic behavior of the solution.

There are some other types of explicit solutions available in the literature. One of them is so-called rational solutions, which is important to be found for integrable equations. It provides us a criterion for integrability as the existence of an infinite sequence of rational solutions appears to be equivalent to the Painlevé property (Newell 1987), and the rational solutions are of, at least, potential value in physical applications. In this regard, of particularly interesting are an important class of what we called nonsingular rational solutions. To the best of our knowledge, the study of nonsingular rational solutions to integrable equations can be traced back to Ames (1967) where N.J. Zabusky found simplest nonsingular rational solution $u=-\frac{4q}{1+4q^2x^2}$ to the Gardner equation

$$u_t+12quu_x+6u^2u_x+u_{xxx}=0.$$

In the literature, there are three types of nonsingular rational solutions: (1) Algebraic solitons; (2) Lump solutions; (3) Rogue wave solutions. There are some examples which exhibit nonsingular rational solutions. In the case of algebraic solitons, a typical example is the Benjamin-Ono (BO) equation

$$u_t+4uu_x+Hu_{xx}=0,\quad Hu(x,t)\equiv\frac{1}{\pi}\text{PV}\int_{-\infty}^{\infty}\frac{u(y,t)}{y-x}dy. \tag{1}$$

In Ono (1975), Ono obtained 1-soliton solution $u=\frac{a}{a^2(x-at-x_0)^2+1}$. Some further results about the algebraic solitons of the BO equation could be found in Matsuno (1982a, b), Case (1979). The second example of algebraic solitons is the mKdV equation $v_t+6v^2v_x+v_{xxx}=0$, whose simplest algebraic solution was also given by Ono (1976) $v=v_0-\frac{4v_0}{4v_0^2(x-6v_0^2t)^2+1}$. Furthermore, N-algebraic solitons were found in Ablowitz and Satsuma (1978). As for lump solutions, the result can be traced to Manakov et al. (1977) where Manakov et al. gave lump solutions to the KPI equation. In particular, in Ablowitz and Satsuma (1978); Satsuma and Ablowitz (1979), Ablowitz and Satsuma developed a new method to seek lump solutions to the KPI equation and DSI equation by taking the “long-wave” limit of the soliton solutions and there have been many results about this topic; please see Feng et al. (1999), Grammaticos et al. (2007), Ablowitz et al. (2000), Villarroel and Ablowitz (1999), Ma (2015), Villarroel and Ablowitz (1994), Gilson and Nimmo (1990), Hu and Willox (1996). The third line of research about nonsingular rational solutions is rogue wave solutions, which is of physical significance. As is known, the NLS equation $iu_t + u_{xx} + 2|u|^2u = 0$ admits the following rogue wave solution $u = \left(1 - \frac{4(1+4it)}{1+4x^2+16t^2}\right) e^{2it}$. Obviously, by taking $u \longrightarrow ue^{-2it}$, we may get a nonsingular rational solution of the equation

$$iu_t + u_{xx} + 2(|u|^2 - 1)u = 0.$$

For more examples, please see, e.g., Kharif et al. (2009), Solli et al. (2007), Peregrine (1983), Dubard et al. (2010), Dubard and Matveev (2011), Gaillard (2011), Guo et al. (2012), Ohta and Yang (2012), Li et al. (2013), Ohta and Yang (2012, 2013) and references therein.

The purpose of this paper is to develop a new procedure to generate lump solutions to several integrable models. Different from those by Ablowitz and Satsuma by taking the “long-wave” limit of the soliton solutions obtained and those by Ablowitz and Villarroel based on inverse scattering transform, the technique we develop here is via Bäcklund transformations and nonlinear superposition formulae in Hirota’s bilinear formalism (Hirota and Satsuma 1978). We will apply our procedure to the some known examples such as KPI equation, two-dimensional Toda equation, BKP equation to show how it works and further to the DJKM equation, Novikov-Veselov equation and negative flow of BKP equation to show its effectiveness.

2 The Lump Solutions of KP Equation

The KP equation takes the form

$$(u_t + 6uu_x + u_{xxx})_x + \alpha u_{yy} = 0. \tag{2}$$

Traditionally, the Eq. (2) with $\alpha = -1$ is called KPI, and the one for $\alpha = 1$ is KPII. The KPI equation does not have stable soliton solutions but has localized solutions that decay algebraically as $x^2 + y^2 \to \infty$ and are called lumps. The lump solutions of KPI have been first obtained by Manakov et al. (1977) and also by Ablowitz and Satsuma (1978). Subsequently, Ablowitz and Satsuma derived the determinant form of the N-lump solution for the KPI equation by taking limits of the corresponding soliton solutions in Satsuma and Ablowitz (1979). In the following, we will use the bilinear Bäcklund transformation and the nonlinear superposition formula to rederive the N-lump solutions of KPI equation.

Through the dependent variable transformation $u = 2(\ln f)_{xx}$, the Eq. (2) can be written in bilinear form

$$(D_x D_t + D_x^4 + \alpha D_y^2) f \cdot f = 0, \tag{3}$$

where the bilinear operator $D_x^m D_t^k$ is defined by Hirota (2004)

$$D_x^m D_t^k a \cdot b \equiv \left(\frac{\partial}{\partial x} - \frac{\partial}{\partial x'}\right)^m \left(\frac{\partial}{\partial t} - \frac{\partial}{\partial t'}\right)^k a(x,t) b(x',t')\Bigg|_{x'=x, t'=t}.$$

A bilinear Bäcklund transformation for Eq. (3) is given by Nakamura (1981), Hu (1997)

$$(a D_y + D_x^2 + \lambda D_x) f \cdot f' = 0, \tag{4}$$
$$(D_t + D_x^3 - 3a\lambda D_y - 3a D_x D_y) f \cdot f' = 0, \tag{5}$$

where $a^2 = \frac{1}{3}\alpha$ and λ is an arbitrary constant. We represent (4)–(5) symbolically by $f \xrightarrow{\lambda} f'$. The associated nonlinear superposition formula for the Eq. (3) is stated in the following proposition (Nakamura 1981; Hu 1997).

Proposition 1 *Let f_0 be a nonzero solution of (3) and suppose that f_1 and f_2 are solutions of (3) such that $f_0 \xrightarrow{\lambda_i} f_i$ $(i = 1, 2)$. Then f_{12} defined by*

$$f_0 f_{12} = c[D_x + \frac{1}{2}(\lambda_2 - \lambda_1)] f_1 \cdot f_2, \quad c \text{ is a nonzero real constant} \tag{6}$$

is a new solution to (3) which is related to f_1 and f_2 under bilinear BT (4)–(5) with parameters λ_2 and λ_1 respectively, i.e.

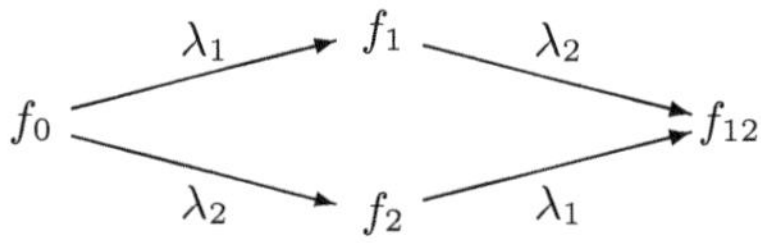

In Hu (1997), it has been shown if we choose $\theta_i = x + p_i y - \alpha p_i^2 t$, then the Bäcklund transformation tells us $1 \xrightarrow{\lambda_i = -ap_i} f_i = \theta_i + \beta_i$ (where β_i is a constant). By using proposition 1, we can obtain the following solution to the KP equation

$$f_{12} = \frac{2}{a(p_1 - p_2)}[f_{1x} f_2 - f_1 f_{2x} + \frac{1}{2}(\lambda_2 - \lambda_1) f_1 f_2] =$$
$$\theta_1\theta_2 + (\beta_1 + \frac{2}{a(p_1 - p_2)})\theta_2 + (\beta_2 - \frac{2}{a(p_1 - p_2)})\theta_1 + \beta_1\beta_2 + \frac{2(\beta_2 - \beta_1)}{a(p_1 - p_2)}, \tag{7}$$

by taking $c = \frac{2}{a(p_1-p_2)}$ in (6). If $\alpha = -1$, $p_2 = p_1^*$, $\beta_1 = -\frac{2}{a(p_1-p_2)}$, $\beta_2 = \frac{2}{a(p_1-p_2)}$ in (7), then we obtain the 1-lump solution

$$f_{12} = \theta_1\theta_1^* - \frac{12}{(p_1 - p_1^*)^2} > 0.$$

Furthermore, we can obtain an N-lump solution of the KP equation by using the nonlinear superposition formula repeatedly. For this purpose, we have the following proposition.

Proposition 2

$$F_N = c_N \begin{vmatrix} f_1 & f_2 & \cdots & f_N \\ (-\partial_x + \frac{\lambda_1}{2})f_1 & (-\partial_x + \frac{\lambda_2}{2})f_2 & \cdots & (-\partial_x + \frac{\lambda_N}{2})f_N \\ \vdots & \vdots & & \vdots \\ (-\partial_x + \frac{\lambda_1}{2})^{N-1}f_1 & (-\partial_x + \frac{\lambda_2}{2})^{N-1}f_2 & \cdots & (-\partial_x + \frac{\lambda_N}{2})^{N-1}f_N \end{vmatrix} \tag{8}$$

is a determinant solution to the KP equation (3), where $f_i (i = 1, 2, \ldots, N)$ *is obtained from the seed solution* f_0 *by using Bäcklund transformation* (4) *and* (5), *i.e.* $f_0 \xrightarrow{\lambda_i} f_i$.

In order to obtain the N-lump solution, we take $f_i = \theta_i + \beta_i, \theta_i = x + p_i y - \alpha p_i^2 t, \lambda_i = -ap_i, \beta_i = \sum_{j\neq i} \frac{2}{\lambda_i-\lambda_j}$ for $i = 1, 2, \ldots, N$ and $c_N = \prod_{1\leq i<j\leq N} \frac{2}{\lambda_j-\lambda_i}$. In this case, from (8), we have

$$F_N = c_N \begin{vmatrix} \theta_1 + \beta_1 & \cdots & \theta_N + \beta_N \\ -1 + \frac{\lambda_1}{2}(\theta_1 + \beta_1) & \cdots & -1 + \frac{\lambda_N}{2}(\theta_N + \beta_N) \\ \vdots & & \vdots \\ (-N+1)(\frac{\lambda_1}{2})^{N-2} + (\frac{\lambda_1}{2})^{N-1}(\theta_1 + \beta_1) & \cdots & (-N+1)(\frac{\lambda_N}{2})^{N-2} + (\frac{\lambda_N}{2})^{N-1}(\theta_N + \beta_N) \end{vmatrix}.$$

It can be verified that the above determinant can be written as the product of the determinants

$$\begin{vmatrix} 1 & \frac{\lambda_1}{2} & \cdots & (\frac{\lambda_1}{2})^{N-1} \\ 1 & \frac{\lambda_2}{2} & \cdots & (\frac{\lambda_2}{2})^{N-1} \\ \vdots & \vdots & & \vdots \\ 1 & \frac{\lambda_N}{2} & \cdots & (\frac{\lambda_N}{2})^{N-1} \end{vmatrix} \times \begin{vmatrix} \theta_1 & \frac{2}{\lambda_1-\lambda_2} & \cdots & \frac{2}{\lambda_1-\lambda_N} \\ -\frac{2}{\lambda_1-\lambda_2} & \theta_2 & \cdots & \frac{2}{\lambda_2-\lambda_N} \\ \vdots & \vdots & \vdots & \vdots \\ -\frac{2}{\lambda_1-\lambda_N} & -\frac{2}{\lambda_2-\lambda_N} & \cdots & \theta_N \end{vmatrix}.$$

By using the basic property of Vandermonde determinant, we know

$$F_N = \begin{vmatrix} \theta_1 & \frac{2}{\lambda_1-\lambda_2} & \cdots & \frac{2}{\lambda_1-\lambda_N} \\ -\frac{2}{\lambda_1-\lambda_2} & \theta_2 & \cdots & \frac{2}{\lambda_2-\lambda_N} \\ \vdots & \vdots & \vdots & \vdots \\ -\frac{2}{\lambda_1-\lambda_N} & -\frac{2}{\lambda_2-\lambda_N} & \cdots & \theta_N \end{vmatrix}. \tag{9}$$

If we choose $N = 2M,\ p_{M+i} = (p_i)^*(i = 1, 2, \ldots, M)$, then F_N gives the M-lump solutions of KPI equation which coincides with those obtained in Satsuma and Ablowitz (1979). The positivity of (9) could be found in Ohta and Yang (2013) for an affirmative answer.

3 The Lump Solutions of the DJKM Equation

The second equation of the KP hierarchy is the DJKM equation which is written as

$$w_{xxxxy} + 2w_{xxx}w_y + 4w_{xxy}w_x + 6w_{xy}w_{xx} - w_{yyy} - 2w_{xxt} = 0. \tag{10}$$

Through the dependent variable transformation $w = 2(\ln f)_x$, the Eq. (10) can be transformed into the multilinear form

$$D_x[(D_x^3D_y - D_xD_t)f \cdot f] \cdot f^2 + \frac{1}{2}D_y[(D_x^4 - 3D_y^2)f \cdot f] \cdot f^2 = 0. \tag{11}$$

A bilinear Bäcklund transformation for Eq. (11) is given by

$$(D_x^2 + iD_y + \lambda + \mu D_x)f \cdot f' = 0, \tag{12a}$$

$$(iD_t + \frac{3}{2}\lambda D_y - \frac{1}{2}D_y^2 - \frac{i}{2}\mu^2 D_y - \frac{i}{2}\mu D_x D_y - \frac{i}{2}D_x^2 D_y)f \cdot f' = 0, \tag{12b}$$

where λ, μ are arbitrary constants. If we take $\lambda = 0$ for simplicity, then Bäcklund transformation (12a) and (12b) can be symbolically written as $f \xrightarrow{\mu} f'$. The associated nonlinear superposition formula for the Eq. (11) is stated in the following proposition.

Proposition 3 *Let f_0 be a nonzero solution of (11) and suppose that f_1 and f_2 are solutions of (11) such that $f_0 \xrightarrow{\mu_i} f_i$ $(i = 1, 2)$. Then f_{12} defined by*

$$f_0 f_{12} = c[D_x + \frac{1}{2}(\mu_2 - \mu_1)]f_1 \cdot f_2, \tag{13}$$

is a new solution to (11) which is related to f_1 and f_2 under bilinear BT with parameters μ_2 and μ_1 respectively. Here c is a nonzero real constant.

Similar with the KP case, we obtain the 1-lump solution to the DJKM equation by using the Bäcklund transformation and nonlinear superposition formula. By setting $\theta_i = x + p_i y - \frac{1}{2}p_i^3 t$, then from bilinear BT, one obtains $1 \overset{\mu_i=-ip_i}{\longrightarrow} f_i = \theta_i + \beta_i$ (where β_i is a constant). Now from the nonlinear superposition formula (13), we obtain the following solution of the DJKM equation

$$f_{12} = \frac{2}{\mu_2 - \mu_1}[f_{1x}f_2 - f_1 f_{2x} + \frac{1}{2}(\mu_2 - \mu_1)f_1 f_2] =$$
$$\theta_1\theta_2 + (\beta_1 + \frac{2}{\mu_2 - \mu_1)})\theta_2 + (\beta_2 - \frac{2}{\mu_2 - \mu_1})\theta_1 + [\beta_1\beta_2 + \frac{2(\beta_2 - \beta_1)}{\mu_2 - \mu_1}] \quad (14)$$

by taking $c = \frac{2}{\mu_2 - \mu_1}$ in (13). If we choose $p_2 = p_1^*$, $\beta_1 = \frac{2}{\mu_1 - \mu_2}$, $\beta_2 = \frac{2}{\mu_2 - \mu_1}$ in (14), then we obtain $\mu_2 = -\mu_1^*$, $\theta_2 = \theta_1^*$ and the 1-lump solution

$$f_{12} = \theta_1\theta_1^* + \frac{4}{(\mu_1 + \mu_1^*)^2} = |\theta_1|^2 + \frac{4}{(\mu_1 + \mu_1^*)^2} > 0. \quad (15)$$

The N-lump solution could be found by using the nonlinear superposition formula repeatedly.

Proposition 4

$$F_N = c_N \begin{vmatrix} f_1 & f_2 & \cdots & f_N \\ (-\partial_x + \frac{\mu_1}{2})f_1 & (-\partial_x + \frac{\mu_2}{2})f_2 & \cdots & (-\partial_x + \frac{\mu_N}{2})f_N \\ \vdots & \vdots & \vdots & \vdots \\ (-\partial_x + \frac{\mu_1}{2})^{N-1}f_1 & (-\partial_x + \frac{\mu_2}{2})^{N-1}f_2 & \cdots & (-\partial_x + \frac{\mu_N}{2})^{N-1}f_N \end{vmatrix}$$

is a determinant solution to the DJKM equation (11), where $f_i (i = 1, 2, \ldots, N)$ are obtained from seed solution f_0 by using BT (12a)–(12b) $f_0 \overset{\mu_i}{\longrightarrow} f_i$.

In order to obtain the multi-lump solution, we take $f_i = \theta_i + \beta_i$, $\theta_i = x + p_i y - \frac{1}{2}p_i^3 t$, $\mu_i = -ip_i$, $\beta_i = \sum\limits_{j \neq i} \frac{2}{\mu_i - \mu_j}$ for $i = 1, 2, \ldots, N$. After the proper choices of parameters, the determinant F_N could be written as

$$F_N = c_N \begin{vmatrix} \theta_1 + \beta_1 & \cdots & \theta_N + \beta_N \\ \vdots & & \vdots \\ (-N+1)(\frac{\mu_1}{2})^{N-2} + (\frac{\mu_1}{2})^{N-1}(\theta_1 + \beta_1) & \cdots & (-N+1)(\frac{\mu_N}{N})^{N-2} + (\frac{\mu_N}{2})^{N-1}(\theta_N + \beta_N) \end{vmatrix}.$$

It can be verified that the above determinant is also a product of determinants

$$\begin{vmatrix} 1 & \frac{\mu_1}{2} & \cdots & (\frac{\mu_1}{2})^{N-1} \\ 1 & \frac{\mu_2}{2} & \cdots & (\frac{\mu_2}{2})^{N-1} \\ \vdots & \vdots & \vdots & \vdots \\ 1 & \frac{\mu_N}{2} & \cdots & (\frac{\mu_N}{2})^{N-1} \end{vmatrix} \times \begin{vmatrix} \theta_1 & \frac{2}{\mu_1 - \mu_2} & \cdots & \frac{2}{\mu_1 - \mu_N} \\ -\frac{2}{\mu_1 - \mu_2} & \theta_2 & \cdots & \frac{2}{\mu_2 - \mu_N} \\ \vdots & \vdots & \vdots & \vdots \\ -\frac{2}{\mu_1 - \mu_N} & -\frac{2}{\mu_2 - \mu_N} & \cdots & \theta_N \end{vmatrix}. \quad (16)$$

The choice of $c_N = \prod_{1\le i<j\le N} \frac{2}{\mu_j-\mu_i}$ gives us

$$F_N = \begin{vmatrix} \theta_1 & \frac{2}{\mu_1-\mu_2} & \cdots & \frac{2}{\mu_1-\mu_N} \\ -\frac{2}{\mu_1-\mu_2} & \theta_2 & \cdots & \frac{2}{\mu_2-\mu_N} \\ \vdots & \vdots & \vdots & \vdots \\ -\frac{2}{\mu_1-\mu_N} & -\frac{2}{\mu_2-\mu_N} & \cdots & \theta_N \end{vmatrix}. \tag{17}$$

For $N = 2M$ and $p_{M+i} = (p_i)^*$ $(i = 1, 2, \ldots, M)$, we could find that $\theta_{M+i} = (\theta_i)^*$, $\mu_{M+i} = -(\mu_i)^*$ and the positivity of F_N is the same with KP case. Therefore, in this case, F_N is the M-lump solution of the DJKM equation.

4 The Lump Solutions of the Elliptic Toda Equation

We now consider the so-called elliptic Toda equation

$$\left(\frac{\partial^2}{\partial x^2} + \frac{\partial^2}{\partial y^2}\right)(\log u_n) = u_{n+1} - 2u_n + u_{n-1}.$$

This equation has been studied in Villarroel (1998); Villarroel and Ablowitz (1994), where the inverse scattering method was applied to obtain lump solutions. By the use of variable transformation $u_n = \frac{f_{n+1} f_{n-1}}{f_n^2}$, we can obtain the following bilinear form

$$(D_x^2 + D_y^2) f_n \cdot f_n = (2e^{D_n} - 2) f_n \cdot f_n \tag{18}$$

which admits a Bäcklund transformation as follows

$$(D_x + iD_y + \lambda^{-1} e^{-D_n} + \mu) f \cdot g = 0, \tag{19a}$$

$$((D_x - iD_y) e^{-\frac{1}{2}D_n} - \lambda e^{\frac{1}{2}D_n} + \gamma e^{-\frac{1}{2}D_n}) f \cdot g = 0. \tag{19b}$$

Furthermore, from the Bäcklund transformation, we may get the following superposition formula.

Proposition 5 *Let $f_0(n)$ be a nonzero solution of Eq. (18) and suppose that $f_1(n)$ and $f_2(n)$ are solutions of (18) such that $f_0(n) \xrightarrow{\lambda_i} f_i(n)$ $(i = 1, 2)$, then there exists the following nonlinear superposition formula*

$$e^{-\frac{1}{2}D_n} f_0 \cdot f_{12} = c(\lambda_1 e^{-\frac{1}{2}D_n} - \lambda_2 e^{\frac{1}{2}D_n}) f_1 \cdot f_2 \tag{20}$$

where f_{12} is a new solution of (18) related to f_1 and f_2 with parameters λ_2 and λ_1 respectively. Here c is a nonzero constant.

In order to get the lump solution, we choose $f_0 = 1$ and $f_i (i = 1, 2)$ as linear functions with respect to x, y and n, i.e. $1 \xrightarrow{\lambda_i} f_i = \theta_i + \beta_i = n + p_i x + q_i y + \beta_i$. Then from the Bäcklund transformation (19a) and (19b), we may get $\mu_i = -\lambda_i^{-1}$, $\gamma_i = \lambda_i$, $p_j = \frac{1}{2}(\lambda_j^{-1} + \lambda_j)$ and $q_j = \frac{1}{2i}(\lambda_j^{-1} - \lambda_j)$. Therefore, we get the seed function of the lump solutions

$$\theta_j = n + \frac{1}{2}(\lambda_j^{-1} + \lambda_j)x + \frac{1}{2i}(\lambda_j^{-1} - \lambda_j)y, \quad j = 1, 2.$$

Therefore the nonlinear superposition formula (20) becomes

$$f_{12}(n) = c(\lambda_1 f_1(n-1) f_2(n) - \lambda_2 f_1(n) f_2(n-1)). \tag{21}$$

In this case, if we take $c = \frac{1}{\lambda_1 - \lambda_2}$ and $f_i = \theta_i + \beta_i$, then (21) can be written as

$$\begin{aligned} f_{12}(n) = \theta_1\theta_2 &+ \frac{1}{\lambda_1 - \lambda_2}(\lambda_1\beta_2 - \lambda_2(\beta_2 - 1))\theta_1 \\ &+ \frac{1}{\lambda_1 - \lambda_2}(\lambda_1(\beta_1 - 1) - \lambda_2\beta_1)\theta_2 + \beta_1\beta_2 + \frac{1}{\lambda_1 - \lambda_2}(\lambda_2\beta_1 - \lambda_1\beta_2). \end{aligned} \tag{22}$$

Furthermore, if we take $\beta_1 = \frac{\lambda_1}{\lambda_1 - \lambda_2}$, $\beta_2 = -\frac{\lambda_2}{\lambda_1 - \lambda_2}$, then we have:

$$f_{12}(n) = \theta_1\theta_2 + A, \tag{23}$$

where $A = \frac{\lambda_1\lambda_2}{(\lambda_1 - \lambda_2)^2}$. Obviously, if we choose $\lambda_1 \neq \lambda_2$, then $\theta_1 = \theta_2^*$, $A > 0$, and therefore we get 1-lump solution of the elliptic Toda equation which is shown in Fig. 1.

Proposition 6 *The elliptic Toda equation admits the general nonlinear superposition formula*

$$e^{-\frac{1}{2}D_n} F_{N-1} \cdot F_{N+1} = c(\lambda_N e^{-\frac{1}{2}D_n} - \lambda_{N+1} e^{\frac{1}{2}D_n}) F_N \cdot \hat{F}_N, \tag{24}$$

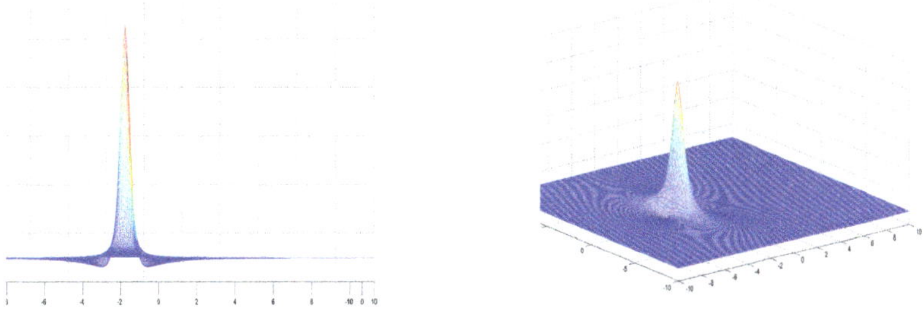

Fig. 1 1-lump solution of the elliptic Toda equation

where

$$F_N(n) = \begin{vmatrix} f_1(n) & \cdots & f_N(n) \\ \vdots & & \vdots \\ (-\lambda_1)^{N-1} f_1(n-N+1) & \cdots & (-\lambda_N)^{N-1} f_N(n-N+1) \end{vmatrix}$$
$$:= |1(n), \ldots, N(n)|.$$
$$\hat{F}_N(n) = |1(n), \ldots, N-1(n), N+1(n)|\,.$$

Here $\{f_j(n,x,y),\ j = 1, 2, \ldots, N+1\}$ *are the seed functions* $f_j(n,x,y) = n + \frac{1}{2}(\lambda_j^{-1} + \lambda_j)x + \frac{1}{2i}(\lambda_j^{-1} - \lambda_j)y + \beta_j$.

Proof It is noted that (24) can be alternatively written as:

$$F_{N-1}(n-1)F_{N+1}(n) = (\lambda_N F_N(n-1)\hat{F}_N(n) - \lambda_{N+1} F_N(n)\hat{F}_N(n-1)) \tag{25}$$

and

$$F_{N-1}(n-1) = |1(n-1), \ldots, N-1(n-1)| = \prod_{i=1}^{N-1}(-\lambda_i) D\begin{bmatrix} 1 \\ N \end{bmatrix}$$

where the determinant D means $F_N(n)$ and $D\begin{bmatrix} j \\ k \end{bmatrix}$ means the $(N-1)$-th minor of D whose j-th row and k-th column are deleted. By taking the explicit forms of F and $\hat{F}$ into the Eq. (25), we may see the nonlinear superposition formula is a Jacobi identity.

Inspired by the 1-lump solution, we now choose $f_j(n) = \theta_j(n) + \beta_j = n + \frac{1}{2}(\lambda_j^{-1} + \lambda_j)x + \frac{1}{2i}(\lambda_j^{-1} - \lambda_j)y + \beta_j$, and therefore the solution $F_N(n)$ can be written as

$$F_N(n) = \begin{vmatrix} \theta_1 + \beta_1 & \cdots & \theta_N + \beta_N \\ -\lambda_1(\theta_1 + \beta_1 - 1) & \cdots & -\lambda_N(\theta_N + \beta_N - 1) \\ \vdots & & \vdots \\ (-\lambda_1)^{N-1}(\theta_1 + \beta_1 - N + 1) & \cdots & (-\lambda_N)^{N-1}(\theta_N + \beta_N - N + 1) \end{vmatrix},$$

from which we see that if and only if we take $\beta_i = \lambda_i \sum_{j \neq i} \frac{1}{\lambda_i - \lambda_j}$, we can get F_{2M} without the odd term. On the other hand, from the determinant identity, we may get

$$F_N(n) = \begin{vmatrix} 1 & -\lambda_1 & \cdots & (-\lambda_1)^{N-1} \\ 1 & -\lambda_2 & \cdots & (-\lambda_2)^{N-1} \\ \vdots & \vdots & & \vdots \\ 1 & -\lambda_N & \cdots & (-\lambda_N)^{N-1} \end{vmatrix} \times \begin{vmatrix} \theta_1 & \frac{-\lambda_1}{\lambda_1 - \lambda_2} & \cdots & \frac{-\lambda_1}{\lambda_1 - \lambda_N} \\ \frac{-\lambda_2}{\lambda_2 - \lambda_1} & \theta_2 & \cdots & \frac{-\lambda_2}{\lambda_2 - \lambda_N} \\ \vdots & \vdots & \ddots & \vdots \\ \frac{-\lambda_N}{\lambda_N - \lambda_1} & \frac{-\lambda_N}{\lambda_N - \lambda_2} & \cdots & \theta_N \end{vmatrix}. \tag{26}$$

In this case, we have the following determinant solution

$$F_N(n) = \begin{vmatrix} \theta_1 & \frac{-\lambda_1}{\lambda_1-\lambda_2} & \cdots & \frac{-\lambda_1}{\lambda_1-\lambda_N} \\ \frac{-\lambda_2}{\lambda_2-\lambda_1} & \theta_2 & \cdots & \frac{-\lambda_2}{\lambda_2-\lambda_N} \\ \vdots & \vdots & \ddots & \vdots \\ \frac{-\lambda_N}{\lambda_N-\lambda_1} & \frac{-\lambda_N}{\lambda_N-\lambda_2} & \cdots & \theta_N \end{vmatrix}. \tag{27}$$

In the following, we want to construct lump solutions from (27). Here we just consider the case of $N = 4$, and set the parameters as $\lambda_3 = \frac{1}{\lambda_1^*}$, $\lambda_4 = \frac{1}{\lambda_2^*}$. In this case, we have

$$\begin{aligned} F_4 = \theta_1\theta_1^*\theta_2\theta_2^* &+ \frac{\lambda_1\lambda_2}{(\lambda_1-\lambda_2)^2}\theta_1^*\theta_2^* + c.c + \frac{\lambda_1^*\lambda_2}{(\lambda_1^*\lambda_2-1)^2}\theta_1\theta_2^* + c.c \\ &+ \frac{\lambda_1\lambda_1^*}{(\lambda_1\lambda_1^*-1)^2}\theta_2\theta_2^* + \frac{\lambda_2\lambda_2^*}{(\lambda_2\lambda_2^*-1)^2}\theta_1\theta_1^* + A, \end{aligned}$$

where c.c means the complex conjugate and A is greater than zero. It means F_4 is 2-lump solution of the Toda equation and Fig. 2 shows 2-lump solution of the Toda equation.

In general, Villarroel has shown in Villarroel (1998) that the F_{2N} given by (27) is always greater than 0 if $\lambda_{N+i}\lambda_i^* = 1$ and $\{\lambda_i, 1 \le i \le 2N\}$ are off the unit circle.

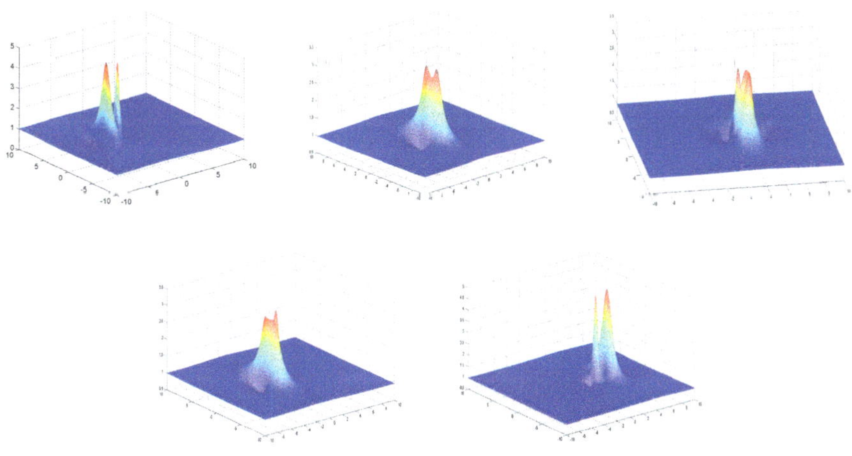

Fig. 2 The interaction of 2-lumps of the Toda equation

5 The Lump Solution of the BKP Equation

In Gilson and Nimmo (1990), the lump solution of the BKP equation has been considered by Claire Gilson and Jon Nimmo. In this part, we would like to show the Bäcklund transformation and nonlinear superposition formula can also provide us a Pfaffian form to the lump solution of BKP, which indicates this technique could also be used for the B_∞-type equations and Pfaffian forms.

Consider the BKP equation

$$(u_t + 15uu_{3x} + 15u_x^3 - 15u_xu_y + u_{5x})_x + 5u_{3x,y} - 5u_{yy} = 0.$$

Through the bilinear transformation $u = 2(\log f)_x$, we obtain the bilinear form for the BKP equation

$$(D_x^6 - 5D_x^3D_y - 5D_y^2 + 9D_xD_t)f \cdot f = 0, \tag{28}$$

whose Bäcklund transformation is indicated as follows (Hirota 2004)

$$(D_x^3 - D_y - 3kD_x^2 + 3k^2D_x)f \cdot g = 0, \tag{29a}$$

$$(-D_x^5 - 5D_x^2D_y + 5kD_x^4 + 5k^2D_x^3 - 10k^2D_y + 10kD_xD_y + 6D_t)f \cdot g = 0. \tag{29b}$$

Furthermore, we have the following nonlinear superposition formula.

Proposition 7 *Let f_0 be a nonzero solution of Eq. (28) and suppose that f_1 and f_2 are solutions such that $f_0 \xrightarrow{\lambda_i} f_i$ $(i = 1, 2)$, then there exists the following nonlinear superposition formula*

$$[D_x - (k_1 + k_2)]f_0 \cdot f_{12} = c[D_x + (k_1 - k_2)]f_1 \cdot f_2 \tag{30}$$

where f_{12} is a new solution related to f_1 and f_2 with parameters λ_2 and λ_1 respectively. Here c is a nonzero constant.

For the Bäcklund transformation (29a) and (29b), if we take $f_0 = 1$ and $f_i (i = 1, 2)$ as the linear functions, then $f_i = \theta_i + \beta_i = x + 3k_i^2y + 5k_i^4t + \beta_i$, $i = 1, 2$. In this case, the nonlinear superposition formula becomes

$$-\frac{d}{dx}f_{12} - (k_1 + k_2)f_{12} = f_2 - f_1 + (k_1 - k_2)f_1 \cdot f_2. \tag{31}$$

By solving this ordinary differential equation, we may obtain the solution of the BKP equation

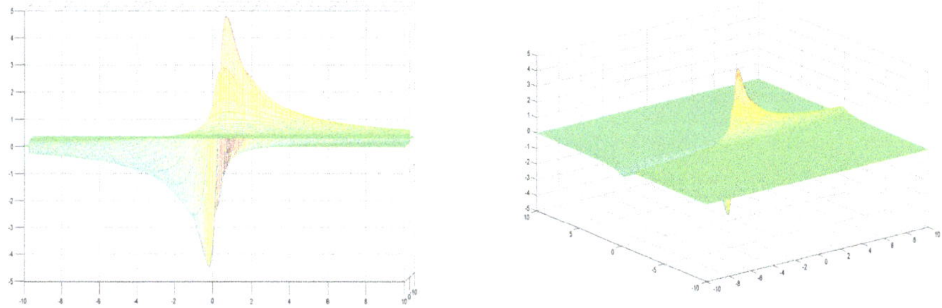

Fig. 3 The figure of 1-lump solution of the BKP equation

$$
\begin{aligned}
f_{12} &= \frac{k_2 - k_1}{k_1 + k_2} f_1 f_2 + \frac{2k_1}{(k_1 + k_2)^2} f_1 - \frac{2k_2}{(k_1 + k_2)^2} f_2 + 2\frac{k_2 - k_1}{(k_1 + k_2)^3} \\
&= \frac{k_2 - k_1}{k_1 + k_2} \theta_1 \theta_2 + \frac{(k_2^2 - k_1^2)\beta_2 + 2k_1}{(k_1 + k_2)^2} \theta_1 + \frac{(k_2^2 - k_1^2)\beta_1 - 2k_2}{(k_1 + k_2)^2} \theta_2 + A,
\end{aligned}
$$

where $A = \frac{k_2-k_1}{k_1+k_2}\beta_1\beta_2 + \frac{\beta_1-\beta_2}{k_1+k_2} - \frac{k_2-k_1}{(k_1+k_2)^2}(\beta_1 + \beta_2) + 2\frac{k_2-k_1}{(k_1+k_2)^3}$. It can be verified that if we take $\beta_1 = \frac{-2k_2}{k_1^2-k_2^2}$, $\beta_2 = \frac{2k_1}{k_1^2-k_2^2}$, $k_2 = k_1^*$ and $|Imk_1| > |Rek_1|$, then the 1-lump solution could be obtained.

Remark 1 Notice that the first order ordinary differential equation (31) may have a general solution, however, in the lump-solution case, we just consider the polynomial solution of f, hence this solution is unique in this sense.

In Fig. 3, the 1-lump solution of the BKP equation is drawn for a particular choice of the parameters.

Proposition 8 *BKP equation has a general nonlinear superposition formula as follows*

$$[D_x - (k_{2n+1} + k_{2n+2})]F_{2n} \cdot F_{2n+2} = [D_x + (k_{2n+1} - k_{2n+2})]\hat{F}_{2n+1} \cdot F_{2n+1}. \quad (32)$$

In particular, the solution F_{2n}, F_{2n+1} and $\hat{F}_{2n+1}$ have the Pfaffian forms

$$F_{2n} = (1, \ldots, 2n),\ F_{2n+1} = (d_0, 1, \ldots, 2n + 1),\ \hat{F}_{2n+1} = (d_0, 1, \ldots, 2n, 2n + 2), \quad (33)$$

in which the Pfaff element satisfies the following relationship

$$
\begin{aligned}
&(d_0, i) = f_i = \theta_i + \beta_i = x + 3k_i^2 y + 5k_i^4 t + \beta_i, \\
&(d_1, i) = \frac{d}{dx}(d_0, i) + k_i(d_0, i) = 1 + k_i f_i, \quad (34) \\
&\frac{d}{dx}(i, j) + (k_i + k_j)(i, j) = (d_0, d_1, i, j),\ (d_0, d_1) = 0.
\end{aligned}
$$

In order to prove the proposition, we need following lemmas.

Lemma 1 *Under the assumption of the Pfaff element (33), we have*

$$\frac{d}{dx}(1,\ldots,2n)+(\sum_{i=1}^{2n}k_i)(1,\ldots,2n)=(d_0,d_1,1,\ldots,2n). \tag{35}$$

Proof We will prove this conclusion by induction. For n=1, it is just the assumption we set in (33). By assumption, it is known that

$$\frac{d}{dx}(2,\ldots,\hat{j},\ldots,2n+2)+\sum_{i=2,i\neq j}^{2n+2}k_i(2,\ldots,\hat{j},\ldots,2n+2)=(d_0,d_1,2,\ldots,\hat{j},\ldots,2n+2)$$

holds for Pfaffian of order n. Then for Pfaffian of order $n+1$, we have

$$\begin{aligned}
&(d_0,d_1,1,\ldots,2n+2)\\
&=\sum_{j=2}^{2n+2}(-1)^j[(d_0,d_1,1,j)(2,\ldots,\hat{j},\ldots,2n+2)+(1,j)(d_0,d_1,2,\ldots,\hat{j},\ldots,2n+2)]\\
&=\sum_{j=2}^{2n+2}(-1)^j\{[\frac{d}{dx}(1,j)+(k_1+k_j)(1,j)](2,\ldots,\hat{j},\ldots,2n+2)\\
&\qquad+(1,j)[\frac{d}{dx}(2,\ldots,\hat{j},\ldots,2n+2)+\sum_{i\neq j}k_i(2,\ldots,\hat{j},\ldots,2n+2)]\}\\
&=\frac{d}{dx}(1,\ldots,2n+2)+\sum_{i=1}^{2n+2}k_i(1,\ldots,2n+2),
\end{aligned}$$

which completes the proof.

Lemma 2 *Under the assumption of the Pfaffian element (33), we also have*

$$(d_1,1,\ldots,2n+1)=\frac{d}{dx}(d_0,1,\ldots,2n+1)+(\sum_{i=1}^{2n+1}k_i)(d_0,1,\ldots,2n+1), \tag{36a}$$

$$(d_1,1,\ldots,2n,2n+2)=\frac{d}{dx}(d_0,1,\ldots,2n,2n+2)+(\sum_{i\neq 2n+1}k_i)(d_0,1,\ldots,2n,2n+2). \tag{36b}$$

Proof We just prove the first equation, and the second one can be verified in a similar way. By expansion of Pfaffian, one has

$$
\begin{aligned}
&(d_1, 1, \ldots, 2n+1)\\
&= \sum_{j=1}^{2n+1} (-1)^j (d_1, j)(1, \ldots, \hat{j}, 2n+1)\\
&= \sum_{j=1}^{2n+1} (-1)^{j-1} [\frac{d}{dx}(d_0, j) + k_j(d_0, j)](1, \ldots, \hat{j}, \ldots, 2n+1)\\
&= \frac{d}{dx}(d_0, 1, \ldots, 2n+1) + \sum_{j=1}^{2n+1} (-1)^{j-1} k_j (d_0, j)(1, \ldots, \hat{j}, \ldots, 2n+1)\\
&- \sum_{j=1}^{2n+1} (-1)^{j-1} (d_0, j)[(d_0, d_1, 1, \ldots, \hat{j}, \ldots, 2n+1) - (\sum_{i=1, i\neq j}^{2n+1} k_i)(1, \ldots, \hat{j}, \ldots, 2n+1)]\\
&= \frac{d}{dx}(d_0, 1, \ldots, 2n+1) + (\sum_{i=1}^{2n+1} k_i)(d_0, \ldots, 2n+1),
\end{aligned}
$$

and the equation is verified.

The Lemma 1 tells us the left side of the nonlinear superposition formula can be written as

$$(d_0, d_1, 1, \ldots, 2n)(1, \ldots, 2n+2) - (1, \ldots, 2n)(d_0, d_1, 1, \ldots, 2n+2), \tag{37}$$

while the Lemma 2 shows the right side can be written as

$$- (d_1, 1, \ldots, 2n+1)(d_0, 1, \ldots, 2n, 2n+2) + (d_0, 1, \ldots, 2n+1)(d_1, 1, \ldots, 2n, 2n+2). \tag{38}$$

Therefore, under these two lemmas, we find that the nonlinear superposition formula of BKP equation (30) can be written as

$$
\begin{aligned}
&(d_0, d_1, 1, \ldots, 2n)(1, \ldots, 2n+2) - (1, \ldots, 2n)(d_0, d_1, 1, \ldots, 2n+2) =\\
&\quad -(d_1, 1, \ldots, 2n+1)(d_0, 1, \ldots, 2n, 2n+2) + (d_0, 1, \ldots, 2n+1)(d_1, 1, \ldots, 2n, 2n+2),
\end{aligned}
$$

which is the Pfaffian identity (Hirota 2004).

And then we would like to prove the F_{2n} given in (33) is always positive or always negative under some constrains. Following the method mentioned in Gilson and Nimmo (1990), we first consider the determinant of $2n \times 2n$ skew-symmetric matrix $A = (a_{i,j})_{1\leq i,j\leq 2n}$ which can be represented as the square of Pfaffian given in (33):

$$F_{2n}^2 = (1, 2, \ldots, 2n)^2 = \det A. \tag{39}$$

Applying Eqs. (33) and (39), we can derive:

$$a_{i,j} = \frac{k_i - k_j}{k_i + k_j}[(f_i + \frac{2k_j}{k_i^2 - k_j^2})(f_j - \frac{2k_i}{k_i^2 - k_j^2}) + \frac{2(k_i^2 + k_j^2)}{(k_i^2 - k_j^2)^2}]. \tag{40}$$

If we set $k_i = k_{n+i}^*, \beta_i = \beta_{n+i}^*$ and $|\mathrm{Im}k_i| > |\mathrm{Re}k_i|$, then the determinant of A can be written as the following form:

$$\det A = \det \begin{vmatrix} \mathrm{C} & \mathrm{B} \\ -\mathrm{B}^* & \mathrm{C}^* \end{vmatrix}, \tag{41}$$

which is always positive. In Eq. (41), $B = (b_{i,j})_{1\le i,j\le n}$, $C = (c_{i,j})_{1\le i,j\le n}$ are two $n \times n$ matrices, whose element $b_{i,j}$, $c_{i,j}$ are given by:

$$b_{i,j} = \frac{k_i - k_j^\star}{k_i + k_j^\star}[(f_i + \frac{2k_j^\star}{k_i^2 - k_j^{\star 2}})(f_j^\star - \frac{2k_i}{k_i^2 - k_j^{\star 2}}) + \frac{2(k_i^2 + k_j^{\star 2})}{(k_i^2 - k_j^{\star 2})^2}],$$

$$c_{i,j} = \frac{k_i - k_j}{k_i + k_j}[(f_i + \frac{2k_j}{k_i^2 - k_j^2})(f_j - \frac{2k_i}{k_i^2 - k_j^2}) + \frac{2(k_i^2 + k_j^2)}{(k_i^2 - k_j^2)^2}].$$

Since $F_{2n}^2 > 0$ by taking $k_i = k_{n+i}^*, \beta_i = \beta_{n+i}^*$ and $|\mathrm{Im}k_i| > |\mathrm{Re}k_i|$ in F_{2n} , and the lump solution is a continuous function, so the F_{2n} is always positive or always negative. Therefore, the solution F_{2n} with $k_i = k_{n+i}^*, \beta_i = \beta_{n+i}^*$ and $|\mathrm{Im}k_i| > |\mathrm{Re}k_i|$ is the nonsingular rational solution of the BKP equation.

6 The Lump Solutions of the Novikov-Veselov Equation

In this part, we want to discuss the lump solution of the Novikov-Veselov equation

$$2u_t + u_{xxx} + u_{yyy} + 3(u\partial_y^{-1}u_x)_x + 3(u\partial_x^{-1}u_y)_y = 0, \tag{42}$$

which can be viewed as an extension the KdV equation in two spatial dimensions and one temporal dimension. Bäcklund Transformation and nonlinear superposition formula and 1,2-lump solutions have been studied in Hu and Willox (1996). Here we revisited some important facts.

Under the dependent variable transformation $u = u_0 + 2(\log f)_{xy}$ with u_0 a constant, the Eq. (42) can be transformed into the multilinear form and enjoys the following Bäcklund transformation

$$(D_xD_y - \mu D_x - \lambda D_y + \lambda\mu + u_0)f \cdot f' = 0, \tag{43a}$$

$$(2D_t + D_x^3 + D_y^3 + 3\lambda^2 D_x - 3\lambda D_x^2 + 3\mu^2 D_y - 3\mu D_y^2)f \cdot f' = 0, \tag{43b}$$

where λ and μ are arbitrary constants. The nonlinear superposition formula can be stated as follows.

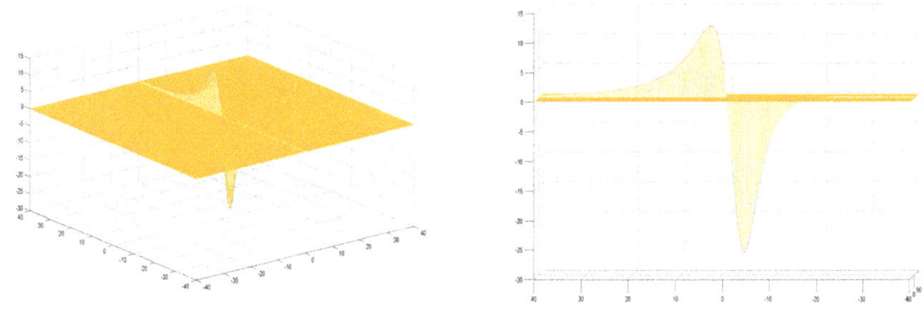

Fig. 4 1-lump solution of the Novikov-Veselov equation

Proposition 9 *Let f_0 be a nonzero solution of (42) and suppose that f_1 and f_2 are solutions of (42) such that $f_0 \xrightarrow{\mu_i} f_i$ $(i = 1, 2)$. Then f_{12} defined by*

$$[D_x - (k_1 + k_2)] f_0 \cdot f_{12} = c[D_x + (k_1 - k_2)] f_1 \cdot f_2 \tag{44}$$

is a new solution to (42) which is related to f_1 and f_2 under bilinear BT (43a) and (43b) with parameters k_2 and k_1 respectively. Here c is a nonzero real constant.

To obtain the lump solutions, we have to take $f_0 = 1$ and $f_i = \theta_i + \beta_i = k_i^2 x + u_0 y - \frac{3}{2(k_i^4 + u_0^3/k_i^2)} t + \beta_i$. For 1-lump solution, if we set $k_2 = k_1^*$, $\beta_1 = \beta_2^*$ and $\mathrm{Im} k_i > \mathrm{Re} k_i$, $(i = 1, 2)$, then

$$f_{12} = (\theta_1 + \frac{2k_1^2 k_2}{k_1^2 - k_2^2}) \times c.c. + 2\frac{k_1^2 k_2^2 (k_1^2 + k_2^2)}{(k_1^2 - k_2^2)^2},$$

where c.c. means the complex conjugate. Obviously, f_{12} is positive and it is a 1-lump solution. We depict the 1-lump solution of the Novikov-Veselov equation in Fig. 4.

Noticing that the nonlinear superposition formula of the Novikov-Veselov equation (44) is the same as the BKP equation (30), the Novikov-Veselov equation (44) possesses the same structure of solution as the BKP equation except the seed function. Hence we have the following proposition.

Proposition 10 *Novikov-Veselov equation owns a general nonlinear superposition formula*

$$[D_x - (k_{2n+1} + k_{2n+2})] F_{2n} \cdot F_{2n+2} = [D_x + (k_{2n+1} - k_{2n+2})] \hat{F}_{2n+1} \cdot F_{2n+1}, \tag{45}$$

where

$$F_{2n} = (1, \ldots, 2n), \; F_{2n+1} = (d_0, 1, \ldots, 2n+1), \; \hat{F}_{2n+1} = (d_0, 1, \ldots, 2n, 2n+2), \tag{46}$$

where the Pfaffian elements satisfy the following relationships

$$
\begin{aligned}
&(d_0, i) = f_i = \theta_i + \beta_i = k_i^2 x + u_0 y - \frac{3}{2(k_i^4 + u_0^3/k_i^2)} t + \beta_i, \\
&(d_1, i) = \frac{d}{dx}(d_0, i) + k_i (d_0, i), \\
&\frac{d}{dx}(i, j) + (k_i + k_j)(i, j) = (d_0, d_1, i, j), \ (d_0, d_1) = 0.
\end{aligned} \tag{47}
$$

Since the proof of this proposition is similar to that of BKP equation, we omit it here. If we set $k_i = k_{n+i}^*$, $\beta_i = \beta_{n+i}^*$ and $|\mathrm{Im}k_i| > |\mathrm{Re}k_i|$, then we can show in a similar way in Sect. 5 that F_{2n} is always positive or always negative. Therefore, we get the N-lump solution of the Novikov-Veselov equation, which has the representation of (46) with $k_i = k_{n+i}^*$, $\beta_i = \beta_{n+i}^*$ and $|\mathrm{Im}k_i| > |\mathrm{Re}k_i|$.

7 The Lump Solutions for Negative Flow of BKP Equation

In Hirota (2004), Sect. 3.3, the author proposed another shallow wave equation, called the negative flow of BKP equation

$$
u_{yt} - u_{xxxy} - 3(u_x u_y)_x + 3u_{xx} = 0. \tag{48}
$$

By the dependent variable transformation $u = 2(\log f)_x$, it can be transformed into a bilinear form

$$
[(D_t - D_x^3)D_y + 3D_x^2] f \cdot f = 0,
$$

which possesses the following Bäcklund transformation

$$
(D_x D_y + \lambda^{-1} D_x + \lambda D_y) f \cdot f' = 0, \tag{49a}
$$

$$
(D_x^3 + 3\lambda D_x^2 + 3\lambda^2 D_x - D_t) f \cdot f' = 0. \tag{49b}
$$

Furthermore, we have the following result.

Proposition 11 *Let f_0 be a nonzero solution of Eq. (48) and suppose that f_1 and f_2 are solutions of (48) such that $f_0 \xrightarrow{\lambda_i} f_i$ $(i = 1, 2)$, then there exists a following nonlinear superposition formula*

$$
(D_x + (k_1 - k_2)) f_0 \cdot f_{12} = c(D_x - (k_1 - k_2)) f_1 \cdot f_2, \tag{50}
$$

where f_{12} is a new solution of (48) related to f_1 and f_2 under bilinear BT (49a) and (49b) with parameters k_2 and k_1 respectively. Here c is a nonzero constant.

A 1-lump solution of the negative flow for BKP equation is derived in the following. Starting with $f_0 = 1$, $f_i = x - k_i^2 y + 3k_i^2 t + \beta_i (i = 1, 2)$, we may obtain the following solution

$$
\begin{aligned}
f_{12} &= \frac{k_2-k_1}{k_1+k_2} f_1 f_2 + \frac{2k_1}{(k_1+k_2)^2} f_1 - \frac{2k_2}{(k_1+k_2)^2} f_2 + 2\frac{k_2-k_1}{(k_1+k_2)^3} \\
&= \frac{k_2-k_1}{k_1+k_2}\theta_1\theta_2 + \frac{(k_2^2-k_1^2)\beta_2 + 2k_1}{(k_1+k_2)^2}\theta_1 + \frac{(k_2^2-k_1^2)\beta_1 - 2k_2}{(k_1+k_2)^2}\theta_2 + A,
\end{aligned}
$$

where $A = \frac{k_2-k_1}{k_1+k_2}\beta_1\beta_2 + \frac{\beta_1-\beta_2}{k_1+k_2} - \frac{k_2-k_1}{(k_1+k_2)^2}(\beta_1+\beta_2) + 2\frac{k_2-k_1}{(k_1+k_2)^3}$. If we take $\beta_1 = \frac{-2k_2}{k_1^2-k_2^2}$, $\beta_2 = \frac{2k_1}{k_1^2-k_2^2}$, $k_2 = k_1^*$ and $|\mathrm{Im}k_1| > |\mathrm{Re}k_1|$, we get the 1-lump solution.

In order to obtain N-lump solutions, we need to establish a general nonlinear superposition formula for the negative flow BKP equation.

Proposition 12 *The negative flow BKP equation owns a general nonlinear superposition formula*

$$
[D_x - (k_{2n+1} + k_{2n+2})]F_{2n} \cdot F_{2n+2} = [D_x + (k_{2n+1} - k_{2n+2})]\hat{F}_{2n+1} \cdot F_{2n+1} \tag{51}
$$

and the solutions F_{2n}, F_{2n+1} and $\hat{F}_{2n+1}$ are expressed as Pfaffians

$$
F_{2n} = (1, \ldots, 2n),\ F_{2n+1} = (d_0, 1, \ldots, 2n+1),\ \hat{F}_{2n+1} = (d_0, 1, \ldots, 2n, 2n+2), \tag{52}
$$

where the Pfaff elements satisfy the following relations

$$
\begin{aligned}
&(d_0, i) = f_i = \theta_i + \beta_i = x - k_i^2 y + 3k_i^2 t + \beta_i, \\
&(d_1, i) = \frac{d}{dx}(d_0, i) + k_i(d_0, i), \\
&\frac{d}{dx}(i, j) + (k_i + k_j)(i, j) = (d_0, d_1, i, j),\ (d_0, d_1) = 0.
\end{aligned} \tag{53}
$$

The proof of Proposition 12 is similar to the case of BKP equation, so we omit it here. Furthermore, we can show in a similar way in Sect. 5 that F_{2n} in (52) with $k_i = k_{n+i}^*$, $\beta_i = \beta_{n+i}^*$ and $|\mathrm{Im}k_i| > |\mathrm{Re}k_i|$ gives the N-lump solution of the negative flow of BKP equation.

8 Conclusion

It is truly remarkable that the lump solutions of several integrable models could be obtained by Bäcklund transformations and nonlinear superposition formulae and the effectiveness presents itself in this paper. It is natural to expect that this technique can be applied to more equations in AKP and BKP type, also for CKP and DKP type equations. The lack of the bilinear Bäcklund transformation of CKP equation brings us essential difficulty to construct the nonlinear superposition formula, as well as the lump solution. In particular, we also expect to develop the similar technique to generate the lump solutions for the discrete integrable lattices.

Acknowledgements The results of this article were previously reported in minisymposium of ICIAM2015 during 10–14 Aug. 2015 and the 13th National Workshop on Solitons and Integrable Systems during 21–24 Aug. 2015 respectively. The work is supported by the National Natural Science Foundation of China (No.11931017, No. 11871336, No. 11601247, No. 11965014).

References

Ablowitz, M.J., Satsuma, J.: Solitons and rational solutions of nonlinear evolution equations. J. Math. Phys. **19**, 2180–2186 (1978)

Ablowitz, M.J., Chakravarty, S., Trubatch, A.D., Villarroel, J.: A novel class of solutions of the non-stationary Schödinger and the Kadomtsev-Petviashvili I equations. Phys. Lett. A **267**, 132–146 (2000)

Ames, W.F.: Nonlinear Partial Differential Equations. Academic Press, New York (1967)

Case, K.M.: Benjamin-Ono-related equations and their solutions. Proc. Nat. Acad. Sci. U.S.A. **76**, 1–3 (1979)

Dubard, P., Gaillard, P., Klein, C., Mateev, V.B.: On multi-rogue wave solutions of the NLS equation and position solutions of the KdV equation. Eur. Phys. J. Spec. Top. **185**, 247–258 (2010)

Dubard, P., Matveev, V.B.: Multi-rogue waves solutions to the focusing NLS equation and the KP-I equation. Nat. Hazards Earth Syst. Sci. **11**, 667–672 (2011)

Feng, B.F., Kawahara, T., Mitsui, T.: A conservative spectral method for several two-dimensional nonlinear wave equations. J. Comput. Phys. **153**, 467–487 (1999)

Gaillard, P.: Families of quasi-rational solutions of the NLS equation and multi-rogue waves. J. Phys. A: Math. Gen. **44**, 435204 (2011)

Gilson, C., Nimmo, J.: Lump solutions of the BKP equation. Phys. Lett. A **147**, 472–476 (1990)

Grammaticos, B., Ramani, A., Papageorgiou, V., Satsuma, J., Willox, R.: Constructing lump-like solutions of the Hirota-Miwa equation. J. Phys. A: Math. Gen. **40**, 12619–12627 (2007)

Guo, B.L., Ling, L.M., Liu, Q.P.: Nonlinear Schrödinger equation: generalized Darboux transformation and rogue wave solutions. Phys. Rev. E **85**, 026607 (2012)

Hirota, R.(Translated by Nagai, A., Nimmo, J., Gilson, C.): Direct Methods in Soliton Theory. Cambridge University Press, Cambridge (2004)

Hirota, R., Satsuma, J.: A simple structure of superposition formula of the Bäcklund transformation. J. Phys. Soc. Jpn. **45**, 1741–1750 (1978)

Hu, X.B., Willox, R.: Some new exact solutions of the Novikov-Veselov equation. J. Phys. A: Math. Gen. **29**, 4589–4592 (1996)

Hu, X.B.: Rational solutions of integrable equations via nonlinear superposition formulae. J. Phys. A: Math. Gen. **30**, 8225–8240 (1997)

Kharif, C., Pelinovsky, E., Slunyaev, A.: Rogue Waves in the Ocean. Springer, Berlin (2009)

Li, L., Wu, Z., Wang, L., He, J.: High-order rogue waves for the Hirota equation. Ann. Phys. **334**, 198–211 (2013)

Ma, W.X.: Lump solutions to the Kadomtsev-Petviashvili equation. Phys. Lett. A **379**, 1975–1978 (2015)

Manakov, S.V., Zahkarov, V.E., Bordag, L.A., Its, A.R., Matveev, V.B.: Two-dimensional solitons of the Kadomtsev-Petviashvili equation and their interaction. Phys. Lett. A **63**, 205–206 (1977)

Matsuno, Y.: Algebra related to the N-soliton solution of the Benjamin-Ono equation. J. Phys. Soc. Jpn. **51**, 2719–2720 (1982)

Matsuno, Y.: On the Benjamin-Ono equation-method for exact solution. J. Phys. Soc. Jpn. **51**, 3734–3739 (1982)

Nakamura, A.: Decay mode solution of the two-dimensional KdV equation and the generalized Bäcklund transformation. J. Math. Phys. **22**, 2456–2462 (1981)

Newell, A.C.: Solitons in Mathematics and Physics. SIAM, Philadelphia (1987)

Ohta, Y., Yang, J.K.: Rogue waves in the Davey-Stewartson I equation. Phys. Rev. E **86**, 036604 (2012)

Ohta, Y., Yang, J.K.: General high-order rogue waves and their dynamics in the nonlinear Schrodinger equation. Proc. R. Soc. A **468**, 1716–1740 (2012)

Ohta, Y., Yang, J.K.: Dynamics of rogue waves in the Davey-Stewartson II equation. J. Phys. A:Math. Gen. **46**, 105202 (2013)

Ono, H.: Algebraic solitary waves in stratified fluids. J. Phys. Soc. Jpn. **39**, 1082–1091 (1975)

Ono, H.: Algebraic soliton of the modified Korteweg-de Vries equation. J. Phys. Soc. Jpn. **41**, 1817–1818 (1976)

Peregrine, D.H.: Water waves, nonlinear Schödinger equations and their solutions. J. Aust. Math. Soc. B **25**, 16–43 (1983)

Satsuma, J., Ablowitz, M.J.: Two-dimensional lumps in nonlinear dispersive systems. J. Math. Phys. **20**, 1496–1503 (1979)

Solli, D.R., Ropers, C., Koonath, P., Jalali, B.: Optical rogue Waves. Nature **450**, 1054–1057 (2007)

Villarroel, J.: On the solution to the inverse problem for the Toda chain. SIAM J. Appl. Math. **59**, 261–285 (1998)

Villarroel, J., Ablowitz, M.J.: Solutions to the 2+1 Toda equation. J. Phys. A: Math Gen. **27**, 931–941 (1994)

Villarroel, J., Ablowitz, M.J.: On the discrete spectrum of the nonstationary Schödinger equation and multipole lumps of the Kadomtsev-Petviashvili I equation. Comm. Math. Phys. **207**, 1–42 (1999)

Stokes Phenomenon Arising in the Confluence of the Gauss Hypergeometric Equation

Calum Horrobin and Marta Mazzocco

Abstract In this paper we study the Gauss and Kummer hypergeometric equations in-depth. In particular, we focus on the confluence of two regular singularities of the Gauss hypergeometric equation to produce the Kummer hypergeometric equation with an irregular singularity at infinity. We show how to pass from solutions with power-like behaviour which are analytic in disks, to solutions with exponential behaviour which are analytic in sectors and have divergent asymptotics. We explicitly calculate the Stokes matrices of the confluent system in terms of the monodromy data, specifically the connection matrices, of the original system around the merging singularities.

Keywords Hypergeometric differential equations · Asymptotic expansions · Confluence · Monodromy data

1 Introduction

This paper studies the Gauss hypergeometric differential equation,

$$x(1-x)\,\frac{d^2y}{dx^2}+(\gamma-(\alpha+\beta+1)x)\,\frac{dy}{dx}-\alpha\beta\,y=0, \tag{1.1}$$

where $x\in\mathbb{C}$, and the Kummer confluent hypergeometric differential equation,

$$z\,\frac{d^2\tilde{y}}{dz^2}+(\gamma-z)\,\frac{d\tilde{y}}{dz}-\beta\,\tilde{y}=0, \tag{1.2}$$

where $z\in\mathbb{C}$.

C. Horrobin · M. Mazzocco (✉)
University of Birmingham, Birmingham, UK
e-mail: m.mazzocco@bham.ac.uk
URL: https://www.birmingham.ac.uk/staff/profiles/maths/mazzocco-marta.aspx

© Springer Nature Switzerland AG 2020
F. Nijhoff et al. (eds.), *Asymptotic, Algebraic and Geometric Aspects of Integrable Systems*, Springer Proceedings in Mathematics & Statistics 338,
https://doi.org/10.1007/978-3-030-57000-2_7

For brevity, in this paper, these equations are simply called Gauss equation and Kummer equation respectively.

The aim of the paper is to give rigour to the confluence of two regular singularities of the Gauss equation to produce the Kummer equation with an irregular singularity at infinity. In particular, the monodromy data of the confluent equation (Kummer), including Stokes data, are produced as limits of the monodromy data of the original equation (Gauss) using explicit formulae.

One of the main difficulties addressed in this paper is how to make sense of the confluence limits by understanding how to pass from the solutions of the original system to the solutions of the confluent system. This is a non-trivial question because it involves passing from a solution with power-like behaviour which converges in a disk to solutions with exponential behaviour which are analytic in a sector and asymptotic to a divergent series.

The procedure of this paper is based on an existence theorem by Glutsyuk (1999). Essentially, this states that there exist certain diagonal matrices K_ε and $K_{-\varepsilon}$ such that the limit,

$$\lim_{\varepsilon\to 0} K_{-\varepsilon}^{-1}\, C\, K_\varepsilon,$$

where C is the connection matrix between the merging simple poles of the original system, exists. Moreover, this limit gives one of the Stokes matrices if $\varepsilon \to 0$ is taken along a certain ray. However, this existence theorem does not prescribe how to calculate the diagonal matrices K_ε and $K_{-\varepsilon}$. The main result of this paper is to calculate such diagonal matrices and thus produce both the Stokes matrices in terms of limits of the connection matrix of the original equation explicitly. In particular calculate how one Stokes matrix is produced as limit along a certain ray and the other one by the limit along the opposite ray.

Despite the fact that the analytic theory of the Gauss and Kummer equations has been developed more than a hundred years ago, the question of producing the Stokes data of the Kummer equation in terms of limits of monodromy data of the Gauss one has only been approached rather recently (Lambert and Rousseau 2008; Watanabe 2007). In particular, in Watanabe (2007), the Mellin-Barnes integral representations of the solutions of Kummer equation are produced as limits of the ones for the Gauss equation, and then the Stokes data are deduced from the Mellin-Barnes integral representations (this last calculation is reported here in Appendix B for completeness). In Lambert and Rousseau (2008), the confluence problem is solved by observing that one of the Fuchsian singularities remains Fuchsian under the confluence, so that the corresponding local fundamental matrix of the Gauss equation admits an analytic limit under the confluence, thus allowing to compute explicitly the monodromy of the Kummer equation around 0. The Stokes matrices are then determined by the fact that loops around 0 are homotopic to loops around ∞ in the Riemann sphere with two punctures.

The approach of the current paper does not require closed form expressions such as Mellin-Barnes integrals. Indeed, in Horrobin et al. (2020), we use this procedure to calculate the Stokes matrices of the linear problem associated to the fifth Painlevé equation (and its higher order analogues) in terms of limits of the connection matrix between 1 and ∞ in the linear problem associated to the sixth Painlevé equation (and its higher order analogues) for which closed form fundamental matrices are unknown.

Another advantage of the procedure of the current paper is that it does not rely on the existence of an additional simple pole which survives the confluence limit, and therefore it can be applied to the confluence from the Bessel differential equation to the Airy one for example, or even more ambitiously, in the confluence from the fifth to the third Painlevé equation—this challenging work is postponed to subsequent publications.

This paper is organised as follows: In Sects. 2 and 3, the authors remind some background on the Gauss and Kummer hypergeometric differential equations respectively. In Sect. 4 the confluence procedure is explained, and the main result of this paper, Theorem 4.9 is proved. In Appendices A and B, the classical derivation of the monodromy data for the Gauss and Kummer hypergeometric differential equations respectively are derived using Mellin-Barnes integrals.

This paper is inspired by some of the facets of Nalini's mathematical taste and style because to tackle a seemingly simple problem it requires an unexpected depth that opens a Pandora's box of beautiful mathematical problems. For this reason, we wish to dedicate this paper to her. [Calum Horrobin and Marta Mazzocco]

I wish to thank Nalini for her friendship of more than twenty years. Throughout her career, Nalini has mentored, supported and sponsored a huge number of early career mathematicians, some formally as her PhD students and post docs, others informally, like myself and many others. [Marta Mazzocco]

2 Gauss Hypergeometric Differential Equation

Throughout the paper we work in the non-resonance assumption: γ, $\gamma-\alpha-\beta$, $\alpha-\beta\notin\mathbb{Z}$.

To define monodromy data, it is easier to deal with a system of first order ODEs by using the following trivial lemma:

Lemma 2.1 *Under the assumptions $\alpha\neq 0$, $\gamma\neq\beta\neq 1$ and $\alpha\neq\beta-1$, the matrix*

$$Y(x)=\begin{pmatrix} y_1(x) & y_2(x)\\ \Psi\left(y_1,y_1';x\right) & \Psi\left(y_2,y_2';x\right)\end{pmatrix}, \tag{2.1}$$

where

$$\Psi\left(y_k,y_k';x\right)=\frac{\alpha\left(\beta-\gamma+(\alpha+1-\beta)x\right)y_k(x)+x(x-1)(\alpha+1-\beta)y_k'(x)}{\alpha(\beta-1)(\beta-\gamma)}, \tag{2.2}$$

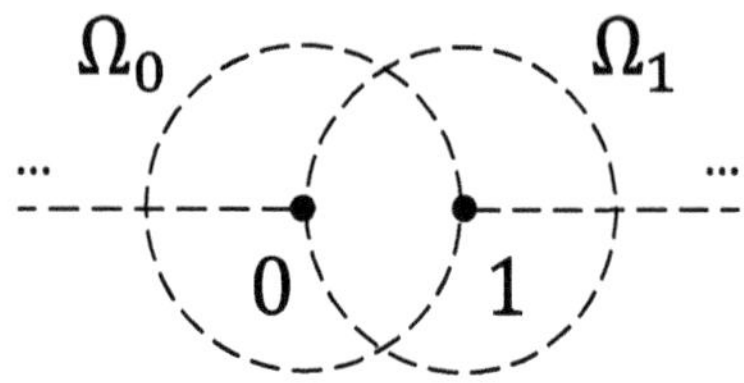

Fig. 1 Chosen disks with branch cuts. Note that Ω_∞ is a disk in the complement of $\overline{\Omega}_0 \cup \overline{\Omega}_1$

is a fundamental solution of the equation

$$\frac{\mathrm{d}Y}{\mathrm{d}x} = \left(\frac{A_0}{x} + \frac{A_1}{x-1}\right) Y, \tag{2.3}$$

$$A_0 = \frac{1}{\alpha+1-\beta}\begin{pmatrix} \alpha(\beta-\gamma) & \alpha(1-\beta)(\beta-\gamma) \\ \alpha+1-\gamma & (1-\beta)(\alpha+1-\gamma) \end{pmatrix},$$
$$A_1 = \frac{1}{\alpha+1-\beta}\begin{pmatrix} \alpha(\gamma-\alpha-1) & \alpha(\beta-1)(\beta-\gamma) \\ \gamma-\alpha-1 & (\beta-1)(\beta-\gamma) \end{pmatrix},$$

if and only if $y_1(x)$ and $y_2(x)$ are linearly independent solutions of Gauss hypergeometric equation (1.1).

So, from now on, we stick to the system of first order ODEs (2.3).
We define the following disks with chosen branches, as illustrated in Fig. 1:

$$\Omega_0 = \{x : |x| < 1,\ -\pi \le \arg(x) < \pi\},$$
$$\Omega_1 = \{x : |x-1| < 1,\ -\pi \le \arg(1-x) < \pi\},$$
$$\Omega_\infty = \{x : |x| > 1,\ -\pi \le \arg(-x) < \pi\},$$

It is well-known that the solutions of Eq. (1.1) are expressible in terms of Gauss hypergeometric ${}_2F_1$ series, in particular the following three pairs of linearly independent local solutions $y_1^{(k)}(x)$ and $y_2^{(k)}(x)$ of (1.1) defined in the neighbourhoods Ω_k form a basis around each singular point:

$$\begin{aligned} y_1^{(0)}(x) &= x^{1-\gamma}{}_2F_1\left(\begin{matrix}\alpha+1-\gamma,\ \beta+1-\gamma \\ 2-\gamma\end{matrix}; x\right), \\ y_2^{(0)}(x) &= {}_2F_1\left(\begin{matrix}\alpha,\ \beta \\ \gamma\end{matrix}; x\right), \end{aligned} \qquad x \in \Omega_0, \tag{2.4}$$

$$\begin{aligned} y_1^{(1)}(x) &= (1-x)^{\gamma-\alpha-\beta}{}_2F_1\left(\begin{matrix}\gamma-\alpha,\ \gamma-\beta \\ \gamma+1-\alpha-\beta\end{matrix}; 1-x\right), \\ y_2^{(1)}(x) &= {}_2F_1\left(\begin{matrix}\alpha,\ \beta \\ \alpha+\beta+1-\gamma\end{matrix}; 1-x\right), \end{aligned} \qquad x \in \Omega_1, \tag{2.5}$$

$$
\left.\begin{aligned}
y_1^{(\infty)}(x) &= (-x)^{-\alpha}{}_2F_1\left(\begin{matrix}\alpha,\ \alpha+1-\gamma\\ \alpha+1-\beta\end{matrix};x^{-1}\right),\\
y_2^{(\infty)}(x) &= (-x)^{-\beta}{}_2F_1\left(\begin{matrix}\beta,\ \beta+1-\gamma\\ \beta+1-\alpha\end{matrix};x^{-1}\right),
\end{aligned}\right.\qquad x\in\Omega_\infty. \tag{2.6}
$$

Lemma 2.2 *The following local fundamental solutions of the matrix hypergeometric equation (2.3) have the following form*

$$
Y^{(0)}(x) = R_0G_0(x)x^{\Theta_0}, \qquad x\in\Omega_0, \tag{2.7}
$$

$$
Y^{(1)}(x) = R_1G_1(x)(1-x)^{\Theta_1}, \qquad x\in\Omega_1, \tag{2.8}
$$

$$
Y^{(\infty)}(x) = R_\infty G_\infty(x)(-x)^{-\Theta_\infty}, \qquad x\in\Omega_\infty, \tag{2.9}
$$

where R_k and Θ_k are the following matrices:

$$
R_0=\begin{pmatrix}1 & 1\\ \frac{\alpha+1-\gamma}{\alpha(\beta-\gamma)} & \frac{1}{\beta-1}\end{pmatrix},\quad R_1=\begin{pmatrix}1 & 1\\ \frac{1}{\alpha} & \frac{\alpha+1-\gamma}{(\beta-1)(\beta-\gamma)}\end{pmatrix},\quad R_\infty=\begin{pmatrix}1 & 0\\ 0 & \frac{(\beta-\alpha)(\alpha+1-\beta)}{\alpha(\beta-1)(\beta-\gamma)}\end{pmatrix},
$$

$$
\Theta_0=\begin{pmatrix}1-\gamma & 0\\ 0 & 0\end{pmatrix},\quad \Theta_1=\begin{pmatrix}\gamma-\alpha-\beta & 0\\ 0 & 0\end{pmatrix},\quad \Theta_\infty=\begin{pmatrix}\alpha & 0\\ 0 & \beta-1\end{pmatrix},
$$

which satisfy $R_k^{-1}A_kR_k=\Theta_k$, and $G_k(x)$ are the following series:

$$
G_0(x)=\begin{pmatrix}
{}_2F_1\left(\begin{matrix}\alpha+1-\gamma,\ \beta-\gamma\\ 1-\gamma\end{matrix};x\right), & \frac{x\alpha(\gamma-\beta)}{\gamma(\gamma-1)}\ {}_2F_1\left(\begin{matrix}\alpha+1,\ \beta\\ \gamma+1\end{matrix};x\right)\\
\frac{x(\alpha+1-\gamma)(1-\beta)}{(1-\gamma)(2-\gamma)}{}_2F_1\left(\begin{matrix}\alpha+2-\gamma,\ \beta+1-\gamma\\ 3-\gamma\end{matrix};x\right), & {}_2F_1\left(\begin{matrix}\alpha,\ \beta-1\\ \gamma-1\end{matrix};x\right)
\end{pmatrix},
$$

$$
G_1(x)=\left(\begin{matrix}
{}_2F_1\left(\begin{matrix}\gamma-\alpha-1,\ \gamma-\beta\\ \gamma-\alpha-\beta\end{matrix};1-x\right),\\
\frac{(1-x)(\beta-1)(\beta-\gamma)}{(\alpha+\beta-\gamma-1)(\alpha+\beta-\gamma)}\ {}_2F_1\left(\begin{matrix}\gamma-\alpha,\ \gamma+1-\beta\\ \gamma+2-\alpha-\beta\end{matrix};1-x\right),
\end{matrix}\right.
$$

$$\left.\begin{matrix} \frac{(1-x)\alpha(\alpha+1-\gamma)}{(\alpha+\beta-\gamma)(\alpha+\beta+1-\gamma)}\,{}_2F_1\left(\begin{matrix}\alpha+1,\ \beta\\ \alpha+\beta+2-\gamma\end{matrix};1-x\right)\\ {}_2F_1\left(\begin{matrix}\alpha,\ \beta-1\\ \alpha+\beta-\gamma\end{matrix};1-x\right)\end{matrix}\right),$$

$$G_\infty(x)=\begin{pmatrix} {}_2F_1\left(\begin{matrix}\alpha,\ \alpha+1-\gamma\\ \alpha+1-\beta\end{matrix};x^{-1}\right), & \frac{\alpha(\beta-1)(\beta-\gamma)(\gamma-\alpha-1)}{(\alpha-\beta)(\alpha+1-\beta)^2(\alpha+2-\beta)}\,\frac{1}{x}\,{}_2F_1\left(\begin{matrix}\alpha+1,\ \alpha+2-\gamma\\ \alpha+3-\beta\end{matrix};x^{-1}\right),\\ -\frac{1}{x}\,{}_2F_1\left(\begin{matrix}\beta,\ \beta+1-\gamma\\ \beta+1-\alpha\end{matrix};x^{-1}\right) & {}_2F_1\left(\begin{matrix}\beta-1,\ \beta-\gamma\\ \beta-\alpha-1\end{matrix};x^{-1}\right)\end{pmatrix}.$$

Proof This result can be proved in two ways: either by reducing Eq. (2.3) to Birkhoff normal form near each singularity and computing the corresponding gauge transformations $R_0G_0(x)$, $R_1G_1(x)$ and $G_\infty(x)$ recursively or by direct substitution of the local solutions (2.4)–(2.6) into expression (2.1) and using Gauss contiguous relations. □

Remark 1 The matrices R_k, $k=0,1$ and ∞, in the above solutions (2.7), (2.8) and (2.9) have been chosen to satisfy $R_k^{-1}A_kR_k=\Theta_k$, where $A_\infty:=-A_0-A_1$. The matrices G_0, G_1, G_∞ have leading term given by the identity.

We now define the monodromy data of Gauss hypergeometric equation (1.1) and recall how to express them in explicit form (Bateman and Erdélyi 2020; Whittaker and Watson 1979). In Appendix A we derive these classical formulae by following the approach of representing solutions using Mellin-Barnes integrals.

When defining local solutions, we have been specific about identifying which sheet of the Riemann surface of the logarithm we are restricting our local solutions to at each singular point. We may extend the definitions of our local fundamental solutions $Y^{(k)}(x)$ to other sheets $e^{2m\pi i}\Omega_k$, $k=0,1,\infty$, by analytically continuing along a closed loop encircling the singularity $x=0,1,\infty$. This action simply means that our solution becomes multiplied by the corresponding exponent $e^{2m\pi i\Theta_k}$, for $k=0,1$ and ∞, $m\in\mathbb{Z}$. Note that, for $k=0$ and 1, the analytic continuation of $Y^{(k)}(x)$ around its singularity in the positive direction means $m>0$ in the previous sentence; while, for $k=\infty$, it means $m<0$. The diagonal matrices $e^{2\pi i\Theta_k}$ are called the local monodromy exponents of the singularities.

We proceed with the global analysis of solutions. Let $Y^{(0)}(x)$, $Y^{(1)}(x)$ and $Y^{(\infty)}(x)$ be the fundamental solutions of the hypergeometric equation as defined in the previous section. Denote by $\gamma_{j,k}\left[Y^{(j)}\right](x)$ the analytic continuation of $Y^{(j)}(x)$ along an orientable curve $\gamma_{j,k}:[0,1]\to\mathbb{C}$ with $\gamma_{j,k}(0)\in\Omega_j$ and $\gamma_{j,k}(1)\in\Omega_k$, for $j,k=0,1,\infty$. We have the following connection formulae (see Appendix A for the detailed derivation of these):

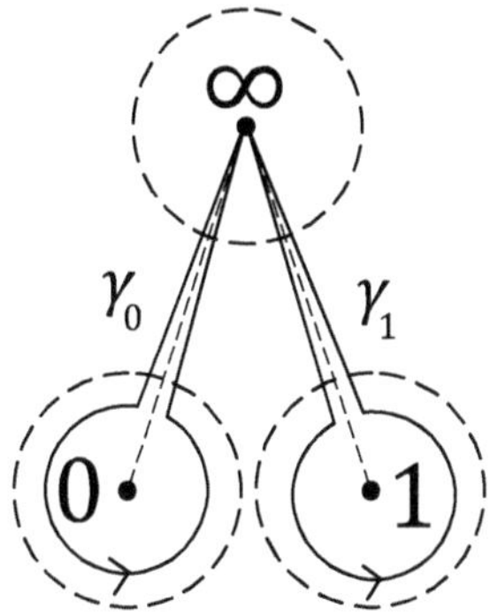

Fig. 2 Curves defining the monodromy matrices M_k of Gauss hypergeometric differential equation

$$\gamma_{j,k}\left[Y^{(j)}\right](x) = Y^{(k)}(x)C^{kj}, \tag{2.10}$$

where:

$$C^{0\infty} = \begin{pmatrix} e^{i\pi(\gamma-1)}\frac{\Gamma(\alpha+1-\beta)\Gamma(\gamma-1)}{\Gamma(\alpha)\Gamma(\gamma-\beta)} & e^{i\pi(\gamma-1)}\frac{\Gamma(\beta+1-\alpha)\Gamma(\gamma-1)}{\Gamma(\beta)\Gamma(\gamma-\alpha)} \\ \frac{\Gamma(\alpha+1-\beta)\Gamma(1-\gamma)}{\Gamma(1-\beta)\Gamma(\alpha+1-\gamma)} & \frac{\Gamma(\beta+1-\alpha)\Gamma(1-\gamma)}{\Gamma(1-\alpha)\Gamma(\beta+1-\gamma)} \end{pmatrix}, \tag{2.11}$$

$$C^{1\infty} = \begin{pmatrix} e^{i\pi(\gamma-\beta)}\frac{\Gamma(\alpha+1-\beta)\Gamma(\alpha+\beta-\gamma)}{\Gamma(\alpha)\Gamma(\alpha+1-\gamma)} & e^{i\pi(\gamma-\alpha)}\frac{\Gamma(\beta+1-\alpha)\Gamma(\alpha+\beta-\gamma)}{\Gamma(\beta)\Gamma(\beta+1-\gamma)} \\ e^{i\pi\alpha}\frac{\Gamma(\alpha+1-\beta)\Gamma(\gamma-\alpha-\beta)}{\Gamma(1-\beta)\Gamma(\gamma-\beta)} & e^{i\pi\beta}\frac{\Gamma(\beta+1-\alpha)\Gamma(\gamma-\alpha-\beta)}{\Gamma(1-\alpha)\Gamma(\gamma-\alpha)} \end{pmatrix}, \tag{2.12}$$

$$C^{01} = \begin{pmatrix} \frac{\Gamma(\gamma+1-\alpha-\beta)\Gamma(\gamma-1)}{\Gamma(\gamma-\alpha)\Gamma(\gamma-\beta)} & \frac{\Gamma(\alpha+\beta+1-\gamma)\Gamma(\gamma-1)}{\Gamma(\alpha)\Gamma(\beta)} \\ \frac{\Gamma(\gamma+1-\alpha-\beta)\Gamma(1-\gamma)}{\Gamma(1-\alpha)\Gamma(1-\beta)} & \frac{\Gamma(\alpha+\beta+1-\gamma)\Gamma(1-\gamma)}{\Gamma(\alpha+1-\gamma)\Gamma(\beta+1-\gamma)} \end{pmatrix}. \tag{2.13}$$

We choose to normalise the monodromy data of Gauss hypergeometric equation with the fundamental solution $Y^{(\infty)}(x)$. Denote by $\gamma_k\left[Y^{(\infty)}\right](x)$ the analytic continuation of $Y^{(\infty)}(x)$ along an orientable, closed curve $\gamma_k : [0, 1] \to \mathbb{C}$ with $\gamma_k(0) = \gamma_k(1) \in \Omega_\infty, k = 0, 1$, which encircles the singularity $x = 0, 1$ respectively in the positive (anti-clockwise) direction. The curves γ_0 and γ_1 are illustrated in Fig. 2, note that $\gamma_\infty := \gamma_1^{-1}\gamma_0^{-1}$. We have:

$$\gamma_k\left[Y^{(\infty)}\right](x) = Y^{(k)}(x)M_k, \qquad k = 0, 1, \infty,$$

where,

$$M_0 = \left(C^{0\infty}\right)^{-1}e^{2\pi i\Theta_0}C^{0\infty}, \quad M_1 = \left(C^{1\infty}\right)^{-1}e^{2\pi i\Theta_1}C^{1\infty}, \quad M_\infty = e^{2\pi i\Theta_\infty}. \tag{2.14}$$

These matrices satisfy the cyclic relation,

$$M_\infty M_1 M_0 = I. \tag{2.15}$$

Definition 2.1 We define the monodromy data of Gauss hypergeometric equation (1.1) as the set,

$$\mathcal{M} := \left\{ (M_0, M_1, M_\infty) \in (\mathrm{GL}_2(\mathbb{C}))^3 \,\middle|\, \begin{array}{c} M_\infty M_1 M_0 = I,\ M_\infty = e^{2\pi i \Theta_\infty} \\ \mathrm{eigenv}(M_k) = e^{2\pi i \Theta_k},\ \text{k=0,1} \end{array} \right\}_{/\mathrm{GL}_2(\mathbb{C})} \tag{2.16}$$

where $\mathrm{eigenv}(M_k) = e^{2\pi i \Theta_k}$ means that the eigenvalues of M_k are given as the elements of the diagonal matrix $e^{2\pi i \Theta_k}$ and the quotient is by global conjugation by a diagonal matrix.

3 Kummer Confluent Hypergeometric Equation

We use z as the variable of Kummer confluent hypergeometric equation, we also write tilde above some of the functions and parameters to distinguish from the Gauss hypergeometric equation. We recall the following,

Lemma 3.1 *Under the assumption* $(\beta - 1)(\beta - \gamma) \neq 0$, *the matrix*

$$\widetilde{Y}(z) = \begin{pmatrix} \tilde{y}_1(z) & \tilde{y}_2(z) \\ \widetilde{\Psi}\left(\tilde{y}_1, \tilde{y}_1'; z\right) & \widetilde{\Psi}\left(\tilde{y}_2, \tilde{y}_2'; z\right) \end{pmatrix}, \tag{3.1}$$

where,

$$\widetilde{\Psi}\left(\tilde{y}_k, \tilde{y}_k'; z\right) = \frac{(z + \beta - \gamma)\, \tilde{y}_k(z) - z \tilde{y}_k'(z)}{(\beta - 1)(\beta - \gamma)},$$

is a fundamental solution of the equation

$$\frac{\partial \widetilde{Y}}{\partial z} = \left(\begin{pmatrix} 1 & 0 \\ 0 & 0 \end{pmatrix} + \frac{\widetilde{A}_0}{z} \right) \widetilde{Y}, \ \textit{where}\ \widetilde{A}_0 = \begin{pmatrix} \beta - \gamma & (1-\beta)(\beta-\gamma) \\ 1 & 1 - \beta \end{pmatrix}, \tag{3.2}$$

if and only if $\tilde{y}_1(z)$ *and* $\tilde{y}_2(z)$ *are linearly independent solutions of Kummer confluent hypergeometric equation (1.2),*

$$z\, \tilde{y}'' + (\gamma - z)\, \tilde{y}' - \beta\, \tilde{y} = 0.$$

Kummer confluent hypergeometric equation (1.2) has one Fuchsian singularity at $z = 0$, since $\frac{\gamma - z}{z}$ and $\frac{-\beta}{z}$ have simple poles at $z = 0$, and an irregular singularity at $z = \infty$ of Poincaré rank one. The exponents of the singularity $z = 0$ are $1 - \gamma$ and 0 and at $z = \infty$ are $\gamma - \beta$ and $\beta - 1$. We make the non-resonance assumption $\gamma \notin \mathbb{Z}$.

3.1 Local Behaviour of the Solutions

Kummer confluent hypergeometric equation has an irregular singularity at $z = \infty$ of Poincaré rank one and, as such, solutions around this point exhibit Stokes phenomenon. In this sub-section, we will state some definitions and theorems which precisely describe fundamental solutions of Kummer equation at the irregular point and the monodromy data, including Stokes matrices.

We first fix the pair of linearly independent local solutions of (1.2) as follows:

$$\begin{aligned} \tilde{y}_1^{(0)}(z) &= z^{1-\gamma}{}_1F_1\left(\begin{matrix}\beta+1-\gamma\\ 2-\gamma\end{matrix};z\right),\\ \tilde{y}_2^{(0)}(z) &= {}_1F_1\left(\begin{matrix}\beta\\ \gamma\end{matrix};z\right),\end{aligned}\qquad z\in\widetilde{\Omega}_0. \tag{3.3}$$

where

$$\widetilde{\Omega}_0 := \left\{z : -\frac{3}{2}\pi \le \arg(z) < \frac{\pi}{2}\right\},$$

is a punctured disk around 0 with branch cut along the positive imaginary axis.

In terms of the linear system (3.1), these solutions correspond to the following local fundamental solution of the matrix hypergeometric equation (3.2):

$$\widetilde{Y}^{(0)}(z) = \widetilde{R}_0 H_0(z) z^{\widetilde{\Theta}_0}, \qquad z\in\widetilde{\Omega}_0, \tag{3.4}$$

where $\widetilde{R}_0$ and $\widetilde{\Theta}_0$ are the following matrices:

$$\widetilde{R}_0 = \begin{pmatrix} 1 & 1\\ \frac{1}{\beta-\gamma} & \frac{1}{\beta-1}\end{pmatrix} \quad \text{and} \quad \widetilde{\Theta}_0 = \begin{pmatrix} 1-\gamma & 0\\ 0 & 0\end{pmatrix},$$

which satisfy $\widetilde{R}_0^{-1}\widetilde{A}_0\widetilde{R}_0 = \widetilde{\Theta}_0$, and $H_0(z)$ is the following series:

$$H_0(z) = \begin{pmatrix} {}_1F_1\left(\begin{matrix}\beta-\gamma\\ 1-\gamma\end{matrix};z\right) & \frac{z(\gamma-\beta)}{\gamma(\gamma-1)}\,{}_1F_1\left(\begin{matrix}\beta\\ \gamma+1\end{matrix};z\right)\\ \frac{z(1-\beta)}{(1-\gamma)(2-\gamma)}\,{}_1F_1\left(\begin{matrix}\beta+1-\gamma\\ 3-\gamma\end{matrix};z\right) & {}_1F_1\left(\begin{matrix}\beta-1\\ \gamma-1\end{matrix};z\right)\end{pmatrix}.$$

We now turn our attention to the irregular singularity $z = \infty$.

Definition 3.1 The rays $\{z : \mathrm{Re}(z) = 0,\ \mathrm{Im}(z) > 0\}$ and $\{z : \mathrm{Re}(z) = 0,\ \mathrm{Im}(z) < 0\}$ are called the Stokes rays of Kummer equation (1.2).

We note that these rays constitute the borderline where the behaviour of e^z changes, as $z \to \infty$; that is to say, on one side of each of these rays we have $e^z \to 0$, whereas on the other side of each ray we have $e^z \to \infty$. This is a key aspect of Stokes phenomenon and plays a role in understanding the following classical theorem.

Theorem 3.2 *Let*

$$\widetilde{\Sigma}_k = \left\{ z : -\frac{\pi}{2} < arg(z) - k\pi < \frac{3\pi}{2} \right\}.$$

For all $k \in \mathbb{Z}$, there exists a solution $\widetilde{Y}^{(\infty,k)}(z)$ of Eq. (3.2) analytic in the sector $\widetilde{\Sigma}_k$ such that,

$$\widetilde{Y}^{(\infty,k)}(z) \sim \widetilde{R}_\infty \left(\sum_{n=0}^{\infty} h_{n,\infty} z^{-n} \right) \begin{pmatrix} e^z z^{\beta-\gamma} & 0 \\ 0 & z^{1-\beta} \end{pmatrix}, \qquad as\ z \to \infty,\ z \in \widetilde{\Sigma}_k, \quad (3.5)$$

where $\widetilde{R}_\infty$ is the following matrix,

$$\widetilde{R}_\infty = \begin{pmatrix} 1 & 0 \\ 0 & \frac{-1}{(\beta-1)(\beta-\gamma)} \end{pmatrix},$$

and $H_\infty(z)$ is the following series

$$H_\infty(z) = \begin{pmatrix} {}_2F_0\left(1-\beta, \gamma-\beta; z^{-1}\right) & \frac{-1}{z}\,{}_2F_0\left(\beta, \beta+1-\gamma; -z^{-1}\right) \\ \frac{(1-\beta)(\beta-\gamma)}{z}\,{}_2F_0\left(2-\beta, \gamma+1-\beta; z^{-1}\right) & {}_2F_0\left(\beta-1, \beta-\gamma; -z^{-1}\right) \end{pmatrix}.$$

Moreover, each solution $\widetilde{Y}^{(\infty,k)}(z)$ is uniquely specified by the relation (3.5).

Proof A proof of the existence of fundamental solutions $\widetilde{Y}^{(\infty,k)}(z)$ which are analytic on sectors $\widetilde{\Sigma}_k$ may be found in Balser et al. (1979). To find the asymptotic behaviour (3.5), we make the following ansatz

$$\widetilde{Y}^{(\infty,k)}(z) \sim \widetilde{R}_\infty H_\infty(z) \exp\left(\int_{-\infty}^{z} \left(\Lambda_0 + \frac{\Lambda_1}{z'} \right) dz' \right), \qquad \text{as } z \to \infty,\ z \in \widetilde{\Sigma}_k,$$

where,

$$\widetilde{R}_\infty = \begin{pmatrix} 1 & 0 \\ 0 & \frac{-1}{(\beta-1)(\beta-\gamma)} \end{pmatrix},$$

Λ_0 and Λ_1 are constant, diagonal matrices to be determined and $H_\infty(z)$ is a formal series

$$H_\infty(z) = \sum_{n=0}^{\infty} h_{n,\infty} z^{-n}.$$

where the coefficients $h_{n,\infty}$ are to be determined.

By substitution in the Eq. (3.2), we obtain

$$-\sum_{n=1}^{\infty} n h_{n,\infty} z^{-n-1} + \left(\sum_{n=0}^{\infty} h_{n,\infty} z^{-n}\right)\left(\Lambda_0 + \frac{\Lambda_1}{z}\right)$$
$$= \widetilde{R}_{\infty}^{-1}\left(\begin{pmatrix} 1 & 0 \\ 0 & 0 \end{pmatrix} + \frac{\widetilde{A}_0}{z}\right)\widetilde{R}_{\infty}\left(\sum_{n=0}^{\infty} h_{n,\infty} z^{-n}\right).$$

By setting $h_{0,\infty} = I$ and equating powers of z^{-n} in this equation, for $n = 0$ and 1, we find:

$$\Lambda_0 = \begin{pmatrix} 1 & 0 \\ 0 & 0 \end{pmatrix} \text{ and } \Lambda_1 = \begin{pmatrix} \beta - \gamma & 0 \\ 0 & 1 - \beta \end{pmatrix},$$

and, for $n \geq 1$, we find the recursion equation,

$$\left[h_{n,\infty}, \begin{pmatrix} 1 & 0 \\ 0 & 0 \end{pmatrix}\right] = (n-1)h_{n-1,\infty} + h_{n-1,\infty}\begin{pmatrix} \gamma - \beta & 0 \\ 0 & \beta - 1 \end{pmatrix} + \widetilde{R}_{\infty}^{-1}\widetilde{A}_0\widetilde{R}_{\infty} h_{n-1,\infty}.$$

It can be verified that the general solution of this equation is,

$$h_{n,\infty} = \begin{pmatrix} \frac{(1-\beta)_n(\gamma-\beta)_n}{n!} & \frac{(\beta)_{n-1}(\beta+1-\gamma)_{n-1}}{(-1)^n(n-1)!} \\ \frac{(1-\beta)(\beta-\gamma)(2-\beta)_{n-1}(\gamma+1-\beta)_{n-1}}{(n-1)!} & \frac{(\beta-1)_n(\beta-\gamma)_n}{(-1)^n n!} \end{pmatrix}, \tag{3.6}$$

which are indeed the coefficients in the asymptotic series given.

To prove uniqueness of solutions, let $\widehat{Y}^{(\infty,k)}(z)$ denote another fundamental solution of Eq. (3.2) which is analytic on the sector $\widetilde{\Sigma}_k$ and has the correct asymptotic behavior, namely,

$$\widehat{Y}^{(\infty,k)}(z) \sim \widetilde{R}_{\infty}\left(\sum_{n=0}^{\infty} h_{n,\infty} z^{-n}\right)\begin{pmatrix} e^z z^{\beta-\gamma} & 0 \\ 0 & z^{1-\beta} \end{pmatrix}, \qquad \text{as } z \to \infty, \ z \in \widetilde{\Sigma}_k. \tag{3.7}$$

Since $\widetilde{Y}^{(\infty,k)}(z)$ and $\widehat{Y}^{(\infty,k)}(z)$ are fundamental solutions defined on the same sector, there exists a constant matrix $C \in \mathrm{GL}_2(\mathbb{C})$ such that,

$$\widetilde{Y}^{(\infty,k)}(z) = \widehat{Y}^{(\infty,k)}(z)C, \qquad z \in \widetilde{\Sigma}_k.$$

Using the asymptotic relations (3.5) and (3.7), we deduce the following,

$$\begin{pmatrix} e^z z^{\beta-\gamma} & 0 \\ 0 & z^{1-\beta} \end{pmatrix} C \begin{pmatrix} e^{-z} z^{\gamma-\beta} & 0 \\ 0 & z^{\beta-1} \end{pmatrix} \sim I, \qquad \text{as } z \to \infty, \ z \in \widetilde{\Sigma}_k.$$

From this relation, we immediately see that $(C)_{1,1} = (C)_{2,2} = 1$. Moreover, since there exists rays belonging to $\widetilde{\Sigma}_k$ along which each exponential, e^z and e^{-z}, explodes as $z \to \infty$, we conclude that $(C)_{1,2} = (C)_{2,1} = 0$. □

Remark 2 The matrices $\widetilde{R}_0$ and $\widetilde{R}_\infty$ in the above solutions (3.4) and (3.5) have been chosen to satisfy $\widetilde{R}_0^{-1}\widetilde{A}_0\widetilde{R}_0 = \widetilde{\Theta}_0$ and,

$$\left[\widetilde{R}_\infty, \begin{pmatrix} 1 & 0 \\ 0 & 0 \end{pmatrix}\right] = 0.$$

We denote the asymptotic behaviour of true solutions at infinity as in (3.5) by,

$$\widetilde{Y}_f^{(\infty)}(z) = \left(\sum_{n=0}^{\infty} h_{n,\infty} z^{-n}\right)\begin{pmatrix} e^z z^{\beta-\gamma} & 0 \\ 0 & z^{1-\beta} \end{pmatrix}, \qquad z \in \widetilde{\Sigma}_k.$$

The series $H_\infty(z) = \sum_{n=0}^{\infty} h_{n,\infty} z^{-n}$ defines a formal gauge transformation which maps equation (3.2) to,

$$\frac{\partial}{\partial z}\widehat{Y}(z) = \left(\begin{pmatrix} 1 & 0 \\ 0 & 0 \end{pmatrix} + \frac{1}{z}\begin{pmatrix} \beta-\gamma & 0 \\ 0 & 1-\beta \end{pmatrix}\right)\widehat{Y}, \tag{3.8}$$

via the transformation $\widetilde{Y}(z) = \widetilde{R}_\infty H_\infty(z)\widehat{Y}(z)$. We define the coefficient of $\frac{1}{z}$ in the new equation to be $-\widetilde{\Theta}_\infty$, namely,

$$\widetilde{\Theta}_\infty := \begin{pmatrix} \gamma-\beta & 0 \\ 0 & \beta-1 \end{pmatrix} \equiv -\text{diag}\left(\widetilde{A}_0\right).$$

In the generic case $a, b \notin \mathbb{Z}^{\leq 0}$, d'Alembert's ratio test shows that the series ${}_2F_0(a, b; z^{-1})$ diverges for all $z \in \mathbb{C}$. In this sense, the asymptotic behaviour $\widetilde{Y}_f^{(\infty)}(z)$ is a formal fundamental solution.

Remark 3 Using expression (3.1) in Lemma 3.1, the formal fundamental solution $\widetilde{Y}_f^{(\infty)}$ of (3.2) corresponds to the following standard formal basis of solutions of (1.2),

$$\begin{aligned} \tilde{y}_{1,f}^{(\infty)}(z) &= e^z z^{\beta-\gamma}{}_2F_0\left(\gamma-\beta,\ 1-\beta; z^{-1}\right), \\ \tilde{y}_{2,f}^{(\infty)}(z) &= -z^{-\beta}{}_2F_0\left(\beta,\ \beta+1-\gamma; -z^{-1}\right). \end{aligned} \tag{3.9}$$

3.2 *Monodromy Data*

We now define the monodromy data, including Stokes data, of Kummer equation (1.2) and recall how to express them in explicit form (Bateman and Erdélyi 2020; Whittaker and Watson 1979). In Appendix B, we derive these classical formulae by representing solutions using Mellin-Barnes integrals.

Definition 3.2 Let $\widetilde{Y}^{(\infty,k)}(z)$ be the fundamental solutions given in Theorem 3.2 and define sectors,

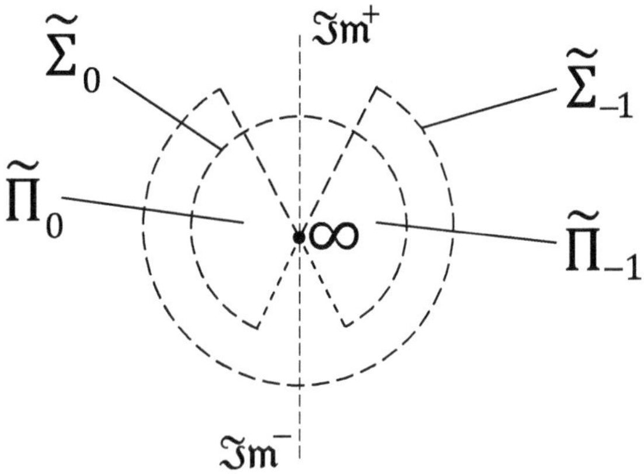

Fig. 3 Sectors $\widetilde{\Pi}_0$, $\widetilde{\Pi}_{-1}$, $\widetilde{\Sigma}_0$ and $\widetilde{\Sigma}_{-1}$ projected onto the plane $\overline{\mathbb{C}}\backslash\{0\}$. The positive and negative imaginary axes are Stokes rays

$$\widetilde{\Pi}_k := \widetilde{\Sigma}_k \cap \widetilde{\Sigma}_{k+1} \equiv \left\{z : |z| > 0,\ \frac{\pi}{2} < \arg(z) - k\pi < \frac{3\pi}{2}\right\},$$

as illustrated in Fig. 3. We define Stokes matrices $\widetilde{S}_k \in \mathrm{SL}_2(\mathbb{C})$ as follows,

$$\widetilde{Y}^{(\infty,k+1)}(z) = \widetilde{Y}^{(\infty,k)}(z)\widetilde{S}_k, \quad z \in \widetilde{\Pi}_k. \tag{3.10}$$

From the asymptotic relation (3.5), it is clear that

$$\widetilde{Y}^{(\infty,k+2)}(z) = \widetilde{Y}^{(\infty,k)}\left(ze^{-2\pi i}\right)e^{-2\pi i\widetilde{\Theta}_\infty}, \qquad z \in \widetilde{\Sigma}_{k+2}. \tag{3.11}$$

due to the fact that these two solutions have the same asymptotic behaviour as $z \to \infty$ in the sector $z \in \widetilde{\Sigma}_{k+2}$. Therefore all solutions $\widetilde{Y}^{(\infty,k)}(z)$ are categorised into two fundamentally distinct cases, namely, when k is even and when k is odd. Combining Definition 3.2 with the relation (3.11), one can show that

$$e^{-2\pi i\widetilde{\Theta}_\infty}\widetilde{S}_{k+1} = \widetilde{S}_{k-1}e^{-2\pi i\widetilde{\Theta}_\infty},$$

which shows that Kummer equation has only two types of Stokes matrices $\widetilde{S}_k$ which are fundamentally different: one with k odd and the other with k even.

Here we select to work with the fundamental solutions $\widetilde{Y}^{(\infty,-1)}(z)$ in the sector $\widetilde{\Sigma}_{-1}$ and $\widetilde{Y}^{(\infty,0)}(z)$ in the sector $\widetilde{\Sigma}_0$ and with the Stokes matrices $\widetilde{S}_0$ and $\widetilde{S}_{-1}$. The explicit form of the Stokes matrices are derived in the Appendix B where the following Lemma is proved:

Lemma 3.3 *We have the following classical formulae:*

$$\widetilde{S}_0 = \begin{pmatrix} 1 & \frac{2\pi i}{\Gamma(\beta)\Gamma(\beta+1-\gamma)}e^{i\pi(\gamma-2\beta)} \\ 0 & 1 \end{pmatrix} \quad \textit{and} \quad \widetilde{S}_{-1} = \begin{pmatrix} 1 & 0 \\ \frac{2\pi i}{\Gamma(1-\beta)\Gamma(\gamma-\beta)} & 1 \end{pmatrix}. \tag{3.12}$$

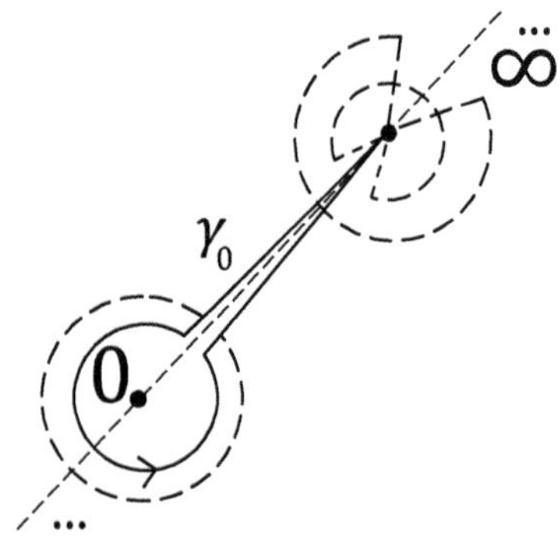

Fig. 4 Curves defining the monodromy matrices $\widetilde{M}_k$ of Kummer hypergeometric differential equation

We choose to normalise our monodromy data with respect to the fundamental solution $\widetilde{Y}^{(\infty,0)}(z)$. Denote by $\gamma_{\infty,0}\left[\widetilde{Y}^{(\infty,0)}\right](z)$ the analytic continuation of $\widetilde{Y}^{(\infty,0)}(z)$ along an orientable curve $\gamma_{\infty,0} : [0,1] \to \mathbb{C}$ with $\gamma_{\infty,0}(0) \in \widetilde{\Sigma}_0$ and $\gamma_{\infty,0}(1) \in \widetilde{\Omega}_0$. We have,

$$\gamma_{\infty 0}\left[\widetilde{Y}^{(\infty,0)}\right](z) = \widetilde{Y}^{(0)}(z)\widetilde{C}^{0\infty},$$

where,

$$\widetilde{C}^{0\infty} = \begin{pmatrix} e^{i\pi(\beta-1)}\frac{\Gamma(\gamma-1)}{\Gamma(\gamma-\beta)} & -\frac{\Gamma(\gamma-1)}{\Gamma(\beta)} \\ e^{i\pi(\beta-\gamma)}\frac{\Gamma(1-\gamma)}{\Gamma(1-\beta)} & -\frac{\Gamma(1-\gamma)}{\Gamma(\beta+1-\gamma)} \end{pmatrix}. \tag{3.13}$$

Denote by $\gamma_0\left[\widetilde{Y}^{(\infty,0)}\right](z)$ the analytic continuation of $\widetilde{Y}^{(\infty,0)}(z)$ along an orientable, closed curve $\gamma_0 : [0,1] \to \mathbb{C}$ with $\gamma_0(0) = \gamma_0(1) \in \widetilde{\Sigma}_0$ which encircles the singularity $z = 0$ in the positive (anti-clockwise) direction. The curve γ_0 is illustrated below, note that $\gamma_\infty := \gamma_0^{-1}$ (Fig. 4).

We have,

$$\gamma_k\left[\widetilde{Y}^{(\infty,0)}\right](z) = Y^{(\infty,k)}(z)\widetilde{M}_k, \qquad k = 0, \infty,$$

where,

$$\widetilde{M}_0 = \left(\widetilde{C}^{0\infty}\right)^{-1} e^{2\pi i\widetilde{\Theta}_0}\widetilde{C}^{0\infty} \quad \text{and} \quad \widetilde{M}_\infty = \widetilde{S}_0 e^{2\pi i\widetilde{\Theta}_\infty}\widetilde{S}_{-1}. \tag{3.14}$$

These matrices satisfy the cyclic relation,

$$\widetilde{M}_\infty\widetilde{M}_0 = I. \tag{3.15}$$

Definition 3.3 We define the monodromy data of Kummer hypergeometric differential equation (1.2) as the set,

$$\widetilde{\mathcal{M}} := \left\{ \begin{array}{c|c} [r]\left(\widetilde{M}_0, \widetilde{S}_0, \widetilde{S}_{-1}\right) & \widetilde{S}_0 \text{ unipotent, upper triangular,} \\ \in (\mathrm{GL}_2(\mathbb{C}))^3 & \widetilde{S}_{-1} \text{unipotent, lower triangular,} \\ & \widetilde{S}_0 e^{2\pi i \widetilde{\Theta}_\infty} \widetilde{S}_{-1} \widetilde{M}_0 = I, \\ & \text{eigenv}\left(\widetilde{M}_0\right) = e^{2\pi i \widetilde{\Theta}_0} \end{array} \right\}_{/GL_2(\mathbb{C})} \tag{3.16}$$

where $\text{eigenv}(\widetilde{M}_0) = e^{2\pi i \widetilde{\Theta}_0}$ means that the eigenvalues of $\widetilde{M}_0$ are given as the elements of the diagonal matrix $e^{2\pi i \widetilde{\Theta}_0}$ and the quotient is by global conjugation by a diagonal matrix.

4 Confluence from Gauss to Kummer Equation

In this Section we analyse the confluence procedure from Gauss equation (1.1) to Kummer equation (1.2). We are primarily concerned with understanding how to produce the monodromy data of the Kummer equation, as defined in Sect. 3.2, from the connection matrices of the Gauss equation (see Sect. 2), under the confluence procedure.

We first explain how the confluence procedure works intuitively. By the substitution $x = \frac{z}{\alpha}$, on the Gauss equation (1.1)

$$\begin{aligned} x(1-x)\, y''(x) + (\gamma - (\alpha+\beta+1)x)\, y'(x) - \alpha\beta\, y(x) = 0, \\ \Leftrightarrow \frac{z}{\alpha}\left(\frac{\alpha - z}{\alpha}\right)\alpha^2\, y_{zz} + \left(\gamma - (\alpha+\beta+1)\frac{z}{\alpha}\right)\alpha\, y_z - \alpha\beta\, y = 0, \\ \Leftrightarrow z\, y_{zz} + (\gamma - z)\, y_z - \beta\, y - \frac{1}{\alpha}\left(z^2 y_{zz} + (\beta+1)y_z\right) = 0. \end{aligned}$$

we produce an differential equation with three Fuchsian singularities at $z = 0$, α and ∞ respectively.

As a heuristic argument, one can see that the final equation becomes Kummer equation (1.2) as $\alpha \to \infty$ so that a double pole is created at $z = \infty$ as the two simple poles $z = \alpha$ and ∞ merge. This derivation does not explain how of obtain solutions of the Kummer equation by taking limits as $\alpha \to \infty$ of certain solutions of Gauss equation under the substitution $x = \frac{z}{\alpha}$. To understand this, we need to use a result by Glutsyuk (1999), which deals with limits of solutions at merging simple poles under a generic confluence procedure. This is explained in the next sub-section.

4.1 A Result by Glutsyuk

Consider the following differential equation,

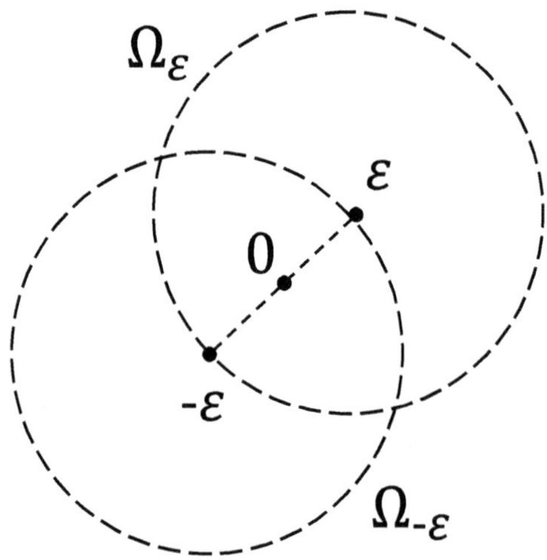

Fig. 5 An illustration of the neighbourhoods $\Omega_{\pm\varepsilon}$ with branch cuts in which we define the fundamental solutions $Y^{(\pm\varepsilon)}(\lambda)$

$$\frac{\partial Y}{\partial \lambda} = \frac{A(\lambda,\varepsilon)}{(\lambda-\varepsilon)(\lambda+\varepsilon)} Y, \quad A(\lambda,\varepsilon) \in \mathrm{GL}_2(\mathbb{C}), \tag{4.1}$$

with $A(\lambda, \varepsilon)$ a holomorphic matrix about $\lambda = \pm\varepsilon$ such that $A(\pm\varepsilon, \varepsilon) \neq 0$ for sufficiently small $\varepsilon \geq 0$ satisfying the following limit,

$$\lim_{\varepsilon\to 0} A(\lambda,\varepsilon) = A(\lambda, 0).$$

Hence, the non-perturbed, or *confluent*, equation,

$$\frac{\partial Y}{\partial \lambda} = \frac{A(\lambda,0)}{\lambda^2} Y, \tag{4.2}$$

has an irregular singularity at $\lambda = 0$ of Poincaré rank one. Moreover, it is assumed that the eigenvalues of the residue matrices $A(\pm\varepsilon, \varepsilon)$ of at $\lambda = \pm\varepsilon$ are non resonant and that the eigenvalues of the leading matrix of $A(\lambda, 0)$ at $\lambda = 0$ are distinct.

We first deal with the perturbed equation (4.1). We define neighbourhoods $\Omega_{\pm\varepsilon}$ of the points $\lambda = \pm\varepsilon$ respectively whose radii are less than $2|\varepsilon|$ and with branch cuts made along the straight line passing through the points $\lambda = -\varepsilon, 0, \varepsilon$, as illustrated in Fig. 5. Equation (4.1) has fundamental solutions $Y^{(\pm\varepsilon)}(\lambda)$ which are analytic in the cut disks $\Omega_\pm(\varepsilon)$ of the following form,

$$Y^{(\pm\varepsilon)}(\lambda) = \left(\sum_{n=0}^{\infty} G_{n,\pm\varepsilon}(\lambda \mp \varepsilon)^n\right)(\lambda \mp \varepsilon)^{\Lambda_{\pm\varepsilon}}, \qquad \lambda \in \Omega_{\pm\varepsilon},$$

where $G_{0,\pm\varepsilon}$ are fixed matrices which diagonalise the residue matrices $A(\pm\varepsilon, \varepsilon)$ and all other terms of the series are determined by certain recursion formulae.

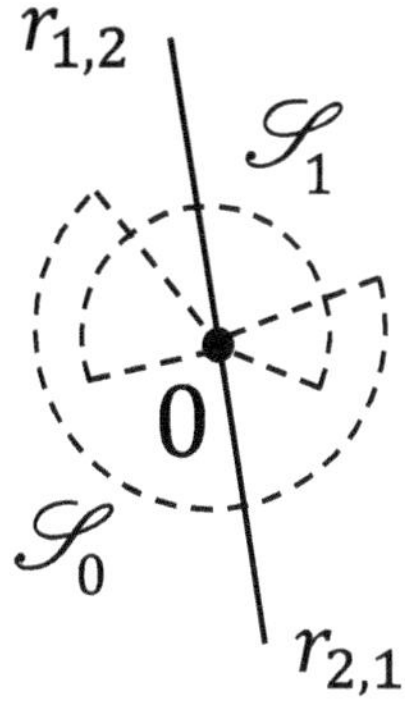

Fig. 6 An illustration of the Stokes rays $r_{i,j}$ and sectors $\mathscr{S}_0$ and $\mathscr{S}_1$

We now turn our attention to the confluent equation (4.2). Denote by μ_1 and μ_2 the eigenvalues of the leading matrix of $A(\lambda, 0)$ at $\lambda = 0$ (by assumption, $\mu_1 \neq \mu_2$) and let,

$$r_{i,j} = \left\{\lambda : \operatorname{Re}\left(\frac{\mu_i - \mu_j}{\lambda}\right) = 0, \ \operatorname{Im}\left(\frac{\mu_i - \mu_j}{\lambda}\right) > 0\right\}, \quad i, j \in \{1, 2\},$$

be the Stokes rays. We denote by $\mathscr{S}_0$ and $\mathscr{S}_1$ open sectors whose union is a punctured neighbourhood of $\lambda = 0$, each of which: has an opening greater than π; contains only one Stokes ray and does not contain the other Stokes ray at its boundary. An illustration of such Stokes rays and sectors is given below (Fig. 6).

We can cover all of the sheets of the Riemann surface of the logarithm at $\lambda = 0$ by extending the notation as follows,

$$\lambda \in \mathscr{S}_{k+2} \Leftrightarrow \lambda e^{-2\pi i} \in \mathscr{S}_k.$$

From the standard theory of linear systems of ordinary differential equations, there exists a number R sufficiently large such that, for all $k \in \mathbb{Z}$, there exist fundamental solutions $Y^{(0,k)}(\lambda)$ of the non-perturbed equation (4.2) analytic in the sectors $\mathscr{S}_k$ such that,

$$Y^{(0,k)}(\lambda) \sim \left(\sum_{n=0}^{\infty} H_n \lambda^n\right) \lambda^{\Theta_0} \exp\left(\lambda^{-1}\begin{pmatrix}\mu_1 & 0\\ 0 & \mu_2\end{pmatrix}\right), \text{ as } \lambda \to 0, \ \lambda \in \mathscr{S}_k,$$

where H_0 is a fixed matrix which diagonalises the leading term of $A(\lambda, 0)$ at $\lambda = 0$, all other terms of the series and the diagonal matrix Θ are uniquely determined by certain recursion relations. Each solution $Y^{(0,k)}(\lambda)$ is uniquely specified by the above asymptotic relation.

We define open sectors $\sigma_{\pm\varepsilon}(\varepsilon) \subset \Omega_\pm$ with base points at $\lambda = \pm\varepsilon$ respectively whose openings do not contain the branch cut between $-\varepsilon$ and ε as illustrated in Fig. 7.

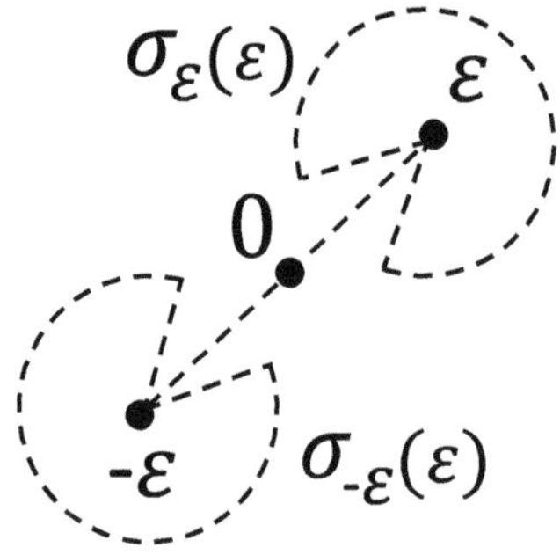

Fig. 7 An illustration of the sectors $\sigma_{\pm\varepsilon}(\varepsilon)$

We impose the condition that, as $\varepsilon \to 0$ along a ray, the sector $\sigma_\varepsilon(\varepsilon)$ (resp. $\sigma_{-\varepsilon}(\varepsilon)$) is translated along a ray to zero and becomes in agreement with the sector $\mathscr{S}_{k+1}$ (resp. $\mathscr{S}_k$), for some $k \in \mathbb{Z}$. We write this condition as follows,

$$\lim_{\varepsilon\to 0} \sigma_\varepsilon(\varepsilon) = \mathscr{S}_{k+1} \quad \text{and} \quad \lim_{\varepsilon\to 0} \sigma_{-\varepsilon}(\varepsilon) = \mathscr{S}_k. \tag{4.3}$$

Theorem 4.1 *Let the fundamental solutions $Y^{(\varepsilon)}(\lambda)$, $Y^{(-\varepsilon)}(\lambda)$ and $Y^{(0,k)}(\lambda)$ and the sectors $\sigma_\varepsilon(\varepsilon)$, $\sigma_{-\varepsilon}(\varepsilon)$ and $\mathscr{S}_k$ be defined as above. There exist diagonal matrices K_ε and $K_{-\varepsilon}$ such that we have the following limits,*

$$\lim_{\varepsilon\to 0} Y^{(\varepsilon)}(\lambda)\big|_{\lambda\in\sigma_\varepsilon(\varepsilon)} K_\varepsilon = Y^{(0,k+1)}(\lambda),$$
$$\lim_{\varepsilon\to 0} Y^{(-\varepsilon)}(\lambda)\big|_{\lambda\in\sigma_{-\varepsilon}(\varepsilon)} K_{-\varepsilon} = Y^{(0,k)}(\lambda),$$

uniformly for $\lambda \in \mathscr{S}_{k+1}$, $\mathscr{S}_k$ respectively, as ε belongs to a fixed ray.

Remark 4 It is well-known that, when solving a linear ordinary differential equation around a Fuchsian singular point, the maximal radius we may take for the neighbourhood on which we can define an analytic solution is the distance to the nearest singularity. For the perturbed equation (4.1), as ε becomes arbitrarily small it is clear from the hypotheses on $A(\lambda, \varepsilon)$ that the closest singularity to $\lambda = \pm\varepsilon$ will be $\lambda = \mp\varepsilon$ respectively. We have illustrated the domains $\Omega_{\pm\varepsilon}$ in Fig. 5 with the maximal radii for which it is possible to define analytic solutions. Observe that the neighbourhoods of analyticity of the fundamental solutions diminish as $\varepsilon \to 0$. The intelligent part of restricting the fundamental solutions $Y^{(\pm\varepsilon)}(\lambda)$ to the sectors $\sigma_{\pm\varepsilon}(\varepsilon)$ as drawn in Fig. 7, rather than the neighbourhoods $\Omega_{\pm\varepsilon}$, is that the radii of these sectors need not be restricted to the distance to the nearest singularity. Indeed, by construction, the singularity $\lambda = \pm\varepsilon$ will not be inside the sector $\sigma_{\mp\varepsilon}(\varepsilon)$ respectively. In particular, this means that the radii of these sectors need not vanish.

By the same reasoning as in the previous remark, it is without loss of generality that we may assume $\sigma_\varepsilon(\varepsilon) \cap \sigma_{-\varepsilon}(\varepsilon) \neq \varnothing$ for ε sufficiently close to zero. Accordingly, since we have two fundamental solutions defined on this intersection, they must be related by multiplication by a constant invertible matrix on the right, namely,

$$Y^{(\varepsilon)}(\lambda) = Y^{(-\varepsilon)}(\lambda)C, \qquad \lambda \in \sigma_\varepsilon(\varepsilon) \cap \sigma_{-\varepsilon}(\varepsilon), \tag{4.4}$$

for some connection matrix $C \in \mathrm{GL}_2(\mathbb{C})$. Similarly, the two fundamental solutions $Y^{(0,0)}(\lambda)$ and $Y^{(0,1)}(\lambda)$ of the confluent equation must be related to each other by multiplication by a constant invertible matrix on the right on the intersection $\mathscr{S}_0$ and $\mathscr{S}_1$, namely,

$$Y^{(0,1)}(\lambda) = Y^{(0,0)}(\lambda)S, \qquad \lambda \in \mathscr{S}_0 \cap \mathscr{S}_1, \tag{4.5}$$

for some Stokes matrix $S \in \mathrm{GL}_2(\mathbb{C})$.

Corollary 4.2 *Let the fundamental solutions $Y^{(\varepsilon)}(\lambda)$, $Y^{(-\varepsilon)}(\lambda)$ and $Y^{(0,k)}(\lambda)$ and the sectors $\sigma_\varepsilon(\varepsilon)$, $\sigma_{-\varepsilon}(\varepsilon)$ and $\mathscr{S}_k$ be defined as above; let $K_{\pm\varepsilon}$ be matrices satisfying Theorem 4.1 and let C and S be the matrices defined by (4.4) and (4.5) respectively. We have the following limit,*

$$\lim_{\varepsilon\to 0} K_{-\varepsilon}^{-1} C K_\varepsilon = S, \tag{4.6}$$

as ε belongs to a fixed ray.

In (4.6) it is clear how to obtain one of the Stokes matrices at the point $\lambda = 0$ of the confluent equation. In order to obtain the second Stokes matrix we take $\varepsilon \to 0$ along the opposite ray to the one already considered. Rather than having the limits in (4.3), we would now have, for example, that $\sigma_\varepsilon(\varepsilon)$ tends to $\mathscr{S}_k$ and $\sigma_{-\varepsilon}(\varepsilon)$ tends to $\mathscr{S}_{k-1}$. In this way, we use the limit in (4.6) to produce the other Stokes matrix. We will explain all of these details and calculate everything explicitly for each of the cases we consider.

4.2 *Limits of Solutions*

As outlined above, our confluence procedure is to introduce the new variable z by the substitution $x = \frac{z}{\alpha}$ and take the limit $\alpha \to \infty$. For the remainder of this chapter we must be careful in which way we are taking α to infinity, for example it would be inconvenient for us if α spiralled towards infinity. We will consider two limits along fixed rays: one with $\arg(\alpha) = \frac{\pi}{2}$ and the other with $\arg(\alpha) = -\frac{\pi}{2}$.

4.2.1 Obtaining the Solutions $\widetilde{Y}^{(\infty,k)}(z)$

We now turn our attention to the main problem of how to obtain fundamental solutions at the double pole of the confluent equation from solutions at the merging simple poles of the original equation. We first examine the behaviour of the fundamental solutions at $x=\infty$, as given in (2.6). Observe that these solutions are expressed using the Gauss ${}_2F_1$ series in the variable $x^{-1}\equiv\frac{\alpha}{z}$, which diverge for $|x^{-1}|>1\Leftrightarrow|z|<|\alpha|$. In this case, we clearly do not have uniform convergence with respect to α and we need to use Glutsyuk's Theorem 4.1.

The fundamental set of solutions (2.6) are written in canonical form. However, we will rewrite the solution $y_1^{(\infty)}(x)$ using one of Kummer relations as follows,

$$\begin{aligned}y_1^{(\infty)}(x)&=(-x)^{-\alpha}{}_2F_1\left(\begin{matrix}\alpha,\ \alpha+1-\gamma\\ \alpha+1-\beta\end{matrix};x^{-1}\right), && x\in\Omega_\infty,\\ &=(-x)^{\beta-\gamma}(1-x)^{\gamma-\alpha-\beta}{}_2F_1\left(\begin{matrix}1-\beta,\ \gamma-\beta\\ \alpha+1-\beta\end{matrix};x^{-1}\right), && x\in\widehat{\Omega}_\infty, \qquad (4.7)\end{aligned}$$

where the new domain $\widehat{\Omega}_\infty$ is defined as,

$$\widehat{\Omega}_\infty=\{x:|x|>1,\ -\pi\le\arg(-x)<\pi,\ -\pi\le\arg(1-x)<\pi\}.$$

There is no need to rewrite the solution $y_2^{(\infty)}(x)$ as given in (2.6) as it is already in a suitable form, this is explained in Lemma 4.3 below. We note that the above two forms of the solution $y_1^{(\infty)}(x)$ are equivalent on the domain $\Omega_\infty\cap\widehat{\Omega}_\infty$. The condition imposed on $\arg(1-x)$ in $\widehat{\Omega}_\infty$ is only necessary to deal with the term $(1-x)^{\gamma-\alpha-\beta}$. After making the substitution $x=\frac{z}{\alpha}$ and taking the limit $\alpha\to\infty$we have

$$\begin{aligned}\left(1-\frac{z}{\alpha}\right)^{\gamma-\alpha-\beta}&=\exp\left((\gamma-\alpha-\beta)\log\left(1-\frac{z}{\alpha}\right)\right),\\ &=\exp\left((\gamma-\alpha-\beta)\left(-\frac{z}{\alpha}+\mathcal{O}\left(\alpha^{-2}\right)\right)\right),\\ &=e^z\left(1+\mathcal{O}\left(\alpha^{-1}\right)\right). \qquad (4.8)\end{aligned}$$

This computation shows how to asymptotically pass from power-like behaviour to exponential behaviour as $\alpha\to\infty$. Moreover, with this new form of $y_1^{(\infty)}(x)$ we are ready to state the following lemma.

Lemma 4.3 *Let $y_2^{(\infty)}(x)$ be given by (2.6) and $y_1^{(\infty)}(x)$ be given in its new form by (4.7). After the substitution $x=\frac{z}{\alpha}$, the terms of these series tend to the terms in the formal series solutions $\tilde{y}_{1,f}^{(\infty)}(z)$ and $\tilde{y}_{2,f}^{(\infty)}(z)$ as given by (3.9), namely we have the following limits:*

$$\begin{aligned}\lim_{\alpha\to\infty}\frac{(1-\beta)_n(\gamma-\beta)_n\alpha^n}{(\alpha+1-\beta)_n n!z^n}&=\frac{(\gamma-\beta)_n(1-\beta)_n}{n!z^n},\\ \lim_{\alpha\to\infty}\frac{(\beta)_n(\beta+1-\gamma)_n\alpha^n}{(\beta+1-\alpha)_n n!z^n}&=(-1)^n\frac{(\beta)_n(\beta+1-\gamma)_n}{n!z^n}.\end{aligned}$$

Proof By direct computation, using

$$\frac{\alpha^n}{(\alpha+1-\beta)_n} = 1 + \mathcal{O}\left(\alpha^{-1}\right) \quad \text{and} \quad \frac{\alpha^n}{(\beta+1-\alpha)_n} = (-1)^n + \mathcal{O}\left(\alpha^{-1}\right).$$

□

Remark 5 Lemma 4.3 is stated in terms of the solutions of the *scalar* hypergeometric equations (1.1) and (1.2). From the viewpoint of working with the (2×2) equations (2.3) and (3.2), we rewrite the solution $Y^{(\infty)}(x)$, as given in (2.9), as follows,

$$\begin{aligned} Y^{(\infty)}(x) &= R_\infty \sum_{n=0}^{\infty} g_{n,\infty} x^{-n} (-x)^{-\Theta_\infty}, && x \in \Omega_\infty, \\ &= R_\infty \sum_{n=0}^{\infty} \widehat{g}_{n,\infty} x^{-n} (-x)^{-\Theta_\infty - \Theta_1} (1-x)^{\Theta_1}, && x \in \widehat{\Omega}_\infty, \end{aligned} \tag{4.9}$$

where $\widehat{g}_{0,\infty} = I$ and we find all other coefficients $\widehat{g}_{n,\infty}$, $n \geq 1$, from the recursive relation,

$$n\widehat{g}_{n,\infty} + \left[\widehat{g}_{n,\infty}, \Theta_{\infty Y}\right] = -R_{\infty Y}^{-1} A_{1Y} R_{\infty Y} \sum_{l=0}^{n-1} \widehat{g}_{l,\infty} + \sum_{l=0}^{n-1} \widehat{g}_{l,\infty} \Theta_1.$$

This recursion equation only differs from that for $g_{n,\infty}$, given in the proof of Lemma 2.2, by the final summation term. We find the solution to this equation is,

$$\widehat{g}_{n,\infty} = \begin{pmatrix} \frac{(1-\beta)_n(\gamma-\beta)_n}{(\alpha+1-\beta)_n n!} & -\frac{(\beta)_{n-1}(\beta+1-\gamma)_{n-1}}{(\beta+1-\alpha)_{n-1}(n-1)!} \\ \frac{\alpha(1-\beta)(\beta-\gamma)(\alpha+1-\gamma)}{(\alpha-\beta)(\alpha+1-\beta)^2(\alpha+2-\beta)} \frac{(2-\beta)_{n-1}(\gamma+1-\beta)_{n-1}}{(\alpha+3-\beta)_{n-1}(n-1)!} & \frac{(\beta-1)_n(\beta-\gamma)_n}{(\beta-\alpha-1)_n n!} \end{pmatrix}. \tag{4.10}$$

The transformation (4.9) is analogous to Kummer relation (4.7). We note that,

$$\begin{aligned} Y^{(\infty)}\left(\frac{z}{\alpha}\right) &= R_\infty \sum_{n=0}^{\infty} \widehat{g}_{n,\infty} \alpha^n z^{-n} \begin{pmatrix} (-\alpha)^{\gamma-\beta} z^{\beta-\gamma} \left(1-\frac{z}{\alpha}\right)^{\gamma-\alpha-\beta} & 0 \\ 0 & (-\alpha)^{\beta-1} z^{1-\beta} \end{pmatrix}, \\ &\equiv R_\infty \begin{pmatrix} 1 & 0 \\ 0 & \alpha^{-1} \end{pmatrix} \begin{pmatrix} 1 & 0 \\ 0 & \alpha \end{pmatrix} \sum_{n=0}^{\infty} \widehat{g}_{n,\infty} \alpha^n z^{-n} \begin{pmatrix} 1 & 0 \\ 0 & \alpha^{-1} \end{pmatrix} \\ &\qquad \begin{pmatrix} z^{\beta-\gamma}\left(1-\frac{z}{\alpha}\right)^{\gamma-\alpha-\beta} & 0 \\ 0 & z^{1-\beta} \end{pmatrix} \begin{pmatrix} (-\alpha)^{\gamma-\beta} & 0 \\ 0 & -(-\alpha)^{\beta} \end{pmatrix}. \end{aligned}$$

The limits analogous to those in Lemma (4.3) are stated as follows: we have the following limit of the leading matrix,

$$\lim_{\alpha\to\infty} R_\infty \begin{pmatrix} 1 & 0 \\ 0 & \alpha^{-1} \end{pmatrix} = \lim_{\alpha\to\infty} \begin{pmatrix} 1 & 0 \\ 0 & \frac{(\beta-\alpha)(\alpha+1-\beta)}{\alpha(\beta-1)(\beta-\gamma)} \end{pmatrix} \begin{pmatrix} 1 & 0 \\ 0 & \alpha^{-1} \end{pmatrix},$$
$$= \begin{pmatrix} 1 & 0 \\ 0 & \frac{-1}{(\beta-1)(\beta-\gamma)} \end{pmatrix} = \widetilde{R}_\infty,$$

and for the terms of the new series,

$$\lim_{\alpha\to\infty} \begin{pmatrix} 1 & 0 \\ 0 & \alpha \end{pmatrix} \alpha^n \widehat{g}_{n,\infty} \begin{pmatrix} 1 & 0 \\ 0 & \alpha^{-1} \end{pmatrix} = h_{n,\infty},$$

where $\widehat{g}_{n,\infty}$ and $h_{n,\infty}$ are given by (4.10) and (3.6) respectively. Hence, we understand that a *term-by-term* limit of the solution,

$$Y^{(\infty)}\left(\frac{z}{\alpha}\right) \begin{pmatrix} (-\alpha)^{\beta-\gamma} & 0 \\ 0 & -(-\alpha)^{-\beta} \end{pmatrix},$$

produces the formal solution $\widetilde{Y}_f^{(\infty)}(z)$, which is analogous to (3.7).

We now turn our attention to the fundamental solutions at $x = 1$, as given in canonical form in (2.5). Observe that these solutions are expressed using Gauss hypergeometric ${}_2F_1$ series in the variable $(1 - x) \equiv (1 - \frac{z}{\alpha})$, which diverge for $|1 - x| > 1 \Leftrightarrow |z - \alpha| > |\alpha|$. As with the fundamental solutions at $x = \infty$, we do not have uniform convergence with respect to α here. Rather than keeping these solutions in canonical form, we use two more of Kummer relations to rewrite them as follows,

$$\begin{aligned} y_1^{(1)}(x) &= (1-x)^{\gamma-\alpha-\beta} {}_2F_1\left(\begin{matrix}\gamma-\alpha,\ \gamma-\beta\\ \gamma+1-\alpha-\beta\end{matrix}; 1-x\right) && x \in \Omega_1, \\ &= x^{\beta-\gamma}(1-x)^{\gamma-\alpha-\beta} {}_2F_1\left(\begin{matrix}\gamma-\beta,\ 1-\beta\\ \gamma+1-\alpha-\beta\end{matrix}; 1-x^{-1}\right), && x \in \widehat{\Omega}_1, \quad (4.11) \\ y_2^{(1)}(x) &= {}_2F_1\left(\begin{matrix}\alpha,\ \beta\\ \alpha+\beta+1-\gamma\end{matrix}; 1-x\right) && x \in \Omega_1, \\ &= x^{-\beta} {}_2F_1\left(\begin{matrix}\beta+1-\gamma,\ \beta\\ \alpha+\beta+1-\gamma\end{matrix}; 1-x^{-1}\right), && x \in \widehat{\Omega}_1, \quad (4.12) \end{aligned}$$

where the new domain $\widehat{\Omega}_1$ is defined as,

$$\widehat{\Omega}_1 = \left\{x : \left|1-x^{-1}\right| < 1,\ -\pi \le \arg(x) < \pi,\ -\pi \le \arg(1-x) < \pi\right\}.$$

We note that the two forms of these solutions are equivalent on the domain $\Omega_1 \cap \widehat{\Omega}_1$. There is a very simple philosophical reason why we rewrite the series in these solutions with $(1 - x^{-1})^n$, rather than $(1 - x)^n$: after the change of variable $x = \frac{z}{\alpha}$, we want to produce a formal series in z^{-n}. Similarly as before, the computations ending in (4.8) show how the solution $y_1^{(1)}(x)$ asymptotically passes from power-like behaviour to exponential behaviour as $\alpha \to \infty$. Moreover, the terms of the series in these new forms of $y_1^{(1)}(x)$ and $y_2^{(1)}(x)$ satisfy the lemma below.

Lemma 4.4 *Let $y_1^{(1)}(x)$ and $y_2^{(1)}(x)$ be given in their new forms by (4.11) and (4.12) respectively. After the substitution $x = \frac{z}{\alpha}$, the terms of these series tend to the terms in the formal series solutions $\tilde{y}_{1,f}^{(\infty)}(z)$ and $\tilde{y}_{2,f}^{(\infty)}(z)$ as given by (3.9), namely we have the following limits:*

$$\lim_{\alpha\to\infty} \frac{(\gamma-\beta)_n(1-\beta)_n(z-\alpha)^n}{(\gamma+1-\alpha-\beta)_n n! z^n} = \frac{(\gamma-\beta)_n(1-\beta)_n}{n! z^n},$$
$$\lim_{\alpha\to\infty} \frac{(\beta+1-\gamma)_n(\beta)_n(z-\alpha)^n}{(\alpha+\beta+1-\gamma)_n n! z^n} = (-1)^n \frac{(\beta)_n(\beta+1-\gamma)_n}{n! z^n}.$$

Proof By direct computation, after expanding the powers of $(z-\alpha)$ and the Pochhammer symbols to find,

$$\frac{(z-\alpha)^n}{(\gamma+1-\alpha-\beta)_n} = 1 + \mathcal{O}\left(\alpha^{-1}\right) \quad \text{and} \quad \frac{(z-\alpha)^n}{(\alpha+\beta+1-\gamma)_n} = (-1)^n + \mathcal{O}\left(\alpha^{-1}\right).$$

□

This lemma shows that *term-by-term* limits of the solutions,

$$y_1^{(1)}(z\alpha^{-1})\,\alpha^{\beta-\gamma} \quad \text{and} \quad -\,y_2^{(1)}(z\alpha^{-1})\,\alpha^{-\beta}, \tag{4.13}$$

produce the formal solutions,

$$\tilde{y}_{1,f}^{(\infty)}(z) \quad \text{and} \quad \tilde{y}_{2,f}^{(\infty)}(z),$$

respectively. The factors $\alpha^{\beta-\gamma}$ and $\alpha^{-\beta}$ in (4.13) are necessary because of the terms,

$$x^{\beta-\gamma} \equiv z^{\beta-\gamma}\alpha^{\gamma-\beta} \quad \text{and} \quad x^{-\beta} \equiv z^{-\beta}\alpha^{\beta},$$

in the solutions $y_1^{(1)}(x)$ and $y_2^{(1)}(x)$ respectively. We note that the direction in which $\alpha \to \infty$ is not yet important for this lemma. The importance of this lemma is shown in the proof of our Main Theorem 4.5.

Remark 6 Similarly as in Remark 5, we may consider the viewpoint of working with the (2×2) equations (2.3) and (3.2) and rewrite the solution $Y^{(1)}(x)$, as given in (2.8), as follows,

$$\begin{aligned} Y^{(1)}(x) &= R_1 \sum_{n=0}^{\infty} g_{n,1}(1-x)^n(1-x)^{\Theta_1}, && x \in \Omega_1, \\ &= R_1 \sum_{n=0}^{\infty} \widehat{g}_{n,1}\left(1-x^{-1}\right)^n x^{-\Theta_\infty-\Theta_1}(1-x)^{\Theta_1}, && x \in \widehat{\Omega}_1, \end{aligned} \tag{4.14}$$

where $\widehat{g}_{0,1} = I$ and we find all other coefficients $\widehat{g}_{n,1}$, $n \geq 1$, from the recursive equation,

$$\left[\widehat{g}_{n,1}, \Theta_1\right] + n\widehat{g}_{n,1} = (n-1)\widehat{g}_{n-1,1} + \widehat{g}_{n-1,1}(\Theta_1 + \Theta_\infty) + R_1^{-1} A_0 R_1 \widehat{g}_{n-1,1}.$$

This recursion equation differs quite significantly from that for $g_{n,1}$, given in the proof of Lemma 2.2. We find the solution to this equation is,

$$\widehat{g}_{n,1} = \begin{pmatrix} \frac{(1-\beta)_n(\gamma-\beta)_n}{(\gamma+1-\alpha-\beta)_n n!} & \frac{(\beta)_n(\beta+1-\gamma)_n}{(\alpha+\beta+1-\gamma)_n n!} - \frac{(\beta)_{n-1}(\beta+1-\gamma)_{n-1}}{(\alpha+\beta+1-\gamma)_{n-1}(n-1)!} \\ \frac{1}{\alpha}\left(\frac{(2-\beta)_n(\gamma+1-\beta)_n}{(\gamma+1-\alpha-\beta)_n n!} - \frac{(2-\beta)_{n-1}(\gamma+1-\beta)_{n-1}}{(\gamma+1-\alpha-\beta)_{n-1}(n-1)!}\right) & \frac{\alpha+1-\gamma}{(\beta-1)(\beta-\gamma)} \frac{(\beta-1)_n(\beta-\gamma)_n}{(\alpha+\beta+1-\gamma)_n n!} \end{pmatrix}. \tag{4.15}$$

The transformation (4.14) is analogous to Kummer relations (4.11) and (4.12). We note that,

$$\begin{aligned} Y^{(1)}\left(\frac{z}{\alpha}\right) &= R_1 \sum_{n=0}^{\infty} \widehat{g}_{n,1} \left(1 - \frac{\alpha}{z}\right)^n \begin{pmatrix} \alpha^{\gamma-\beta} z^{\beta-\gamma} \left(1-\frac{z}{\alpha}\right)^{\gamma-\alpha-\beta} & 0 \\ 0 & \alpha^{\beta-1} z^{1-\beta} \end{pmatrix}, \\ &\equiv R_1 \begin{pmatrix} 1 & 0 \\ 0 & -\alpha^{-1} \end{pmatrix} \begin{pmatrix} 1 & 0 \\ 0 & -\alpha \end{pmatrix}, \sum_{n=0}^{\infty} \widehat{g}_{n,1} \left(1 - \frac{\alpha}{z}\right)^n \begin{pmatrix} 1 & 0 \\ 0 & -\alpha^{-1} \end{pmatrix} \\ &\qquad \begin{pmatrix} z^{\beta-\gamma}\left(1-\frac{z}{\alpha}\right)^{\gamma-\alpha-\beta} & 0 \\ 0 & z^{1-\beta} \end{pmatrix} \begin{pmatrix} \alpha^{\gamma-\beta} & 0 \\ 0 & -\alpha^{\beta} \end{pmatrix}. \end{aligned}$$

The limits analogous to those in Lemma 4.4 are stated as follows: we have the following limit of the leading matrix,

$$\begin{aligned} \lim_{\alpha\to\infty} R_1 \begin{pmatrix} 1 & 0 \\ 0 & -\alpha^{-1} \end{pmatrix} &= \lim_{\alpha\to\infty} \begin{pmatrix} 1 & 1 \\ \frac{1}{\alpha} & \frac{\alpha+1-\gamma}{(\beta-1)(\beta-\gamma)} \end{pmatrix} \begin{pmatrix} 1 & 0 \\ 0 & -\alpha^{-1} \end{pmatrix}, \\ &= \begin{pmatrix} 1 & 0 \\ 0 & \frac{-1}{(\beta-1)(\beta-\gamma)} \end{pmatrix} = \widetilde{R}_\infty, \end{aligned}$$

and for the terms of the new series,

$$\lim_{\alpha\to\infty} \begin{pmatrix} 1 & 0 \\ 0 & -\alpha \end{pmatrix} (-\alpha)^n \widehat{g}_{n,1} \begin{pmatrix} 1 & 0 \\ 0 & -\alpha^{-1} \end{pmatrix} = h_{n,\infty},$$

where $\widehat{g}_{n,1}$ and $h_{n,\infty}$ are given by (4.15) and (3.6) respectively. Hence, we understand that a *term-by-term* limit of the solution,

$$Y^{(1)}\left(\frac{z}{\alpha}\right) \begin{pmatrix} \alpha^{\beta-\gamma} & 0 \\ 0 & -\alpha^{-\beta} \end{pmatrix},$$

produces the formal solution $\widetilde{Y}_f^{(\infty)}(z)$, which is analogous to (4.13).

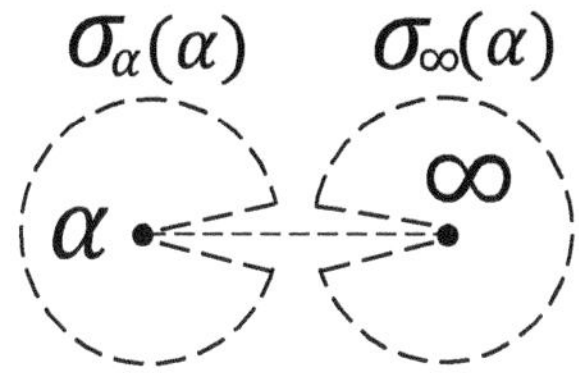

Fig. 8 Sectors $\sigma_\alpha(\alpha)$ and $\sigma_\infty(\alpha)$

Having understood how to take term-by-term limits of the series solutions of Gauss equation around $x = 1$ and ∞ to produce the formal solutions of Kummer equation around $z = \infty$, we now show how to apply Glutsyuk's Theorem 4.1 to Gauss hypergeometric equation. Let $\eta \in \left(0, \frac{\pi}{2}\right)$ be some fixed value. We define the following sectors,

$$\widetilde{\mathscr{S}_k} := \left\{ z : \arg(z) - k\pi \in \left(\eta - \frac{\pi}{2}, \frac{3\pi}{2} - \eta\right)\right\}, \tag{4.16}$$

we note that if $z \in \widetilde{\mathscr{S}_k}$ then $z \in \widetilde{\Sigma}_k$. The presence of η is to ensure that the boundaries of the sectors $\widetilde{\mathscr{S}_k}$ do not contain a Stokes ray, as is necessary in the hypothesis of Glutsyuk's Theorem 4.1. We note that this condition is not satisfied by the sectors $\widetilde{\Sigma}_k$ defined in Theorem 3.2, which are the maximal sectors on which we can define single-valued analytic fundamental solutions.

We also define the following sectors,

$$\sigma_\alpha(\alpha) := \left\{ z : \begin{array}{c} \left|1 - \frac{\alpha}{z}\right| < |\alpha|^2, \ \arg\left(\frac{z}{\alpha}\right) \in (\eta - \pi, \pi - \eta), \\ \arg\left(1 - \frac{z}{\alpha}\right) \in (\eta - \pi, \pi - \eta) \end{array} \right\}, \tag{4.17}$$

$$\sigma_\infty(\alpha) := \left\{ z : \begin{array}{c} \arg\left(-z\alpha^{-1}\right) \in (\eta - \pi, \pi - \eta), \\ \arg\left(1 - \frac{z}{\alpha}\right) \in (\eta - \pi, \pi - \eta) \end{array} \right\}. \tag{4.18}$$

We note that if z is sufficiently close to α with $z \in \sigma_\alpha(\alpha)$ then $x = \frac{z}{\alpha} \in \widehat{\Omega}_1$ and if z is sufficiently large with $z \in \sigma_\infty(\alpha)$ then $x = \frac{z}{\alpha} \in \widehat{\Omega}_\infty$. These sectors will be the new domains of our solutions $y_1^{(1)}(z\alpha^{-1})$, $y_2^{(1)}(z\alpha^{-1})$ and $y_1^{(\infty)}(z\alpha^{-1})$, $y_2^{(\infty)}(z\alpha^{-1})$ respectively, they are illustrated below.

Compared with the domains $\widehat{\Omega}_1$ and $\widehat{\Omega}_\infty$, which are disks with branch cuts, the sectors $\sigma_\alpha(\alpha)$ and $\sigma_\infty(\alpha)$ have larger radii and do not contain any part of the branch cut between α and ∞. We can analytically extend our solutions $y_k^{(1)}(z\alpha^{-1})$ and $y_k^{(\infty)}(z\alpha^{-1})$, $k = 1, 2$, to these larger domains because the singularity $z = \infty$ (resp. $z = \alpha$) can never lie inside the sector $\sigma_\alpha(\alpha)$ (resp. $\sigma_\infty(\alpha)$) or on its boundary. That is the key reason to restrict our solutions to sectors rather than disks.

We examine the sector $\sigma_\alpha(\alpha)$ more closely. From the first condition,

$$\left|1 - \frac{\alpha}{z}\right| < |\alpha|^2 \quad \Leftrightarrow \quad \left|\frac{1}{\alpha} - \frac{1}{z}\right| < |\alpha|,$$

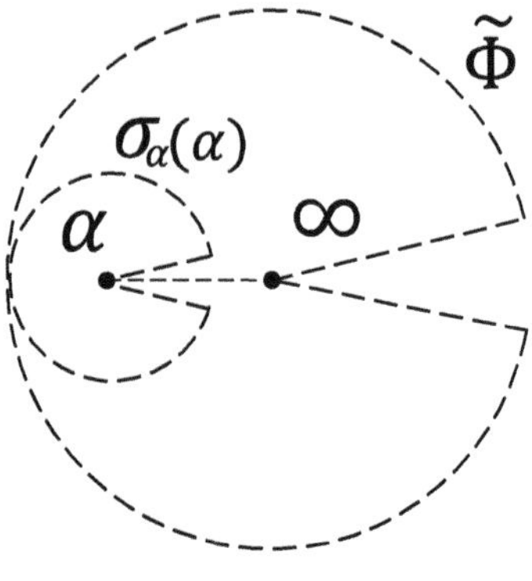

Fig. 9 As $\alpha \to \infty$ along a ray, the sector $\sigma_\alpha(\alpha)$ is translated along the branch cut and becomes in agreement with the sector $\widetilde{\Phi} := \left\{z : \left|\arg\left(\frac{z}{\alpha}\right)\right| < \pi - \eta\right\}$

observe that as $\alpha \to \infty$ the radius of this sector becomes infinite, indeed the above inequality becomes simply $|z| > 0$. Furthermore, as $\alpha \to \infty$ along a ray, the base point of the sector $\sigma_\alpha(\alpha)$ is translated along that ray, tending to infinity. We illustrate this phenomenon in Fig. 9.

In the two limit directions we are concerned with, for $\arg(\alpha) = \pm\frac{\pi}{2}$, we have,

$$\arg\left(\frac{z}{\alpha}\right) \in (\eta - \pi, \pi - \eta) \quad \Leftrightarrow \quad \arg(z) \in \left(\eta - \pi \pm \frac{\pi}{2}, \pi \pm \frac{\pi}{2} - \eta\right),$$

for the sector $\sigma_\infty(\alpha)$, whose base point is already fixed at infinity, we have,

$$\arg\left(-\frac{z}{\alpha}\right) \in (\eta - \pi, \pi - \eta) \quad \Leftrightarrow \quad \arg(z) \in \left(\eta \pm \frac{\pi}{2}, 2\pi \pm \frac{\pi}{2} - \eta\right),$$

recall from (4.8) that the condition on $\arg\left(1 - \frac{z}{\alpha}\right)$ in $\sigma_\infty(\alpha)$ does not play a role after taking the limit. With these considerations in mind, we write,

$$\lim_{\substack{\alpha\to\infty \\ \arg(\alpha)=-\frac{\pi}{2}}} \sigma_\alpha(\alpha) = \widetilde{\mathscr{S}}_{-1}, \qquad \lim_{\substack{\alpha\to\infty \\ \arg(\alpha)=-\frac{\pi}{2}}} \sigma_\infty(\alpha) = \widetilde{\mathscr{S}}_0,$$

$$\lim_{\substack{\alpha\to\infty \\ \arg(\alpha)=\frac{\pi}{2}}} \sigma_\alpha(\alpha) = \widetilde{\mathscr{S}}_0, \qquad \lim_{\substack{\alpha\to\infty \\ \arg(\alpha)=\frac{\pi}{2}}} \sigma_\infty(\alpha) = \widetilde{\mathscr{S}}_1.$$

We now apply Glutsyuk's Theorem 4.1 with the (2×2) hypergeometric equation (2.3) in place of the perturbed equation and the confluent hypergeometric equation (3.2) in place of the non-perturbed equation. Glutsyuk's theorem asserts the existence of invertible diagonal matrices $K_\infty^\pm(\alpha)$ and $K_1^\pm(\alpha)$ such that:

$$\lim_{\substack{\alpha\to\infty \\ \arg(\alpha)=-\frac{\pi}{2}}} Y^{(1)}\left(z\alpha^{-1}\right)\Big|_{z\in\sigma_\alpha(\alpha)} K_1^-(\alpha) = \widetilde{Y}^{(\infty,-1)}(z), \tag{4.19}$$

$$\lim_{\substack{\alpha\to\infty \\ \arg(\alpha)=-\frac{\pi}{2}}} Y^{(\infty)}\left(z\alpha^{-1}\right)\Big|_{z\in\sigma_\infty(\alpha)} K_\infty^-(\alpha) = \widetilde{Y}^{(\infty,0)}(z), \tag{4.20}$$

uniformly for $z \in \widetilde{\mathscr{S}}_{-1}$ and $z \in \widetilde{\mathscr{S}}_0$ respectively, and:

$$\lim_{\substack{\alpha\to\infty\\ \arg(\alpha)=\frac{\pi}{2}}} Y^{(1)}\left(z\alpha^{-1}\right)\Big|_{z\in\sigma_\alpha(\alpha)} K_1^+(\alpha) = \widetilde{Y}^{(\infty,0)}(z), \tag{4.21}$$

$$\lim_{\substack{\alpha\to\infty\\ \arg(\alpha)=\frac{\pi}{2}}} Y^{(\infty)}\left(z\alpha^{-1}\right)\Big|_{z\in\sigma_\infty(\alpha)} K_\infty^+(\alpha) = \widetilde{Y}^{(\infty,1)}(z), \tag{4.22}$$

uniformly for $z \in \widetilde{\mathscr{S}}_0$ and $z \in \widetilde{\mathscr{S}}_1$ respectively. We note that since we are considering two limits, namely one with $\arg(\alpha) = \frac{\pi}{2}$ and another with $\arg(\alpha) = -\frac{\pi}{2}$, we have distinguished the diagonal matrices in each case with a superscript $+$ or $-$ respectively. Due to the asymptotics of the fundamental solutions of Kummer equation as given in Theorem 3.2, each of these four limits is asymptotic to the formal fundamental solution $\widetilde{Y}_f^{(\infty)}(z)$ as $z \to \infty$ with z belonging to the corresponding sector.

Equivalently, from the viewpoint of studying the classical scalar hypergeometric equations (1.1) and (1.2), Glutsyuk's Theorem 4.1 asserts the existence of scalars $k_{1,\infty}^{\pm}(\alpha)$, $k_{2,\infty}^{\pm}(\alpha)$, $k_{1,1}^{\pm}(\alpha)$ and $k_{2,1}^{\pm}(\alpha)$ such that, for $j \in \{1, 2\}$:

$$\lim_{\substack{\alpha\to\infty\\ \arg(\alpha)=-\frac{\pi}{2}}} y_j^{(1)}(z\alpha^{-1})\Big|_{z\in\sigma_\alpha(\alpha)} k_{j,1}^-(\alpha) = \tilde{y}_j^{(\infty,-1)}(z), \tag{4.23}$$

$$\lim_{\substack{\alpha\to\infty\\ \arg(\alpha)=-\frac{\pi}{2}}} y_j^{(\infty)}(z\alpha^{-1})\Big|_{z\in\sigma_\infty(\alpha)} k_{j,\infty}^-(\alpha) = \tilde{y}_j^{(\infty,0)}(z), \tag{4.24}$$

uniformly for $z \in \widetilde{\mathscr{S}}_{-1}$ and $\widetilde{\mathscr{S}}_0$ respectively, and:

$$\lim_{\substack{\alpha\to\infty\\ \arg(\alpha)=\frac{\pi}{2}}} y_j^{(1)}(z\alpha^{-1})\Big|_{z\in\sigma_\alpha(\alpha)} k_{j,1}^+(\alpha) = \tilde{y}_j^{(\infty,0)}(z), \tag{4.25}$$

$$\lim_{\substack{\alpha\to\infty\\ \arg(\alpha)=\frac{\pi}{2}}} y_j^{(\infty)}(z\alpha^{-1})\Big|_{z\in\sigma_\infty(\alpha)} k_{j,\infty}^+(\alpha) = \tilde{y}_j^{(\infty,1)}(z), \tag{4.26}$$

uniformly $z \in \widetilde{\mathscr{S}}_0$ and $\widetilde{\mathscr{S}}_1$ respectively.

Having applied Glutsyuk's theorem to our confluence of the hypergeometric equation, we now focus on understanding what we can deduce about these scalars $k_{j,\infty}^{\pm}(\alpha)$ and $k_{j,1}^{\pm}(\alpha)$, $j = 1, 2$. We are ready to state our first main theorem.

Theorem 4.5 *If $k_{j,\infty}^{\pm}(\alpha)$ and $k_{j,1}^{\pm}(\alpha)$ are scalars satisfying (4.23)–(4.26), then these numbers satisfy the following limits,*

$$\lim_{\substack{\alpha\to\infty\\ arg(\alpha)=\pm\frac{\pi}{2}}} k_{1,\infty}^{\pm}(\alpha)\,(-\alpha)^{\gamma-\beta} = 1, \tag{4.27}$$

$$\lim_{\substack{\alpha\to\infty\\ arg(\alpha)=\pm\frac{\pi}{2}}} -k_{2,\infty}^{\pm}(\alpha)\,(-\alpha)^{\beta} = 1, \tag{4.28}$$

$$\lim_{\substack{\alpha\to\infty\\ arg(\alpha)=\pm\frac{\pi}{2}}} k_{1,1}^{\pm}(\alpha)\,\alpha^{\gamma-\beta}=1, \tag{4.29}$$

$$\lim_{\substack{\alpha\to\infty\\ arg(\alpha)=\pm\frac{\pi}{2}}} -k_{2,1}^{\pm}(\alpha)\,\alpha^{\beta}=1. \tag{4.30}$$

Proof In either case $\arg(\alpha)=\frac{\pi}{2}$ or $-\frac{\pi}{2}$, let $\mathscr{S}^*$ be a closed, proper subsector of $\widetilde{\mathscr{S}_1}$ or $\widetilde{\mathscr{S}_0}$ respectively. Combining the statements (4.24) and (4.26), together with the asymptotic behaviour (3.5), we have,

$$\lim_{\substack{\alpha\to\infty\\ \arg(\alpha)=\pm\frac{\pi}{2}}} y_1^{(\infty)}(z\alpha^{-1})\Big|_{z\in\sigma_\infty(\alpha)} k_{1,\infty}^{\pm}(\alpha)\sim \tilde{y}_{1,f}^{(\infty)}(z),\quad \text{as } z\to\infty,\ z\in\mathscr{S}^*. \tag{4.31}$$

We now re-write $y_1^{(\infty)}(z\alpha^{-1})$ using Kummer transformation as in (4.7),

$$y_1^{(\infty)}\left(z\alpha^{-1}\right)\Big|_{z\in\sigma_\infty(\alpha)}=z^{\beta-\gamma}(-\alpha)^{\gamma-\beta}\left(1-\frac{z}{\alpha}\right)^{\gamma-\alpha-\beta}\sum_{n=0}^{\infty}\frac{(1-\beta)_n(\gamma-\beta)_n\alpha^n}{(\alpha+1-\beta)_n n!z^n}\Bigg|_{z\in\sigma_\infty(\alpha)}.$$

We therefore deduce,

$$\begin{aligned}
&\lim_{\substack{\alpha\to\infty\\ \arg(\alpha)=\pm\frac{\pi}{2}}} z^{\beta-\gamma}(-\alpha)^{\gamma-\beta}\left(1-\frac{z}{\alpha}\right)^{\gamma-\alpha-\beta}\sum_{n=0}^{\infty}\frac{(1-\beta)_n(\gamma-\beta)_n\alpha^n}{(\alpha+1-\beta)_n n!z^n}\Bigg|_{z\in\sigma_\infty(\alpha)} k_{1,\infty}^{\pm}(\alpha)\\
&=\lim_{\substack{\alpha\to\infty\\ \arg(\alpha)=\pm\frac{\pi}{2}}} z^{\beta-\gamma}(-\alpha)^{\gamma-\beta}e^{z}\sum_{n=0}^{\infty}\frac{(1-\beta)_n(\gamma-\beta)_n\alpha^n}{(\alpha+1-\beta)_n n!z^n}\Bigg|_{z\in\sigma_\infty(\alpha)} k_{1,\infty}^{\pm}(\alpha).
\end{aligned}$$

Combining this with (4.31) and writing $\tilde{y}_{1,f}^{(\infty)}(z)$ as in (3.9), we have,

$$\lim_{\substack{\alpha\to\infty\\ \arg(\alpha)=\pm\frac{\pi}{2}}} \sum_{n=0}^{\infty}\frac{(1-\beta)_n(\gamma-\beta)_n\alpha^n}{(\alpha+1-\beta)_n n!z^n}\Bigg|_{z\in\sigma_\infty(\alpha)}(-\alpha)^{\gamma-\beta}k_{1,\infty}^{\pm}\sim\sum_{n=0}^{\infty}\frac{(\gamma-\beta)_n(1-\beta)_n}{n!z^n},$$

as $z\to\infty$ for $z\in\mathscr{S}^*$.

We now define $w=z^{-1}$ so that $w\to 0\Leftrightarrow z\to\infty$ and we can apply the following classical result (Wasow 1965):

Lemma 4.6 *Let $f(w)$ be holomorphic in an open sector σ at $w=0$ and let σ^* be a closed, proper sub-sector of σ. If,*

$$f(w)\sim\sum_{n=0}^{\infty}a_n w^n,\qquad \textit{as } w\to 0,\ w\in\sigma,$$

then:

$$a_n = \frac{1}{n!} \lim_{\substack{w \to 0 \\ w \in \sigma^*}} f^{(n)}(z),$$

where $f^{(n)}(w)$ denotes the nth derivative of $f(w)$,

to find,

$$\frac{(\gamma-\beta)_n(1-\beta)_n}{n!} =$$
$$\frac{1}{n!} \lim_{\substack{w \to 0 \\ w^{-1} \in \mathscr{S}^*}} \frac{d^n}{dw^n} \lim_{\substack{\alpha \to \infty \\ \arg(\alpha)=\pm\frac{\pi}{2}}} \sum_{l=0}^{\infty} \frac{(1-\beta)_l(\gamma-\beta)_l \alpha^l w^l}{(\alpha+1-\beta)_l l!}\Bigg|_{w^{-1}\in\sigma_\infty(\alpha)} (-\alpha)^{\gamma-\beta} k^{\pm}_{1,\infty}(\alpha).$$

We proceed to treat the limits on the right hand side with special care. We first note that, due to the uniformity of the limits (4.24) and (4.26), we may interchange the limit in α with the derivative and the limit in w as follows,

$$\frac{(\gamma-\beta)_n(1-\beta)_n}{n!} =$$
$$\frac{1}{n!} \lim_{\alpha \to \infty} \lim_{\substack{w \to 0 \\ w^{-1} \in \mathscr{S}^*}} \frac{d^n}{dw^n} \sum_{l=0}^{\infty} \frac{(1-\beta)_l(\gamma-\beta)_l \alpha^l w^l}{(\alpha+1-\beta)_l l!}\Bigg|_{w^{-1}\in\sigma_\infty(\alpha)} (-\alpha)^{\gamma-\beta} k^{\pm}_{1,\infty}(\alpha).$$

The next step is to notice that the series inside the limits on the right hand side represents an analytic function (or at least its analytic extension to the sector $\sigma_\infty(\varepsilon)$ does). We may therefore interchange the derivative and series as follows,

$$\frac{(\gamma-\beta)_n(1-\beta)_n}{n!} =$$
$$\frac{1}{n!} \lim_{\substack{\alpha \to \infty \\ \arg(\alpha)=\pm\frac{\pi}{2}}} \lim_{\substack{w \to 0 \\ w^{-1} \in \mathscr{S}^*}} \sum_{l=0}^{\infty} \frac{d^n}{dw^n} \frac{(1\ \ \beta)_l(\gamma\ \ \beta)_l \alpha^l w^l}{(\alpha+1-\beta)_l l!}\Bigg|_{w^{-1}\in\sigma_\infty(\alpha)} (-\alpha)^{\gamma-\beta} k^{\pm}_{1,\infty}(\alpha) =$$
$$\frac{1}{n!} \lim_{\substack{\alpha \to \infty \\ \arg(\alpha)=\pm\frac{\pi}{2}}} \lim_{\substack{w \to 0 \\ w^{-1} \in \mathscr{S}^*}} \sum_{l=0}^{\infty} \frac{(l+n)!}{l!} \frac{(1-\beta)_{l+n}(\gamma-\beta)_{l+n} \alpha^{l+n} w^l}{(\alpha+1-\beta)_{l+n}(l+n)!}\Bigg|_{w^{-1}\in\sigma_\infty(\alpha)} (-\alpha)^{\gamma-\beta} k^{\pm}_{1,\infty}(\alpha).$$

Furthermore, due to the analyticity of the series on the right hand side, its limit as $w \to 0$ certainly exists and is simply equal to the first term of the series. We finally deduce,

$$\frac{(\gamma-\beta)_n(1-\beta)_n}{n!} = \frac{1}{n!} \lim_{\substack{\alpha \to \infty \\ \arg(\alpha)=\pm\frac{\pi}{2}}} n! \frac{(1-\beta)_n(\gamma-\beta)_n \alpha^n}{(\alpha+1-\beta)_n n!} (-\alpha)^{\gamma-\beta} k^{\pm}_{1,\infty}(\alpha). \tag{4.32}$$

Therefore

$$\lim_{\substack{\alpha\to\infty\\ \arg(\alpha)=\pm\frac{\pi}{2}}} \frac{(1-\beta)_n(\gamma-\beta)_n\alpha^n}{(\alpha+1-\beta)_n n!}(-\alpha)^{\gamma-\beta}k^{\pm}_{1,\infty}(\alpha)$$
$$= \frac{(1-\beta)_n(\gamma-\beta)_n}{n!} \lim_{\substack{\alpha\to\infty\\ \arg(\alpha)=\pm\frac{\pi}{2}}} (-\alpha)^{\gamma-\beta}k^{\pm}_{1,\infty}(\alpha).$$

Comparing with the left hand side of (4.32) we deduce the desired result (4.27). The limit (4.28) can be proved by using $y_2^{(\infty)}(z\alpha^{-1})$ as given by (2.6). The limits (4.29) and (4.30) can be proved using $y_1^{(1)}(z\alpha^{-1})$ and $y_2^{(1)}(z\alpha^{-1})$ as given by (4.11) and (4.12) and using Lemma 4.4 in place of Lemma 4.3. □

Remark 7 Returning to the point of view of studying the hypergeometric equations as the (2×2) equations (2.3) and (3.2), our Main Theorem 4.5 may be equivalently stated as follows. If $K_1^{\pm}(\alpha)$ and $K_\infty^{\pm}(\alpha)$ are diagonal matrices satisfying (4.19)–(4.22), then they satisfy the following:

$$\lim_{\substack{\alpha\to\infty\\ \arg(\alpha)=\pm\frac{\pi}{2}}} K_\infty^{\pm}(\alpha)\begin{pmatrix}(-\alpha)^{\gamma-\beta} & 0\\ 0 & -(-\alpha)^{\beta}\end{pmatrix} = I, \tag{4.33}$$

$$\lim_{\substack{\alpha\to\infty\\ \arg(\alpha)=\pm\frac{\pi}{2}}} K_1^{\pm}(\alpha)\begin{pmatrix}\alpha^{\gamma-\beta} & 0\\ 0 & -\alpha^{\beta}\end{pmatrix} = I. \tag{4.34}$$

These limits can be proved in an analogous way to the limits in our Main Theorem 4.5 by using Remarks 5 and 6 in place of Lemmas 4.3 and 4.4 respectively.

4.2.2 Obtaining $\widetilde{Y}^{(0)}(z)$ from $Y^{(0)}(z)$

Since the substitution $x = \frac{z}{\alpha}$ and limit $\alpha \to \infty$ do not interfere with the nature of the Fuchsian singularity $x = 0$, corresponding to $z = 0$, this limit is much easier. We will only consider the limit along $\arg(\alpha) = -\frac{\pi}{2}$, the other case is completely analogous even though it requires to change the branch cut in $\widetilde{\Omega}_0$.

Lemma 4.7 *We have the following limit,*

$$\lim_{\alpha\to\infty} {}_2F_1\left(\begin{matrix}\alpha,\ \beta\\ \gamma\end{matrix};\frac{z}{\alpha}\right) = {}_1F_1\left(\begin{matrix}\beta\\ \gamma\end{matrix};z\right).$$

Proof By taking the term by term limit in the series for ${}_2F_1$ we obtain a uniformly convergent series that coincides with ${}_1F_1$. We conclude by uniqueness of Taylor series expansion for analytic functions. □

Theorem 4.8 *Let* $y_k^{(0)}(x)$ *and* $\tilde{y}_k^{(0)}(z)$*,* $k = 1, 2$*, be defined as in (2.4) and (3.3) respectively. For* $\arg(\alpha) = -\frac{\pi}{2}$*, we have the following limits,*

$$\begin{aligned} &\lim_{\substack{\alpha\to\infty \\ z\in\omega_0(\alpha)}} y_1^{(0)}\left(z\alpha^{-1}\right)\alpha^{1-\gamma} = \tilde{y}_1^{(0)}(z), \\ &\lim_{\substack{\alpha\to\infty \\ z\in\omega_0(\alpha)}} y_2^{(0)}\left(z\alpha^{-1}\right) = \tilde{y}_2^{(0)}(z), \end{aligned} \qquad z \in \widetilde{\Omega}_0. \tag{4.35}$$

where

$$\omega_0(\alpha) = \left\{ z : |z| < |\alpha|,\ -\frac{3}{2}\pi \leq arg(z) < \frac{\pi}{2} \right\}.$$

Proof Notice that for $\arg(\alpha) = \frac{\pi}{2}$, $x \in \Omega_0 \Leftrightarrow z \in \omega_0(\alpha)$. Since the radius of this neighbourhood clearly becomes infinite as $\alpha \to \infty$, if $z \in \omega_0(\alpha)$ for all $|\alpha|$ sufficiently large, then the domain ω_0 tends to the domain $\widetilde{\Omega}_0$ (i.e. the domain in our definition of the fundamental solutions of Kummer equation around $z = 0$ as given in Sect. 3.1).

Using Lemma 4.7, we compute the limits as follows,

$$\begin{aligned} \lim_{\alpha\to\infty} y_1^{(0)}\left(z\alpha^{-1}\right)\alpha^{1-\gamma} &= \lim_{\alpha\to\infty} z^{1-\gamma}{}_2F_1\left(\begin{matrix}\alpha+1-\gamma,\ \beta+1-\gamma \\ 2-\gamma\end{matrix};\frac{z}{\alpha}\right) \\ &= z^{1-\gamma}{}_1F_1\left(\begin{matrix}\beta+1-\gamma \\ 2-\gamma\end{matrix};z\right) = \tilde{y}_1^{(0)}(z), \qquad z \in \widetilde{\Omega}_0, \end{aligned}$$

$$\begin{aligned} \text{and } \lim_{\alpha\to\infty} y_2^{(0)}\left(z\alpha^{-1}\right) &= \lim_{\alpha\to\infty} {}_2F_1\left(\begin{matrix}\alpha,\ \beta \\ \gamma\end{matrix};\frac{z}{\alpha}\right) \\ &= {}_1F_1\left(\begin{matrix}\beta \\ \gamma\end{matrix};z\right) = \tilde{y}_2^{(0)}(z), \qquad z \in \widetilde{\Omega}_0, \end{aligned}$$

as required. □

Remark 8 The factor $\alpha^{1-\gamma}$ in the first limit of Theorem 4.8 is necessary because of the term,

$$x^{1-\gamma} \equiv z^{1-\gamma}\alpha^{\gamma-1},$$

in the solution $y_1^{(0)}(x)$, as given in (2.4).

Remark 9 We have stated Theorem 4.8 in terms of the solutions of the *scalar* hypergeometric equations (1.1) and (1.2). The limits (4.35) can be equivalently stated in terms of the solutions of the (2×2) equations (2.3) and (3.2): for $\arg(\alpha) = \pm\frac{\pi}{2}$,

$$\lim_{\substack{\alpha\to\infty \\ z\in\omega_0(\alpha)}} Y^{(0)}\left(\frac{z}{\alpha}\right)\alpha^{\Theta_0} = \widetilde{Y}^{(0)}(z), \qquad z \in \widetilde{\Omega}_0. \tag{4.36}$$

To see how this is equivalent to (4.35), we observe that for the diagonalising matrices we have

$$\lim_{\alpha\to\infty} R_0 = \lim_{\alpha\to\infty}\begin{pmatrix} 1 & 1 \\ \frac{\alpha+1-\gamma}{\alpha(\beta-\gamma)} & \frac{1}{\beta-1}\end{pmatrix} = \begin{pmatrix} 1 & 1 \\ \frac{1}{\beta-\gamma} & \frac{1}{\beta-1}\end{pmatrix} = \widetilde{R}_0,$$

and for the series, using Lemma 4.7,

$$\lim_{\alpha\to\infty} G_0\left(z\alpha^{-1}\right) = \lim_{\alpha\to\infty}\begin{pmatrix} {}_2F_1\left(\begin{matrix}\alpha+1-\gamma,\ \beta-\gamma\\ 1-\gamma\end{matrix};\frac{z}{\alpha}\right), & \frac{z(\gamma-\beta)}{\gamma(\gamma-1)}\,{}_2F_1\left(\begin{matrix}\alpha+1,\ \beta\\ \gamma+1\end{matrix};\frac{z}{\alpha}\right)\\ \frac{z(\alpha+1-\gamma)(1-\beta)}{\alpha(1-\gamma)(2-\gamma)}\,{}_2F_1\left(\begin{matrix}\alpha+2-\gamma,\ \beta+1-\gamma\\ 3-\gamma\end{matrix};\frac{z}{\alpha}\right), & {}_2F_1\left(\begin{matrix}\alpha,\ \beta-1\\ \gamma-1\end{matrix};\frac{z}{\alpha}\right)\end{pmatrix},$$

$$=\begin{pmatrix} {}_1F_1\left(\begin{matrix}\beta-\gamma\\ 1-\gamma\end{matrix};z\right), & \frac{z(\gamma-\beta)}{\gamma(\gamma-1)}\,{}_1F_1\left(\begin{matrix}\beta\\ \gamma+1\end{matrix};z\right)\\ \frac{z(1-\beta)}{(1-\gamma)(2-\gamma)}\,{}_1F_1\left(\begin{matrix}\beta+1-\gamma\\ 3-\gamma\end{matrix};z\right), & {}_1F_1\left(\begin{matrix}\beta-1\\ \gamma-1\end{matrix};z\right)\end{pmatrix} = H_0(z).$$

4.3 Limits of Monodromy Data

Summarising the results so far, in Sect. 4.2 we showed how *term-by-term* limits of the solutions of Gauss equation around $x=\infty$ and $x=1$ produce the formal solutions of Kummer equation around $z=\infty$. We then explained how Glutsyuk's Theorem 4.1 asserts the existence of certain scalars which multiply Gauss solutions so that their true limits exist and are equal to the solutions of Kummer equation analytic in sectors at $z=\infty$. We have also proved our Main Theorem 4.5, which establishes some important limits which these factors must satisfy. We now bring these results together to prove our second main theorem, concerned with explicitly producing the set of monodromy data $\widetilde{\mathcal{M}}$ from the set $\mathcal{M}$.

Theorem 4.9 *Define the monodromy data of Gauss equation as given in (2.11)–(2.16) and of Kummer equation as in (3.12)–(3.16). We have the following limits,*

$$\lim_{\substack{\alpha\to\infty\\ arg(\alpha)=\frac{\pi}{2}}}\begin{pmatrix}\alpha^{\gamma-\beta} & 0\\ 0 & -\alpha^{\beta}\end{pmatrix} C^{1\infty}\begin{pmatrix}(-\alpha)^{\beta-\gamma} & 0\\ 0 & -(-\alpha)^{-\beta}\end{pmatrix} = \widetilde{S}_0, \tag{4.37}$$

$$\lim_{\substack{\alpha\to\infty\\ arg(\alpha)=-\frac{\pi}{2}}}\begin{pmatrix}\alpha^{\gamma-\beta} & 0\\ 0 & -\alpha^{\beta}\end{pmatrix} C^{1\infty}\begin{pmatrix}(-\alpha)^{\beta-\gamma} & 0\\ 0 & -(-\alpha)^{-\beta}\end{pmatrix} = \widetilde{S}_{-1}, \tag{4.38}$$

$$\lim_{\substack{\alpha\to\infty\\ arg(\alpha)=-\frac{\pi}{2}}}\begin{pmatrix}\alpha^{\gamma-1} & 0\\ 0 & 1\end{pmatrix} C^{0\infty}\begin{pmatrix}(-\alpha)^{\beta-\gamma} & 0\\ 0 & -(-\alpha)^{-\beta}\end{pmatrix} = \widetilde{C}^{0\infty} \tag{4.39}$$

Furthermore, as immediate consequences of the above limits of connection matrices, we have the following limits of monodromy matrices,

$$\lim_{\substack{\alpha\to\infty\\ arg(\alpha)=-\frac{\pi}{2}}} \begin{pmatrix} (-\alpha)^{\gamma-\beta} & 0 \\ 0 & -(-\alpha)^{\beta} \end{pmatrix} M_0 \begin{pmatrix} (-\alpha)^{\beta-\gamma} & 0 \\ 0 & -(-\alpha)^{-\beta} \end{pmatrix} = \widetilde{M}_0, \tag{4.40}$$

$$\lim_{\substack{\alpha\to\infty\\ arg(\alpha)=-\frac{\pi}{2}}} \begin{pmatrix} (-\alpha)^{\gamma-\beta} & 0 \\ 0 & -(-\alpha)^{\beta} \end{pmatrix} M_\infty M_1 \begin{pmatrix} (-\alpha)^{\beta-\gamma} & 0 \\ 0 & -(-\alpha)^{-\beta} \end{pmatrix} = \widetilde{M}_\infty. \tag{4.41}$$

Proof As part of the proof of this theorem, we will use the following elementary lemma.

Lemma 4.10 *Let* $f(\alpha)$ *and* $g(\alpha)$ *be matrices such that* $\lim_{\alpha\to\infty} f(\alpha)g(\alpha)$ *exists.*
(i) *If* $\lim_{\alpha\to\infty} det(f(\alpha))$ *exists and is non-zero and* $det(f(\alpha)) \neq 0$ *for all* α *sufficiently large and if the limit* $\lim_{\alpha\to\infty} f(\alpha)$ *exists and is invertible, then the limit* $\lim_{\alpha\to\infty} g(\alpha)$ *exists.*
(ii) *If* $\lim_{\alpha\to\infty} det(g(\alpha))$ *exists and is non-zero and* $det(g(\alpha)) \neq 0$ *for all* α *sufficiently large and if the limit* $\lim_{\alpha\to\infty} g(\alpha)$ *exists, then the limit* $\lim_{\alpha\to\infty} f(\alpha)$ *exists.*

Let $\sigma_\alpha(\alpha)$ and $\sigma_\infty(\alpha)$ be the sectors defined in (4.17) and (4.18) respectively. As mentioned previously, if $z \in \sigma_\alpha(\alpha)$ then $x \in \Omega_1$ and if $z \in \sigma_\infty(\alpha)$ then $x \in \Omega_\infty$, so that the connection matrix $C^{1\infty}$ remains valid for the solutions $Y^{(1)}(z\alpha^{-1})$ and $Y^{(\infty)}(z\alpha^{-1})$ restricted to the sectors $\sigma_\alpha(\alpha)$ and $\sigma_\infty(\alpha)$ respectively. Since the radii of these sectors do not diminish as $\alpha \to \infty$, for $|\alpha|$ sufficiently large we must have,

$$\sigma_\alpha(\alpha) \cap \sigma_\infty(\alpha) \neq \varnothing,$$

recall Fig. 8. Therefore, for $|\alpha|$ sufficiently large, we have,

$$Y^{(\infty)}\left(z\alpha^{-1}\right) = Y^{(1)}\left(z\alpha^{-1}\right) C^{1\infty}, \qquad z \in \sigma_\alpha(\alpha) \cap \sigma_\infty(\alpha). \tag{4.42}$$

Let $\widetilde{\mathscr{S}}_k$ be the sectors defined in (4.16). To prove the first limit (4.37), we first give a proof of Glutsyuk's Corollary 4.2 in our case. We multiply by the matrices $K_\infty^+(\alpha)$ and $K_1^+(\alpha)$ and take the limit $\alpha \to \infty$, with $\arg(\alpha) = \frac{\pi}{2}$, so that (4.42) becomes,

$$\begin{aligned} &\lim_{\substack{\alpha\to\infty\\ \arg(\alpha)=\frac{\pi}{2}}} Y^{(\infty)}(z\alpha^{-1})\big|_{z\in\sigma_\infty(\alpha)} K_\infty^+(\alpha) \\ &= \lim_{\substack{\alpha\to\infty\\ \arg(\alpha)=\frac{\pi}{2}}} Y^{(1)}(z\alpha^{-1})\big|_{z\in\sigma_\alpha(\alpha)} K_1^+(\alpha) \left(K_1^+(\alpha)\right)^{-1} C^{1\infty} K_\infty^+(\alpha), \end{aligned} \tag{4.43}$$

for $z \in \widetilde{\mathscr{S}}_0 \cap \widetilde{\mathscr{S}}_1$. We apply Lemma 4.10 **(i)** with,

$$f(\alpha) = Y^{(1)}(z\alpha^{-1})\big|_{z\in\sigma_\alpha(\alpha)} K_1^+(\alpha) \quad \text{and} \quad g(\alpha) = \left(K_1^+(\alpha)\right)^{-1} C^{1\infty} K_\infty^+(\alpha).$$

Observe that the hypotheses of Lemma 4.10 hold: the limit,

$$\lim_{\substack{\alpha\to\infty\\ \arg(\alpha)=\frac{\pi}{2}}} f(\alpha)g(\alpha),$$

exists and equals $\widetilde{Y}^{(\infty,1)}(z)$, by (4.22), and the limit,

$$\lim_{\substack{\alpha\to\infty\\ \arg(\alpha)=\frac{\pi}{2}}} f(\alpha),$$

exists and equals $\widetilde{Y}^{(\infty,0)}(z)$, by (4.21), which is clearly invertible because it is a fundamental solution. For all α, $f(\alpha)$ is also clearly invertible because it is a fundamental solution. The limit,

$$\lim_{\substack{\alpha\to\infty\\ \arg(\alpha)=\frac{\pi}{2}}} g(\alpha) = \lim_{\substack{\alpha\to\infty\\ \arg(\alpha)=\frac{\pi}{2}}} \left(K_1^+(\alpha)\right)^{-1} C^{1\infty} K_\infty^+(\alpha),$$

therefore exists and, from (4.43),

$$\widetilde{Y}^{(\infty,1)}(z) = \widetilde{Y}^{(\infty,0)}(z) \lim_{\substack{\alpha\to\infty\\ \arg(\alpha)=\frac{\pi}{2}}} \left(K_1^+(\alpha)\right)^{-1} C^{1\infty} K_\infty^+(\alpha), \qquad z \in \widetilde{\mathscr{S}_0} \cap \widetilde{\mathscr{S}_1}.$$

Recall that if $z \in \widetilde{\mathscr{S}_k}$ then $z \in \widetilde{\Sigma}_k$ and recall Definition 3.2 of Stokes matrices, namely we have,

$$\widetilde{Y}^{(\infty,1)}(z) = \widetilde{Y}^{(\infty,0)}(z)\widetilde{S}_0, \qquad z \in \widetilde{\Sigma}_0 \cap \widetilde{\Sigma}_1.$$

We conclude that,

$$\lim_{\substack{\alpha\to\infty\\ \arg(\alpha)=\frac{\pi}{2}}} \left(K_1^+(\alpha)\right)^{-1} C^{1\infty} K_\infty^+(\alpha) = \widetilde{S}_0,$$

which is precisely Glutsyuk's Corollary 4.2 in our case. Combining this with (4.33) and (4.34), we compute,

$$\begin{aligned}
\widetilde{S}_0 &= \lim_{\substack{\alpha\to\infty\\ \arg(\alpha)=\frac{\pi}{2}}} \left(K_1^+(\alpha)\right)^{-1} C^{1\infty} K_\infty^+(\alpha),\\
&= \lim_{\substack{\alpha\to\infty\\ \arg(\alpha)=\frac{\pi}{2}}} \left(K_1^+(\alpha)\begin{pmatrix}\alpha^{\gamma-\beta} & 0\\ 0 & -\alpha^{\beta}\end{pmatrix}\begin{pmatrix}\alpha^{\beta-\gamma} & 0\\ 0 & -\alpha^{-\beta}\end{pmatrix}\right)^{-1}\\
&\qquad C^{1\infty} K_\infty^+(\alpha)\begin{pmatrix}(-\alpha)^{\gamma-\beta} & 0\\ 0 & -(-\alpha)^{\beta}\end{pmatrix}\begin{pmatrix}(-\alpha)^{\beta-\gamma} & 0\\ 0 & -(-\alpha)^{-\beta}\end{pmatrix},\\
&= \lim_{\substack{\alpha\to\infty\\ \arg(\alpha)=\frac{\pi}{2}}} \begin{pmatrix}\alpha^{\gamma-\beta} & 0\\ 0 & -\alpha^{\beta}\end{pmatrix} C^{1\infty} \begin{pmatrix}(-\alpha)^{\beta-\gamma} & 0\\ 0 & -(-\alpha)^{-\beta}\end{pmatrix},
\end{aligned}$$

where we have implicitly used Lemma 4.10 again, this proves the first limit (4.37) of the theorem. To prove the second limit (4.38), we multiply by the matrices $K_\infty^-(\alpha)$ and $K_1^-(\alpha)$ and take the limit $\alpha \to \infty$, with $\arg(\alpha) = -\frac{\pi}{2}$, so that (4.42) becomes,

$$
\begin{aligned}
&\lim_{\substack{\alpha\to\infty\\ \arg(\alpha)=-\frac{\pi}{2}}} Y^{(\infty)}(z\alpha^{-1})\big|_{z\in\sigma_\infty(\alpha)} K_\infty^-(\alpha)\\
&= \lim_{\substack{\alpha\to\infty\\ \arg(\alpha)=-\frac{\pi}{2}}} Y^{(1)}(z\alpha^{-1})\big|_{z\in\sigma_\alpha(\alpha)} K_1^-(\alpha)\left(K_1^-(\alpha)\right)^{-1} C^{1\infty} K_\infty^-(\alpha), \qquad (4.44)
\end{aligned}
$$

for $z\in\widetilde{\mathscr{S}}_{-1}\cap\widetilde{\mathscr{S}}_0$. By following a similar procedure as above, using Lemma 4.10 and the relations (4.19) and (4.20), we deduce,

$$
\lim_{\substack{\alpha\to\infty\\ \arg(\alpha)=-\frac{\pi}{2}}} \left(K_1^-(\alpha)\right)^{-1} C^{1\infty} K_\infty^-(\alpha) = \widetilde{S}_{-1}.
$$

Combining this with (4.33) and (4.34), we compute,

$$
\begin{aligned}
\widetilde{S}_{-1} &= \lim_{\substack{\alpha\to\infty\\ \arg(\alpha)=-\frac{\pi}{2}}} \left(K_1^-(\alpha)\right)^{-1} C^{1\infty} K_\infty^-(\alpha),\\
&= \lim_{\substack{\alpha\to\infty\\ \arg(\alpha)=-\frac{\pi}{2}}} \left(K_1^-(\alpha)\begin{pmatrix}\alpha^{\gamma-\beta} & 0\\ 0 & -\alpha^{\beta}\end{pmatrix}\begin{pmatrix}\alpha^{\beta-\gamma} & 0\\ 0 & -\alpha^{-\beta}\end{pmatrix}\right)^{-1}\\
&\qquad C^{1\infty}K_\infty^-(\alpha)\begin{pmatrix}(-\alpha)^{\gamma-\beta} & 0\\ 0 & -(-\alpha)^{\beta}\end{pmatrix}\begin{pmatrix}(-\alpha)^{\beta-\gamma} & 0\\ 0 & -(-\alpha)^{-\beta}\end{pmatrix},\\
&= \lim_{\substack{\alpha\to\infty\\ \arg(\alpha)=-\frac{\pi}{2}}} \begin{pmatrix}\alpha^{\gamma-\beta} & 0\\ 0 & -\alpha^{\beta}\end{pmatrix} C^{1\infty}\begin{pmatrix}(-\alpha)^{\beta-\gamma} & 0\\ 0 & -(-\alpha)^{-\beta}\end{pmatrix},
\end{aligned}
$$

where we have implicitly used Lemma 4.10, this proves the second limit (4.38) of the theorem.

To prove the third limit (4.39) we first note that the curve $\gamma_{\infty 0}$ which defines the connection matrix $C^{0\infty}$ survives the confluence limit. In other words, after the substitution $x=\frac{z}{\alpha}$, the curve does not diminish or become broken under the limit $\alpha\to\infty$. This fact is expressed as follows,

$$
\lim_{\substack{\alpha\to\infty\\ \arg(\alpha)=-\frac{\pi}{2}}} \gamma_{\infty 0}\left[Y^{(\infty)}K_\infty^-(\alpha)\right](z\alpha^{-1}) = \gamma_{\infty 0}\left[\widetilde{Y}^{(\infty,0)}\right](z),
$$

or equivalently, using the domains $\omega_0^-(\alpha)$ and $\widetilde{\Omega}_0^-$ defined in Sects. 4.2 and 3.1 respectively,

$$
\lim_{\substack{\alpha\to\infty\\ \arg(\alpha)=-\frac{\pi}{2}}} Y^{(0)}\left(z\alpha^{-1}\right)\big|_{z\in\omega_0^-(\alpha)} C^{0\infty}\left(C^{1\infty}\right)^{-1} K_\infty^-(\alpha) = \widetilde{Y}^{(0)}(z)\widetilde{C}^{0\infty}, \qquad z\in\widetilde{\Omega}_0^-.
$$

Combining this with the limits (4.36) and (4.33), we deduce the required result (4.39) as follows,

$$\lim_{\substack{\alpha\to\infty\\ \arg(\alpha)=-\frac{\pi}{2}}} Y^{(0)}\left(z\alpha^{-1}\right)\Big|_{z\in\omega_0^-(\alpha)} \begin{pmatrix}\alpha^{1-\gamma} & 0\\ 0 & 1\end{pmatrix}\begin{pmatrix}\alpha^{\gamma-1} & 0\\ 0 & 1\end{pmatrix} C^{0\infty}$$

$$K_\infty^-(\alpha)\begin{pmatrix}(-\alpha)^{\gamma-\beta} & 0\\ 0 & -(-\alpha)^{\beta}\end{pmatrix}\begin{pmatrix}(-\alpha)^{\beta-\gamma} & 0\\ 0 & -(-\alpha)^{-\beta}\end{pmatrix}\Bigg)$$

$$= \widetilde{Y}^{(0)}(z) \lim_{\substack{\alpha\to\infty\\ \arg(\alpha)=-\frac{\pi}{2}}} \begin{pmatrix}\alpha^{\gamma-1} & 0\\ 0 & 1\end{pmatrix} C^{0\infty}\begin{pmatrix}(-\alpha)^{\beta-\gamma} & 0\\ 0 & -(-\alpha)^{-\beta}\end{pmatrix}, \qquad z\in\widetilde{\Omega}_0^-,$$

$$= \widetilde{Y}^{(0)}(z)\widetilde{C}^{0\infty}, \qquad z\in\widetilde{\Omega}_0^-,$$

$$\Leftrightarrow \lim_{\substack{\alpha\to\infty\\ \arg(\alpha)=-\frac{\pi}{2}}} \begin{pmatrix}\alpha^{\gamma-1} & 0\\ 0 & 1\end{pmatrix} C^{0\infty}\begin{pmatrix}(-\alpha)^{\beta-\gamma} & 0\\ 0 & -(-\alpha)^{-\beta}\end{pmatrix} = \widetilde{C}^{0\infty},$$

where we have implicitly used Lemma 4.10.

Having deduced the limit (4.39) of the connection matrix, the limit (4.40) follows directly since $M_0 = \left(C^{0\infty}\right)^{-1} e^{2\pi i\Theta_0} C^{0\infty}$ and $\Theta_0 \equiv \widetilde{\Theta}_0$. For (4.40), we have,

$$\lim_{\substack{\alpha\to\infty\\ \arg(\alpha)=-\frac{\pi}{2}}} \begin{pmatrix}(-\alpha)^{\gamma-\beta} & 0\\ 0 & -(-\alpha)^{\beta}\end{pmatrix} M_0 \begin{pmatrix}(-\alpha)^{\beta-\gamma} & 0\\ 0 & -(-\alpha)^{-\beta}\end{pmatrix}$$

$$= \lim_{\substack{\alpha\to\infty\\ \arg(\alpha)=-\frac{\pi}{2}}} \begin{pmatrix}(-\alpha)^{\gamma-\beta} & 0\\ 0 & -(-\alpha)^{\beta}\end{pmatrix} \left(C^{0\infty}\right)^{-1} e^{2\pi i\Theta_0} C^{0\infty}\begin{pmatrix}(-\alpha)^{\beta-\gamma} & 0\\ 0 & -(-\alpha)^{-\beta}\end{pmatrix},$$

$$= \lim_{\substack{\alpha\to\infty\\ \arg(\alpha)=-\frac{\pi}{2}}} \begin{pmatrix}(-\alpha)^{\gamma-\beta} & 0\\ 0 & -(-\alpha)^{\beta}\end{pmatrix} \left(C^{0\infty}\right)^{-1}\begin{pmatrix}\alpha^{\gamma-1} & 0\\ 0 & 1\end{pmatrix} e^{2\pi i\Theta_0}$$

$$\begin{pmatrix}\alpha^{1-\gamma} & 0\\ 0 & 1\end{pmatrix} C^{0\infty}\begin{pmatrix}(-\alpha)^{\beta-\gamma} & 0\\ 0 & -(-\alpha)^{-\beta}\end{pmatrix},$$

$$= \left(\widetilde{C}^{0\infty}\right)^{-1} e^{2\pi i\widetilde{\Theta}_0}\widetilde{C}^{0\infty} = \widetilde{M}_0,$$

as required. □

4.3.1 Explicit Computations of Limits of Monodromy Data

Here we apply Theorem 4.9 to calculate explicitly the Stokes' matrices. We will use the following classical facts:

$$\lim_{\alpha\to\infty} a^{c-b}\frac{\Gamma(a+b)}{\Gamma(a+c)} = 1, \text{ as } a\to\infty, |\arg(a)|<\pi, \tag{4.45}$$

$$\Gamma(a) \equiv \frac{\pi}{\sin(\pi a)\Gamma(1-a)}, \tag{4.46}$$

$$\lim_{a\to\infty} e^{i\pi a}\csc(\pi a) = 2i \text{ for } \mathrm{Im}(a)<0. \tag{4.47}$$

The proof of (4.47) is elementary, the proofs of (4.45) and (4.46) can be found in Whittaker and Watson (1979) and Bateman and Erdélyi (2020).

Let $C^{1\infty}$ be given by (2.12). Using $(-\alpha) \equiv \alpha e^{i\pi}$, we calculate,

$$\begin{aligned}
&\begin{pmatrix} \alpha^{\gamma-\beta} & 0 \\ 0 & -\alpha^{-\beta} \end{pmatrix} C^{1\infty} \begin{pmatrix} (-\alpha)^{\beta-\gamma} & 0 \\ 0 & -(-\alpha)^{\beta} \end{pmatrix} \\
&= \begin{pmatrix} \alpha^{\gamma-\beta} & 0 \\ 0 & -\alpha^{-\beta} \end{pmatrix} \\
&\quad \begin{pmatrix} e^{i\pi(\gamma-\beta)} \frac{\Gamma(\alpha+1-\beta)\Gamma(\alpha+\beta-\gamma)}{\Gamma(\alpha)\Gamma(\alpha+1-\gamma)} & e^{i\pi(\gamma-\alpha)} \frac{\Gamma(\beta+1-\alpha)\Gamma(\alpha+\beta-\gamma)}{\Gamma(\beta)\Gamma(\beta+1-\gamma)} \\ e^{i\pi\alpha} \frac{\Gamma(\alpha+1-\beta)\Gamma(\gamma-\alpha-\beta)}{\Gamma(1-\beta)\Gamma(\gamma-\beta)} & e^{i\pi\beta} \frac{\Gamma(\beta+1-\alpha)\Gamma(\gamma-\alpha-\beta)}{\Gamma(1-\alpha)\Gamma(\gamma-\alpha)} \end{pmatrix} \begin{pmatrix} (-\alpha)^{\beta-\gamma} & 0 \\ 0 & -(-\alpha)^{\beta} \end{pmatrix}, \\
&= \begin{pmatrix} \frac{\Gamma(\alpha+1-\beta)\Gamma(\alpha+\beta-\gamma)}{\Gamma(\alpha)\Gamma(\alpha+1-\gamma)} & -e^{\pi i(\gamma-\alpha-\beta)} \alpha^{\gamma-2\beta} \frac{\Gamma(\beta+1-\alpha)\Gamma(\alpha+\beta-\gamma)}{\Gamma(\beta)\Gamma(\beta+1-\gamma)} \\ -e^{\pi i(\alpha+\beta-\gamma)} \alpha^{2\beta-\gamma} \frac{\Gamma(\alpha+1-\beta)\Gamma(\gamma-\alpha-\beta)}{\Gamma(1-\beta)\Gamma(\gamma-\beta)} & \frac{\Gamma(\beta+1-\alpha)\Gamma(\gamma-\alpha-\beta)}{\Gamma(1-\alpha)\Gamma(\gamma-\alpha)} \end{pmatrix}.
\end{aligned}$$

Using (4.45), we find for the (1,1) and (2,2) elements:

$$\lim_{\substack{\alpha\to\infty \\ \arg(\alpha)=\pm\frac{\pi}{2}}} \frac{\Gamma(\alpha+1-\beta)\Gamma(\alpha+\beta-\gamma)}{\Gamma(\alpha)\Gamma(\alpha+1-\gamma)} = 1,$$

$$\text{and} \quad \lim_{\substack{\alpha\to\infty \\ \arg(\alpha)=\pm\frac{\pi}{2}}} \frac{\Gamma(\beta+1-\alpha)\Gamma(\gamma-\alpha-\beta)}{\Gamma(1-\alpha)\Gamma(\gamma-\alpha)} = 1,$$

respectively, as required. We rewrite the (1,2) and (2,1) elements using (4.46) as follows:

$$\begin{aligned}
&-e^{\pi i(\gamma-\alpha-\beta)} \alpha^{\gamma-2\beta} \frac{\Gamma(\beta+1-\alpha)\Gamma(\alpha+\beta-\gamma)}{\Gamma(\beta)\Gamma(\beta+1-\gamma)} \\
&\qquad = \frac{-e^{i\pi(\gamma-\alpha-\beta)}}{\sin(\pi(\alpha+\beta-\gamma))} \alpha^{\gamma-2\beta} \frac{\Gamma(\beta+1-\alpha)}{\Gamma(\gamma+1-\alpha-\beta)} \frac{\pi}{\Gamma(\beta)\Gamma(\beta+1-\gamma)},
\end{aligned}$$

and,

$$\begin{aligned}
&-e^{i\pi(\alpha+\beta-\gamma)} \alpha^{2\beta-\gamma} \frac{\Gamma(\alpha+1-\beta)\Gamma(\gamma-\alpha-\beta)}{\Gamma(1-\beta)\Gamma(\gamma-\beta)} \\
&\qquad = \frac{-e^{i\pi(\alpha+\beta-\gamma)}}{\sin(\pi(\gamma-\alpha-\beta))} \alpha^{2\beta-\gamma} \frac{\Gamma(\alpha+1-\beta)}{\Gamma(\alpha+\beta+1-\gamma)} \frac{\pi}{\Gamma(1-\beta)\Gamma(\gamma-\beta)},
\end{aligned}$$

respectively. As $\alpha \to \infty$, the dominant terms in these expressions are $e^{\mp i\pi\alpha}$ respectively; observe that, if $\arg(\alpha) = \pm\frac{\pi}{2}$ then $e^{\pm i\pi\alpha} \to 0$ as $\alpha \to \infty$, as required. Finally, for the most important computations, we have:

$$\lim_{\substack{\alpha\to\infty\\ \arg(\alpha)=-\frac{\pi}{2}}} \underbrace{\frac{-e^{i\pi(\alpha+\beta-\gamma)}}{\sin(\pi(\gamma-\alpha-\beta))}}_{\to 2i \text{ by } (4.47)} \underbrace{\alpha^{2\beta-\gamma}\frac{\Gamma(\alpha+1-\beta)}{\Gamma(\alpha+\beta+1-\gamma)}}_{\to 1 \text{ by } (4.45)} \frac{\pi}{\Gamma(1-\beta)\Gamma(\gamma-\beta)},$$

$$= \frac{2\pi i}{\Gamma(1-\beta)\Gamma(\gamma-\beta)} \equiv (S_{-1})_{2,1},$$

and,

$$\lim_{\substack{\alpha\to\infty\\ \arg(\alpha)=\frac{\pi}{2}}} \underbrace{\frac{-e^{i\pi(\gamma-\alpha-\beta)}}{\sin(\pi(\alpha+\beta-\gamma))}}_{\to 2i \text{ by } (4.47)} \underbrace{\alpha^{\gamma-2\beta}\frac{\Gamma(\beta+1-\alpha)}{\Gamma(\gamma+1-\alpha-\beta)}}_{\to e^{i\pi(\gamma-2\beta)} \text{ by } (4.45)} \frac{\pi}{\Gamma(\beta)\Gamma(\beta+1-\gamma)},$$

$$= \frac{2\pi i e^{i\pi(\gamma-2\beta)}}{\Gamma(\beta)\Gamma(\beta+1-\gamma)} \equiv (S_0)_{1,2},$$

as required by formulae (3.12).

Acknowledgements We thank D. Guzzetti for many helpful conversations. This research was supported by the EPSRC Research Grant $EP/P021913/1$ and EPSRC DTA allocation to the Mathematical Sciences Department at Loughborough University.

Appendix A: Gauss Monodromy Data and Mellin-Barnes Integral

Here, following Bateman and Erdélyi (2020), Whittaker and Watson (1979) and Andrews et al. (1999), we re-derive the classical formulae (2.11)–(2.13). This is a worthwhile exercise as it gives a greater understanding of how to analytically continue solutions and compute their monodromy data.

We will work with the following Mellin-Barnes integral,

$$\frac{1}{2\pi i}\int_{-i\infty}^{+i\infty} I(s,x)\,ds \quad \text{where} \quad I(s,x) = \frac{\Gamma(\alpha+s)\Gamma(\beta+s)\Gamma(-s)}{\Gamma(c+s)}(-x)^s, \tag{A.48}$$

with $|\arg(-x)| < \pi$ and whose path of integration is along the imaginary axis with indentations as necessary so that the poles of $\Gamma(\alpha+s)\Gamma(\beta+s)$ lie on its left and the poles of $\Gamma(-s)$ lie on its right, as shown in Fig. 10. It is always possible to construct such a path as long as α and $\beta \notin \mathbb{Z}^{\leq 0}$, which is a general assumption since the case in which α or $\beta \in \mathbb{Z}^{\leq 0}$ corresponds to some of the solutions in (2.4)–(2.6) being polynomials.

We will prove the following proposition, which is sufficient to derive the connection formulae (2.11)–(2.13).

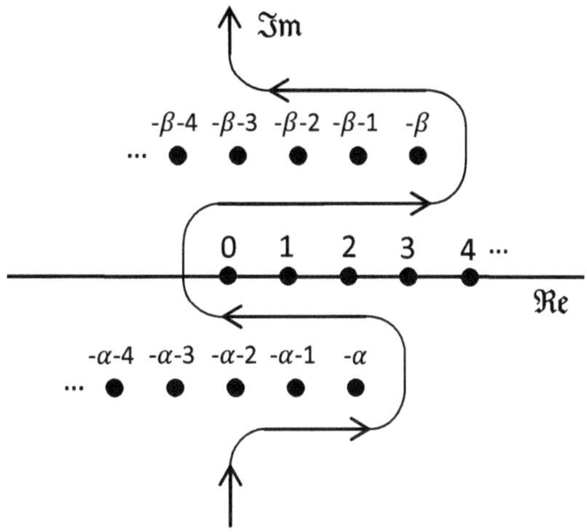

Fig. 10 Path of integration with indentations as in (A.48)

Proposition A.11 *The integral as given by (A.48) satisfies the following properties:*

1. *for* $|arg(-x)| < \pi$,

$$\frac{1}{2\pi i}\int_{-i\infty}^{+i\infty} I(s,x)\,ds,$$

defines an analytic function of x*;*

2. *for* $|arg(-x)| < \pi$ *and* $|x| < 1$,

$$\frac{1}{2\pi i}\int_{-i\infty}^{+i\infty} I(s,x)\,ds = \frac{\Gamma(\alpha)\Gamma(\beta)}{\Gamma(\gamma)}\,y_2^{(0)}(x),$$

where $y_2^{(0)}(x)$ *is the solution of Gauss equation as given by (2.4).*

3. *for* $|arg(-x)| < \pi$ *and* $|x| > 1$,

$$\frac{1}{2\pi i}\int_{-i\infty}^{+i\infty} I(s,x)\,ds = \frac{\Gamma(\alpha)\Gamma(\beta-\alpha)}{\Gamma(\gamma-\alpha)}y_1^{(\infty)}(x) + \frac{\Gamma(\beta)\Gamma(\alpha-\beta)}{\Gamma(\gamma-\beta)}y_2^{(\infty)}(x),$$

where $y_1^{(\infty)}(x)$ *and* $y_2^{(\infty)}(x)$ *are the solutions of Gauss equation as given by (2.6).*

Proof This proof is organised into three parts to prove each statement consecutively.

We start by proving the analyticity of the integral. We use Euler's reflection formula $\Gamma(-s)\Gamma(s+1) = -\pi\csc(\pi s)$ to re-write the integrand,

$$I(s,x) = -\frac{\Gamma(\alpha+s)\Gamma(\beta+s)}{\Gamma(c+s)\Gamma(s+1)}\frac{\pi}{\sin(\pi s)}(-x)^s. \tag{A.49}$$

Using the following asymptotic expansion of the Gamma function (Whittaker and Watson 1979) Sect. 13.6,

$$\Gamma(s+a) = s^{s+a-\frac{1}{2}} e^{-s} \sqrt{2\pi}(1+o(1)), \quad \text{with } |s|, \tag{A.50}$$

which is valid for $|\arg(s+a)| < \pi$, we deduce,

$$\frac{\Gamma(\alpha+s)\Gamma(\beta+s)}{\Gamma(c+s)\Gamma(s+1)} = \mathcal{O}\left(|s|^{\alpha+\beta-\gamma-1}\right), \quad \text{as } |s| \to \infty. \tag{A.51}$$

Writing $\sin(\pi s) = \frac{1}{2i}(e^{i\pi s} - e^{-i\pi s})$ we also deduce,

$$\sin(\pi s) = \mathcal{O}\left(e^{|s|\pi}\right), \quad \text{as } |s| \to \infty, \tag{A.52}$$

along the contour of integration (the imaginary axis). Combining (A.51) and (A.52), the integrand has the following asymptotic behavior,

$$I(s,x) = \mathcal{O}\left(|s|^{\alpha+\beta-\gamma-1} e^{-|s|\pi}(-x)^s\right), \quad \text{as } |s| \to \infty,$$

along the contour of integration, we therefore need only consider the analyticity of the following integral,

$$\begin{aligned} &\int_{-i\infty}^{+i\infty} e^{-|s|\pi}(-x)^s \, ds \\ &\equiv i \int_0^\infty e^{-\sigma\pi} e^{i\sigma(\log|x|+i\arg(-x))} \, d\sigma - i \int_0^\infty e^{-\sigma\pi} e^{-i\sigma(\log|x|+i\arg(-x))} \, d\sigma. \end{aligned} \tag{A.53}$$

We recall the following lemma, see for instance Whittaker and Watson (1979) Sect. 5.32.

Lemma A.12 *If $f : \mathbb{R} \to \mathbb{R}$ is a continuous function such that $|f(t)| \leq K e^{rt}$ for constants K and r, then the integral $\int_0^\infty f(t) e^{-\lambda t} \, dt$ defines an analytic function of λ for $r < Re(\lambda)$.*

Applying this lemma to the first integral in (A.53), with $r = -\pi$, $K = 1$ and $\lambda = \arg(-x)$, we find an analytic function for $-\pi < \arg(-x)$. Applying this lemma to the second integral in (A.53), with $r = -\pi$, $K = 1$ and $\lambda = -\arg(-x)$, we find an analytic function for $\arg(-x) < \pi$. This concludes the proof that the integral (A.48) defines an analytic function for $-\pi < \arg(-x) < \pi$.

We now represent $y_2^{(0)}(x)$ using a Mellin-Barnes integral. We write $I(s,x)$ as in (A.49) and consider the following integral,

$$\frac{1}{2\pi i} \int_{C_N} I(s,x) \, ds,$$

for $N \in \mathbb{N}^{\geq 0}$, where C_N is the following semicircle,

$$C_N = \left\{ s = \left(N + \frac{1}{2}\right) e^{i\theta} : \theta \in \left[-\frac{\pi}{2}, \frac{\pi}{2}\right] \right\}.$$

Let $s \in C_N$, using formula (A.50) from above, we deduce the following asymptotic behavior,

$$\frac{\Gamma(\alpha+s)\Gamma(\beta+s)}{\Gamma(\gamma+s)\Gamma(s+1)} = \mathcal{O}\left(N^{\alpha+\beta-\gamma-1}\right), \quad \text{as } N \to \infty, \tag{A.54}$$

and, using $\sin(\pi s) = \frac{1}{2i}(e^{i\pi s} - e^{-i\pi s})$,

$$\frac{(-x)^s}{\sin(\pi s)} = \mathcal{O}\left(e^{(N+\frac{1}{2})(\cos(\theta)\log|x|-\sin(\theta)\arg(-x)-\pi|\sin(\theta)|)}\right), \quad \text{as } N \to \infty. \tag{A.55}$$

Since $|\arg(-x)| < \pi$, we write $|\arg(-x)| \leq \pi - \delta$ for some $\delta > 0$, so that,

$$\begin{aligned} \pm\arg(-x) + \pi \geq \delta \quad &\Leftrightarrow \quad \sin(\theta)\arg(-x) + |\sin(\theta)|\pi \geq |\sin(\theta)|\delta, \\ &\Leftrightarrow \quad e^{-\sin(\theta)\arg(-x)-\pi|\sin(\theta)|} \leq e^{-|\sin(\theta)|\delta}. \end{aligned} \tag{A.56}$$

Combining (A.54)–(A.56), the integrand has the following asymptotic behaviour for $s \in C_N$,

$$I(s,x) = \mathcal{O}\left(N^{\alpha+\beta-\gamma-1}e^{(N+\frac{1}{2})(\cos(\theta)\log|x|-|\sin(\theta)|\delta)}\right), \quad \text{as } N \to \infty.$$

Since $\cos(\theta)$ and $|\sin(\theta)|$ are even functions, we need only consider $\theta \in \left[0, \frac{\pi}{2}\right]$. For $\theta \in \left[0, \frac{\pi}{4}\right]$, $\cos(\theta) \geq \frac{1}{\sqrt{2}}$ and for $\theta \in \left[\frac{\pi}{4}, \frac{\pi}{2}\right]$, $\sin(\theta) \geq \frac{1}{\sqrt{2}}$. Henceforth, we impose the condition that $|x| < 1$, or equivalently $\log|x| < 0$. For $s \in C_N$ we deduce:

$$I(s,x) = \begin{cases} \mathcal{O}\left(N^{\alpha+\beta-\gamma-1}e^{(N+\frac{1}{2})\frac{1}{\sqrt{2}}\log|x|}\right), & \theta \in \left[0, \frac{\pi}{4}\right), \\ \mathcal{O}\left(N^{\alpha+\beta-\gamma-1}e^{(N+\frac{1}{2})\frac{1}{\sqrt{2}}(\log|x|-\delta)}\right), & \theta = \frac{\pi}{4}, \\ \mathcal{O}\left(N^{\alpha+\beta-\gamma-1}e^{-(N+\frac{1}{2})\frac{1}{\sqrt{2}}\delta}\right), & \theta \in \left(\frac{\pi}{4}, \frac{\pi}{2}\right], \end{cases}$$

as $N \to \infty$. This shows that the integral of $I(s,x)$ along the semicircle C_N tends to zero as N tends to infinity, for $|x| < 1$ and $|\arg(-x)| < \pi$. Due to Cauchy's theorem, we have,

$$\frac{1}{2\pi i}\left(\int_{-i\infty}^{+i\infty} - \int_{(N+\frac{1}{2})i}^{+i\infty} - \int_{C_N} - \int_{-i\infty}^{-(N+\frac{1}{2})i}\right) I(s,x)\,ds = -\sum_{n=0}^{N} \operatorname*{Res}_{s=n} I(s,x). \tag{A.57}$$

We note that there is a minus sign since the path of integration is a contour oriented clockwise, see Fig. 11.

Using $\operatorname*{Res}_{\lambda=-n} \Gamma(\lambda) = \frac{(-1)^n}{n!}$, for $n \geq 0$, we compute the residues to find,

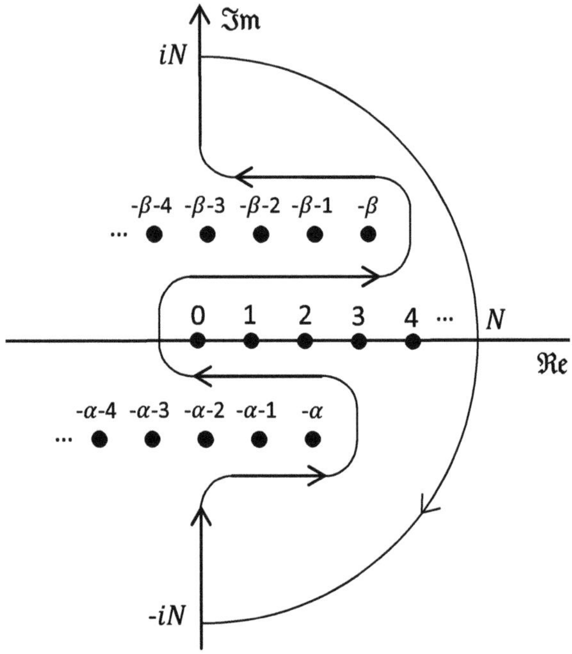

Fig. 11 Paths of integration along the imaginary axis and the semicircle C_N as in (A.57)

$$\frac{1}{2\pi i}\int_{-i\infty}^{+i\infty} I(s,x)\,ds = \lim_{N\to\infty}\sum_{n=0}^{N}\frac{\Gamma(\alpha+n)\Gamma(\beta+n)}{\Gamma(\gamma+n)\Gamma(n+1)}x^n,$$

for $|x|<1$ and $|\arg(-x)|<\pi$ and the desired result is proved after noting $\frac{\Gamma(\alpha+n)}{\Gamma(\alpha)}\equiv(\alpha)_n$.

Finally, we carry out the analytic continuation of $y_2^{(0)}(x)$ for $|x|>1$. The technique to derive the connection formulae is similar to that already used in the second part of this proof, the main difference being that we will now consider taking an integral on the left hand side of the imaginary axis. For $N\in\mathbb{N}$ consider the integral,

$$\frac{1}{2\pi i}\int_{C_N'} I(s,x)\,ds,$$

where C_N' is the semicircle,

$$C_N' = \left\{s = Ne^{i\theta} : \theta\in\left[-\frac{3\pi}{2},-\frac{\pi}{2}\right]\right\}.$$

We summarise the results, following a similar procedure as before. Using (A.50) we deduce,

$$\frac{\Gamma(\alpha+s)\Gamma(\beta+s)\Gamma(-s)}{\Gamma(\gamma+s)} = \mathcal{O}\left(N^{\alpha+\beta-\gamma-1}e^{-N\pi|\sin(\theta)|}\right),$$

for $s \in C'_N$ as $N \to \infty$, and hence,

$$I(s,x) = \mathcal{O}\left(N^{\alpha+\beta-\gamma-1} e^{N(\cos(\theta)\log|x| - \sin(\theta)\arg(-x) - \pi|\sin(\theta)|)}\right),$$
$$= \mathcal{O}\left(N^{\alpha+\beta-\gamma-1} e^{N(\cos(\theta)\log|x| - |\sin(\theta)|\delta)}\right),$$

where δ is a small positive number such that $|\arg(-x)| \leq \pi - \delta$. Clearly $\cos(\theta)$ and $-|\sin(\theta)|$ are both non-positive for $\theta \in \left[-\frac{3\pi}{2}, -\frac{\pi}{2}\right]$ and they are never both simultaneously zero. Furthermore, for $|x| > 1$ we have $\log|x| > 0$, so that the integral of $I(s,x)$ along the semicircle C'_N tends to zero as N tends to infinity, for $|x| > 1$ and $|\arg(-x)| < \pi$. Due to Cauchy's theorem, we have,

$$\frac{1}{2\pi i}\left(\int_{-i\infty}^{+i\infty} - \int_{Ni}^{+i\infty} - \int_{C'_N} - \int_{-i\infty}^{-Ni}\right) I(s,x)\, ds$$
$$= \sum_{n=0}^{M_1(N)} \operatorname*{Res}_{s=\alpha-n} I(s,x) + \sum_{n=0}^{M_2(N)} \operatorname*{Res}_{s=\beta-n} I(s,x), \tag{A.58}$$

where $M_1(N)$ and $M_2(N)$ are the number of poles $-\alpha, -\alpha-1, \ldots$ and $-\beta, -\beta-1, \ldots$ which lie to the right of the semicircle respectively. Clearly $M_1(N)$ and $M_2(N)$ become infinite as N tends to infinity, see Fig. 12.

We compute the residues to find,

$$\frac{1}{2\pi i}\int_{-i\infty}^{+i\infty} I(s,x)\, ds = \frac{\Gamma(\alpha)\Gamma(\beta-\alpha)}{\Gamma(\gamma-\alpha)}(-x)^{-\alpha} \lim_{N\to\infty} \sum_{n=0}^{M_1(N)} \frac{(\alpha)_n(\alpha+1-\gamma)_n}{(\alpha+1-\beta)_n n! x^n}$$
$$+ \frac{\Gamma(\beta)\Gamma(\alpha-\beta)}{\Gamma(\gamma-\beta)}(-x)^{-\beta} \lim_{N\to\infty} \sum_{n=0}^{M_2(N)} \frac{(\beta)_n(\beta+1-\gamma)_n}{(\beta+1-\alpha)_n n! x^n},$$

for $|x| > 1$ and $|\arg(-x)| < \pi$ and the desired result is proved. □

We conclude these computations by explaining how Proposition A.11 leads to the formulae (2.11)–(2.13). Let $\gamma_{j,k}$ be a curve as described at the beginning of this subsection. The second statement in proposition A.11 shows how to represent Gauss ${}_2F_1$ series using a Mellin-Barnes integral. Due to the analyticity of this integral, as shown in the first statement, the third statement provides the formula for the analytic continuation of Gauss hypergeometric series beyond its radius of convergence. That is to say,

$$\gamma_{0,\infty}\left[y_2^{(0)}\right](x) = \frac{\Gamma(\alpha-\beta)\Gamma(\gamma)}{\Gamma(\alpha-\gamma)\Gamma(\beta)}\, y_1^{(\infty)}(x) + \frac{\Gamma(\beta-\alpha)\Gamma(\gamma)}{\Gamma(\beta-\gamma)\Gamma(\alpha)}\, y_2^{(\infty)}(x).$$

By manipulating the parameters as follows: $\alpha \mapsto \alpha+1-\gamma$, $\beta \mapsto \beta+1-\gamma$, $\gamma \mapsto 2-\gamma$ and multiplying through by $x^{1-\gamma}$ we also deduce,

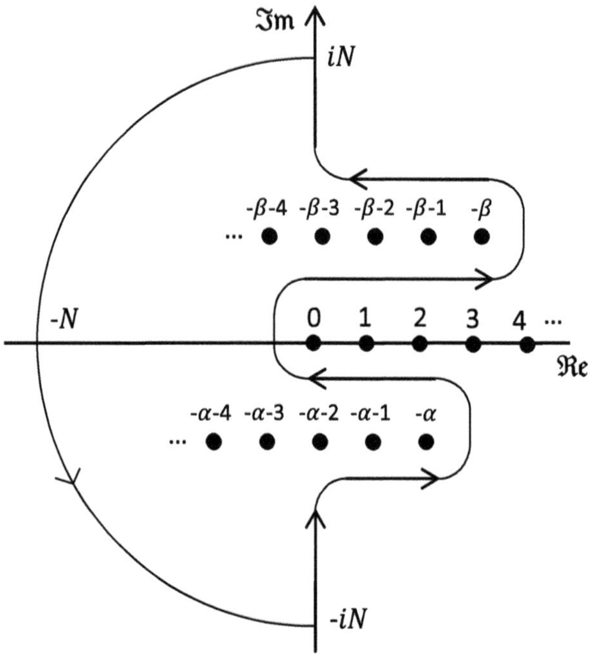

Fig. 12 Paths of integration along the imaginary axis and the semicircle C'_N as in (A.58)

$$\gamma_{0,\infty}\left[y_1^{(0)}\right](x) = -e^{-i\pi\gamma}\frac{\Gamma(\beta-\alpha)\Gamma(2-\gamma)}{\Gamma(1-\alpha)\Gamma(\beta+1-\gamma)}\, y_1^{(\infty)}(x) - e^{-i\pi\gamma}\frac{\Gamma(\alpha-\beta)\Gamma(2-\gamma)}{\Gamma(1-\beta)\Gamma(\alpha+1-\gamma)}\, y_2^{(\infty)}(x),$$

recall that we have selected a branch of $\log(x)$ in the definition of our solutions (2.4) around zero so $x^{1-\gamma}$ is well-defined. These factors constitute the entries of the connection matrix,

$$\left(\gamma_{0,\infty}\left[y_1^{(0)}\right](x),\ \gamma_{0,\infty}\left[y_2^{(0)}\right](x)\right) = \left(y_1^{(\infty)}(x),\ y_2^{(\infty)}(x)\right)\, C^{\infty 0},$$

where,

$$C^{\infty 0} = \begin{pmatrix} -e^{-i\pi\gamma}\frac{\Gamma(\beta-\alpha)\Gamma(2-\gamma)}{\Gamma(1-\alpha)\Gamma(\beta+1-\gamma)} & \frac{\Gamma(\alpha-\beta)\Gamma(\gamma)}{\Gamma(\alpha-\gamma)\Gamma(\beta)} \\ -e^{-i\pi\gamma}\frac{\Gamma(\alpha-\beta)\Gamma(2-\gamma)}{\Gamma(1-\beta)\Gamma(\alpha+1-\gamma)} & \frac{\Gamma(\beta-\alpha)\Gamma(\gamma)}{\Gamma(\beta-\gamma)\Gamma(\alpha)} \end{pmatrix},$$

which is indeed the inverse of the connection matrix $C^{0\infty}$ as given by (2.11). To find the analytic continuation of the solutions around $x = 1$ we manipulate the variable x as well as the parameters. From the transformations $\alpha \mapsto \alpha$, $\beta \mapsto \beta$, $\gamma \mapsto \alpha+\beta+1-\gamma$ and $x \mapsto 1-x$, we have,

$$\gamma_{1,\infty}\left[y_2^{(1)}\right](x) =$$
$$e^{-i\pi\alpha}\frac{\Gamma(\beta-\alpha)\Gamma(\alpha+\beta+1-\gamma)}{\Gamma(\beta)\Gamma(\beta+1-\gamma)}(1-x)^{-\alpha}{}_2F_1\left(\begin{matrix}\alpha,\ \gamma-\beta\\ \alpha+1-\beta\end{matrix};(1-x)^{-1}\right)$$
$$+e^{-i\pi\beta}\frac{\Gamma(\alpha-\beta)\Gamma(\alpha+\beta+1-\gamma)}{\Gamma(\alpha)\Gamma(\alpha+1-\gamma)}(1-x)^{-\beta}{}_2F_1\left(\begin{matrix}\beta,\ \gamma-\alpha\\ \beta+1-\alpha\end{matrix};(1-x)^{-1}\right),$$

and from the transformations $\alpha\mapsto\gamma-\alpha$, $\beta\mapsto\gamma-\beta$, $\gamma\mapsto\gamma+1-\alpha-\beta$ and $x\mapsto 1-x$,

$$\gamma_{1,\infty}\left[y_1^{(1)}\right](x) =$$
$$e^{i\pi(\beta-\gamma)}\frac{\Gamma(\beta-\alpha)\Gamma(\gamma+1-\alpha-\beta)}{\Gamma(1-\alpha)\Gamma(\gamma-\alpha)}(1-x)^{-\alpha}{}_2F_1\left(\begin{matrix}\alpha,\ \gamma-\beta\\ \alpha+1-\beta\end{matrix};(1-x)^{-1}\right)$$
$$+e^{i\pi(\alpha-\gamma)}\frac{\Gamma(\alpha-\beta)\Gamma(\gamma+1-\alpha-\beta)}{\Gamma(1-\beta)\Gamma(\gamma-\beta)}(1-x)^{-\beta}{}_2F_1\left(\begin{matrix}\beta,\ \gamma-\alpha\\ \beta+1-\alpha\end{matrix};(1-x)^{-1}\right),$$

both for $|\arg(x-1)|<\pi$ and $|x-1|>1$. After applying Kummer transformation,

$$(1-x)^{-a}{}_2F_1\left(\begin{matrix}a,\ c-b\\ a+1-b\end{matrix};(1-x)^{-1}\right)=(-x)^{-a}{}_2F_1\left(\begin{matrix}a,\ a+1-c\\ a+1-b\end{matrix};x^{-1}\right),$$

which is valid for $|\arg(x-1)|<\pi$, $|\arg(-x)|<\pi$, $|x-1|>1$ and $|x|>1$, we deduce the connection matrix,

$$\left(\gamma_{0,\infty}\left[y_1^{(1)}\right](x),\ \gamma_{0,\infty}\left[y_2^{(1)}\right](x)\right)=\left(y_1^{(\infty)}(x),\ y_2^{(\infty)}(x)\right)\,C^{\infty 1},$$

where,

$$C^{\infty 1}=\begin{pmatrix} e^{i\pi(\beta-\gamma)}\frac{\Gamma(\beta-\alpha)\Gamma(\gamma+1-\alpha-\beta)}{\Gamma(1-\alpha)\Gamma(\gamma-\alpha)} & e^{-i\pi\alpha}\frac{\Gamma(\beta-\alpha)\Gamma(\alpha+\beta+1-\gamma)}{\Gamma(\beta)\Gamma(\beta+1-\gamma)}\\ e^{i\pi(\alpha-\gamma)}\frac{\Gamma(\alpha-\beta)\Gamma(\gamma+1-\alpha-\beta)}{\Gamma(1-\beta)\Gamma(\gamma-\beta)} & e^{-i\pi\beta}\frac{\Gamma(\alpha-\beta)\Gamma(\alpha+\beta+1-\gamma)}{\Gamma(\alpha)\Gamma(\alpha+1-\gamma)}\end{pmatrix},$$

which is indeed the inverse of the connection matrix $C^{1\infty}$ as given by (2.12). The connection matrix C^{01} as in (2.13) can be deduced from the relation,

$$C^{01}=C^{\infty 1}C^{0\infty}.$$

Appendix B: Mellin-Barnes Integral for Kummer Equation

In Appendix B, we follow the classical approach to show that these solutions can be expressed in closed form by certain Mellin-Barnes integrals and thus derive the connection matrices. This analysis allows us to explicitly compute the monodromy data, including Stokes matrices, of Kummer equation in the following section and thus obtain a richer understanding of Stokes phenomenon.

The remainder of this subsection is dedicated to deriving the classical formulae (3.12)–(3.13). This is a valuable exercise in its own right as it gives us a richer understanding of Stokes phenomenon using a concrete example. Our approach is to use Mellin-Barnes integrals to represent the fundamental solutions $\widetilde{Y}^{(\infty,k)}(z)$, as defined in Theorem 3.2, for which we are able to compute their analytic continuations. Our analysis of Mellin-Barnes integrals is based on Whittaker and Watson's (1979) Sect. 16, who study a different form of the confluent hypergeometric differential equation but is equivalent to ours using analytic transformations.

Define the following functions,

$$\begin{aligned} \tilde{y}_1^{(\infty,-1)}(z) &= e^{-i\pi(\beta-\gamma)} e^z \varphi\left(\gamma-\beta,\gamma; e^{i\pi} z\right), \\ \tilde{y}_2^{(\infty,-1)}(z) &= -\varphi(\beta,\gamma;z), \end{aligned} \qquad z \in \widetilde{\Sigma}_{-1}, \tag{B.59}$$

$$\begin{aligned} \tilde{y}_1^{(\infty,0)}(z) &= e^{i\pi(\beta-\gamma)} e^z \varphi\left(\gamma-\beta,\gamma; e^{-i\pi} z\right), \\ \tilde{y}_2^{(\infty,0)}(z) &= -\varphi(\beta,\gamma;z), \end{aligned} \qquad z \in \widetilde{\Sigma}_{0}, \tag{B.60}$$

where φ is the Mellin-Barnes integral,

$$\varphi(\beta,\gamma;z) = \frac{1}{2\pi i}\int_{-i\infty}^{+i\infty} \frac{\Gamma(s)\Gamma(\beta-s)\Gamma(\beta+1-\gamma-s)}{\Gamma(\beta)\Gamma(\beta+1-\gamma)} z^{s-\beta}\, ds, \tag{B.61}$$

whose path of integration is along the imaginary axis with indentations as necessary so that the poles of $\Gamma(s)$ lie on its left and the poles of $\Gamma(\beta-s)\Gamma(\beta+1-\gamma-s)$ lie on its right, as shown in Fig. 13. When dealing with $\varphi(\beta,\gamma;z)$ it is to be understood that $\arg(z)$ belongs to an interval of length at most 2π, as in (B.59) and (B.60), so that we have a well-defined function.

Proposition B.13 *Let $\widetilde{Y}^{(\infty,k)}(z)$ be the fundamental solutions defined in Theorem 3.2. Also let $\tilde{y}_1^{(\infty,k)}(z)$ and $\tilde{y}_2^{(\infty,k)}(z)$, $k=-1,0$, be the functions defined in (B.59) and (B.60) and denote by $\widetilde{Y}(\tilde{y}_1,\tilde{y}_2;z)$ the matrix function given by (3.1). We have,*

$$\widetilde{Y}\left(\tilde{y}_1^{(\infty,k)}, \tilde{y}_2^{(\infty,k)}; z\right) = \widetilde{Y}^{(\infty,k)}(z), \qquad z \in \widetilde{\Sigma}_k, \tag{B.62}$$

for $k=-1,0$.

Proof We prove this proposition in three steps: we first show that the functions $\tilde{y}_1^{(\infty,k)}(z)$ and $\tilde{y}_2^{(\infty,k)}(z)$ are analytic on their respective sectors; using this fact, we secondly show that these functions satisfy Kummer equation (1.2); finally, we show that these functions have the correct asymptotic behaviour (3.5). By the uniqueness statement of Theorem 3.2, these conditions are sufficient to conclude (B.62).

First step: analyticity of $\tilde{y}_1^{(\infty,k)}(z)$ and $\tilde{y}_2^{(\infty,k)}(z)$.

We require formula (A.50) and Lemma A.12, as used in the derivation of Gauss monodromy data formulae. Using (A.50), we have the following behaviour in the

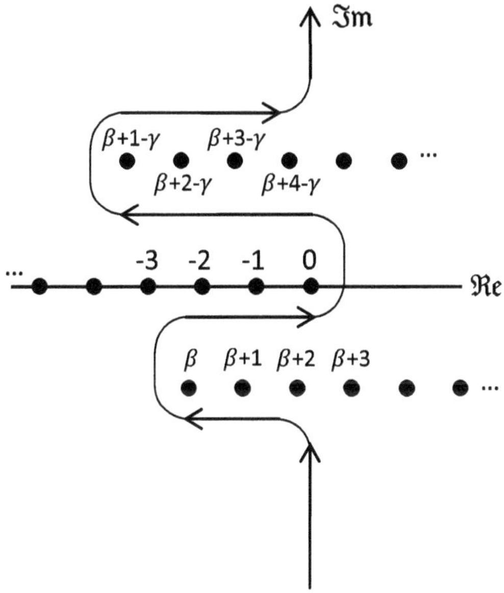

Fig. 13 Path of integration in the Mellin-Barnes integral $\varphi(\beta, \gamma; z)$, the dots represent the poles of the integrand

integrand of $\varphi(a, c; z)$,

$$\Gamma(s)\Gamma(\beta - s)\Gamma(\beta + 1 - \gamma - s) = \mathcal{O}\left(e^{-\frac{3\pi}{2}|s|}|s|^{2\beta-\gamma-\frac{1}{2}}\right), \quad \text{as } |s| \to \infty \quad \text{(B.63)}$$

along the contour of integration. We therefore need only consider the analyticity of the following integral,

$$\int_{-i\infty}^{+i\infty} e^{-\frac{3\pi}{2}|s|} z^{s-\beta}\, ds$$
$$\equiv i\int_0^\infty e^{-\frac{3\pi}{2}|\sigma|} z^{-\beta} e^{i\sigma(\log|z|+i\arg(z))}\, d\sigma - i\int_0^\infty e^{-\frac{3\pi}{2}|\sigma|} z^{-\beta} e^{-i\sigma(\log|z|+i\arg(z))}\, d\sigma.$$

Applying Lemma A.12 to the first integral, with $r = -\frac{3\pi}{2}$, $K = 1$ and $\lambda = \arg(z)$, we find an analytic function for $-\frac{3\pi}{2} < \arg(z)$. Applying Lemma A.12 to the second integral, with $r = -\frac{3\pi}{2}$, $K = 1$ and $\lambda = -\arg(z)$, we find an analytic function for $\arg(z) < \frac{3\pi}{2}$. We conclude that $\varphi(\beta, \gamma; z)$ defines analytic functions $\tilde{y}_2^{(\infty,-1)}(z)$ and $\tilde{y}_2^{(\infty,0)}(z)$ on their respective sectors $\widetilde{\Sigma}_{-1}$ and $\widetilde{\Sigma}_0$. It therefore follows that $\tilde{y}_1^{(\infty,-1)}(z)$ and $\tilde{y}_1^{(\infty,0)}(z)$ are also analytic functions, since $\varphi\left(\gamma - \beta - 1, \gamma; e^{i\pi z}\right)$ must be analytic on $z \in \widetilde{\Sigma}_{-1}$ and $\varphi\left(\gamma - \beta - 1, \gamma; e^{-i\pi z}\right)$ must be analytic on $z \in \widetilde{\Sigma}_0$.

Second step: Showing $\tilde{y}_1^{(\infty,k)}(z)$ and $\tilde{y}_2^{(\infty,k)}(z)$ satisfy the Kummer equation (1.2).
We will now substitute $\varphi(\beta, \gamma; z)$ for $\tilde{y}(z)$ into the left hand side of Kummer equation (1.2) and show that the result is zero. Having established the analyticity

of $\varphi(\beta, \gamma; z)$ on the sectors $\widetilde{\Sigma}_{-1}$ and $\widetilde{\Sigma}_0$, we can compute the derivatives of this integral by taking the derivatives inside the integral. After multiplying through by $2\pi i\Gamma(\beta)\Gamma(\beta+1-\gamma)$ to cancel all multiplicative constant terms, we find,

$$\begin{aligned}
&\left(z\,\varphi''(\beta,\gamma;z) + (\gamma - z)\,\varphi'(\beta,\gamma;z) - \beta\,\varphi(\beta,\gamma;z)\right) 2\pi i\Gamma(\beta)\Gamma(\beta+1-\gamma)\\
&= \int_{-i\infty}^{+i\infty} \Gamma(s)\Gamma(\beta+2-s)\Gamma(\beta+1-\gamma-s)z^{s-\beta-1}\,ds\\
&\quad - \int_{-i\infty}^{+i\infty} \gamma\Gamma(s)\Gamma(\beta+1-s)\Gamma(\beta+1-\gamma-s)z^{s-\beta-1}\,ds\\
&\quad + \int_{-\infty}^{+\infty} \Gamma(s)\Gamma(\beta+1-s)\Gamma(\beta+1-\gamma-s)z^{s-\beta}\,ds\\
&\quad - \int_{-\infty}^{+\infty} (\beta)\Gamma(s)\Gamma(\beta-s)\Gamma(\beta+1-\gamma-s)z^{s-\beta}\,ds\\
&= \int_{-1-i\infty}^{-1+i\infty} \Gamma(s+1)\Gamma(\beta-\gamma-s)z^{s-\beta}\left(\Gamma(\beta+1-s) - \gamma\Gamma(\beta-s)\right)\,ds\\
&\quad - \int_{-i\infty}^{+i\infty} \Gamma(s)\Gamma(\beta+1-\gamma-s)z^{s-\beta}\left((\beta)\Gamma(\beta-s) - \Gamma(\beta+1-s)\right)\,ds\\
&= \left(\int_{-1-i\infty}^{-1+i\infty} - \int_{-i\infty}^{+i\infty}\right)\Gamma(s+1)\Gamma(\beta-s)\Gamma(\beta+1-\gamma-s)z^{s-\beta}\,ds. \qquad \text{(B.64)}
\end{aligned}$$

Due to the choice of the path of integration, the final integrand has no poles between the contours of integration, see Fig. 14. Therefore, due to Cauchy's theorem, the expression equals zero and we have shown that $\varphi(\beta, \gamma; z)$ satisfies Kummer confluent hypergeometric equation (1.2) on $z \in \widetilde{\Sigma}_{-1}$ and $\widetilde{\Sigma}_0$.

Observe the following differential identity,

$$\begin{aligned}
&z\,\frac{d^2}{dz^2}\left(e^z f(-z)\right) + (\gamma - z)\,\frac{d}{dz}\left(e^z f(-z)\right) - \beta\, e^z f(-z)\\
&\equiv e^z\left(z\,\frac{d^2}{dz^2} f(z) - (\gamma - (-z))\,\frac{d}{dz} f(z) - (\gamma - \beta)\, f(z)\right).
\end{aligned}$$

Given that $\varphi(\beta, \gamma; z)$ satisfies Kummer equation (1.2), it follows that the right hand side of this identity equals zero for $f(-z) = \varphi(\gamma - \beta, \gamma; -z)$. Looking at the left hand side of the identity, we deduce that $e^z\varphi(\gamma - \beta, \gamma; -z)$ also satisfies Eq. (1.2).

Third step: Asymptotic behaviour of $\tilde{y}_1^{(\infty,k)}(z)$ and $\tilde{y}_2^{(\infty,k)}(z)$ for large $|z|$.

Recalling the formal solutions given in Remark 3, we will deduce the following asymptotics, for $j \in \{0, -1\}$:

$$y_1^{(\infty,j)}(z) \sim e^z z^{\beta-\gamma}{}_2F_0\left(\gamma - \beta,\ 1 - \beta; z^{-1}\right), \quad \text{as } z \to \infty,\ z \in \widetilde{\Sigma}_j, \qquad \text{(B.65)}$$

$$y_2^{(\infty,j)}(z) \sim -z^{-\beta}{}_2F_0\left(\beta,\ \beta + 1 - \gamma; -z^{-1}\right), \quad \text{as } z \to \infty,\ z \in \widetilde{\Sigma}_j. \qquad \text{(B.66)}$$

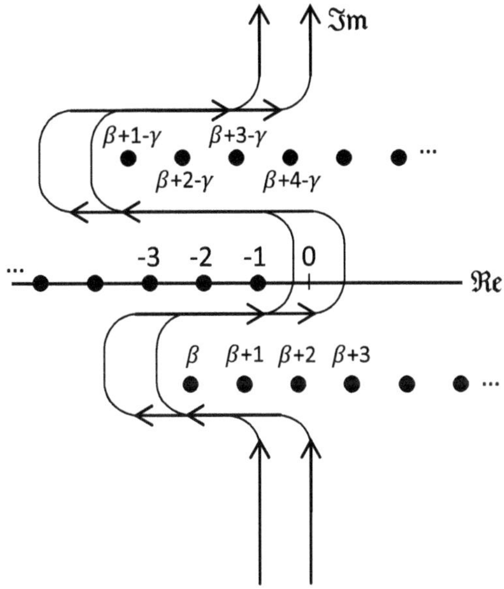

Fig. 14 Paths of integration in (B.64), the dots represent poles of the integrand. Note the crucial detail that $s = 0$ is not a pole of the integrand, so there are no singularities between the two paths

Denote the integrand of $\varphi(\beta, \gamma; z)$ by,

$$I(s, z) = \frac{\Gamma(s)\Gamma(\beta - s)\Gamma(\beta + 1 - \gamma - s)}{\Gamma(\beta)\Gamma(\beta + 1 - \gamma)} z^{s-\beta}, \tag{B.67}$$

and let τ be a large, positive real number. For $N \geq 0$, consider the path of integration along the rectangle R with vertices at $\pm i\tau$ and $-N - \frac{1}{2} \pm i\tau$, with indentations so that the poles of the integrand are separated as usual and with a positive orientation as shown in Fig. 15.

By Cauchy's theorem, we have,

$$\begin{aligned}\frac{1}{2\pi i} \int_R I(s, z)\, ds &= \frac{1}{2\pi i} \left(\int_{-N-\frac{1}{2}-i\tau}^{-i\tau} + \int_{-i\tau}^{+i\tau} + \int_{+i\tau}^{-N-\frac{1}{2}+i\tau} + \int_{-N-\frac{1}{2}+i\tau}^{-N-\frac{1}{2}-i\tau} \right) I(s, z)\, ds.\\ &= \sum_{n=0}^{N} \operatorname*{Res}_{s=-n} I(s, z),\end{aligned}$$

We examine these integrals in the limit $\tau \to \infty^+$ one-by-one, using the asymptotics (A.50) of the Gamma function:

1. By writing $s = x - i\tau$ in the first integral we obtain,

$$e^{\tau\left(\arg(z) - \frac{3\pi}{2}\right)} \int_{-N-\frac{1}{2}}^{0} \mathcal{O}\left(|z|^x \tau^{\operatorname{Re}(2\beta-\gamma)-x-\frac{1}{2}} \right)\, dx,$$

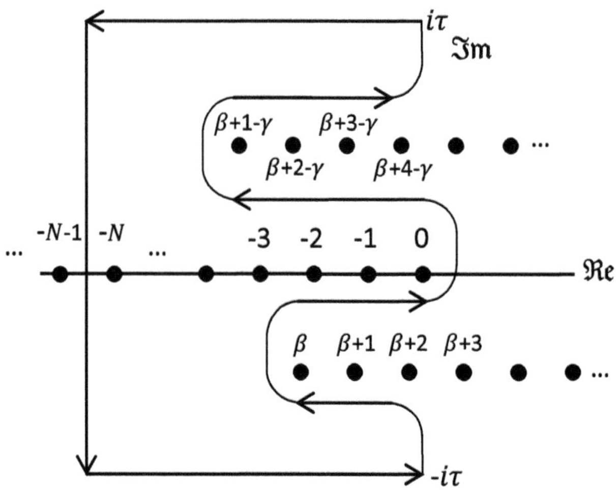

Fig. 15 Path of integration around the rectangle R, the dots represent the poles of the integrand of $\varphi(\beta, \gamma; z)$

which tends to zero as $\tau \to \infty^+$, thanks to $\arg(z) < \frac{3\pi}{2}$.

2. In the limit $\tau \to \infty^+$, the second integral becomes $\varphi(\beta, \gamma; z)$, by definition.

3. Similarly to the first integral, by writing $s = x + i\tau$ in the third integral, we obtain,

$$e^{-\tau\left(\arg(z)+\frac{3\pi}{2}\right)} \int_0^{-N-\frac{1}{2}} \mathcal{O}\left(|z|^x \tau^{\mathrm{Re}(2\beta-\gamma)-x-\frac{1}{2}}\right) dx,$$

which also tends to zero as $\tau \to \infty^+$, thanks to $\arg(z) > -\frac{3\pi}{2}$.

4. We write $s = -N - \frac{1}{2} + iy$ in the fourth integral to obtain,

$$\begin{aligned}\int_{-N-\frac{1}{2}+i\tau}^{-N-\frac{1}{2}-i\tau} I(s,z)\, ds &= i\left(\int_0^{-\tau} + \int_\tau^0\right) I\left(-N - \frac{1}{2} + iy, z\right) dy \\ &= iz^{-N-\frac{1}{2}-\beta}\left(\int_0^\tau \mathcal{O}\left(|y|^{N+\mathrm{Re}(2\beta-\gamma)-1} e^{-y\left(\frac{3\pi}{2}-\arg(z)\right)}\right) dy\right. \\ &\quad \left. - \int_0^\tau \mathcal{O}\left(|y|^{N+\mathrm{Re}(2\beta-\gamma)-1} e^{-y\left(\arg(z)+\frac{3\pi}{2}\right)}\right) dy\right). \quad \text{(B.68)}\end{aligned}$$

Using the fact that $\lim_{\tau\to\infty^+} \int_0^\tau e^{-ky}\, dy$ for $k > 0$ exists, the limit as $\tau \to \infty^+$ of the fourth integral exists and is of order $\mathcal{O}\left(|z|^{-N-\frac{1}{2}-\beta}\right)$ as $\tau \to \infty^+$, thanks to $|\arg(z)| < \frac{3\pi}{2}$.

Summarising the above analysis, we have shown that for large τ,

$$\varphi(\beta,\gamma;z) = \sum_{n=0}^{N} \operatorname*{Res}_{s=-n} I(s,z) + \mathcal{O}\left(|z|^{-N-\frac{1}{2}-\beta}\right), \tag{B.69}$$

$$= z^{-\beta} \sum_{n=0}^{N} \frac{(\beta)_n(\beta+1-\gamma)_n}{(-z)^n n!} + \mathcal{O}\left(|z|^{-N-\frac{1}{2}-\beta}\right),$$

where we have used the formula $\operatorname*{Res}_{\lambda=-n} \Gamma(\lambda) = \frac{(-1)^n}{n!}$, for $n \geq 0$, to calculate the residues. This proves (B.66). Moreover, for $N \geq 0$, we can immediately deduce,

$$e^{\mp i\pi(\beta-\gamma)} e^z \varphi\left(\gamma-\beta,\gamma; e^{\pm i\pi} z\right) = e^z z^{\beta-\gamma} \sum_{n=0}^{N} \frac{(\gamma-\beta)_n(1-\beta)_n}{z^n n!} + \mathcal{O}\left(e^z |z|^{-N-\frac{1}{2}+\beta-\gamma}\right),$$

which proves (B.65). □

Remark 10 The expression (B.69) is valid for all finite N. In order to take the limit as $N \to \infty$ it is important to understand that (B.69) becomes an asymptotic result. This is because the integrals in (B.68) depend on N and, in particular, they diverge as $N \to \infty^+$, hence the interchange between limits $\lim_{N\to\infty^+}$ and $\lim_{\tau\to\infty^+}$ is not justified here.

Having established how to represent the fundamental solutions $\widetilde{Y}^{(\infty,k)}(z)$ using Mellin-Barnes integrals, we now show how to analytically continue them to $z = 0$. We will prove the following proposition, which is sufficient to deduce the monodromy data formulae (3.12)–(3.13).

Proposition B.14 *Let $\tilde{y}_1^{(0)}(z)$ and $\tilde{y}_2^{(0)}(z)$ be the solutions as given in (3.3). For $-\pi \pm \frac{\pi}{2} < arg(z) < \pi \pm \frac{\pi}{2}$, the integral as given by (B.61) satisfies,*

$$\varphi(\beta,\gamma;z) = \frac{\Gamma(\gamma-1)}{\Gamma(\beta)} \tilde{y}_1^{(0)}(z) + \frac{\Gamma(1-\gamma)}{\Gamma(\beta+1-\gamma)} \tilde{y}_2^{(0)}(z).$$

Proof Let $I(s,z)$ be the integrand of $\varphi(\beta,\gamma;z)$ as given by (B.67). For large $\tau > 0$ and an integer $N > 0$, we now consider the integral around the rectangle R' with vertices $\pm i\tau$ and $N + \frac{1}{2} \pm i\tau$, with indentations along the imaginary axis as usual and with a negative orientation as shown in Fig. 16. Our analysis of this integral is analogous to that of the integral around the rectangle R, which lies to the left of the imaginary axis.

By Cauchy's theorem, we have,

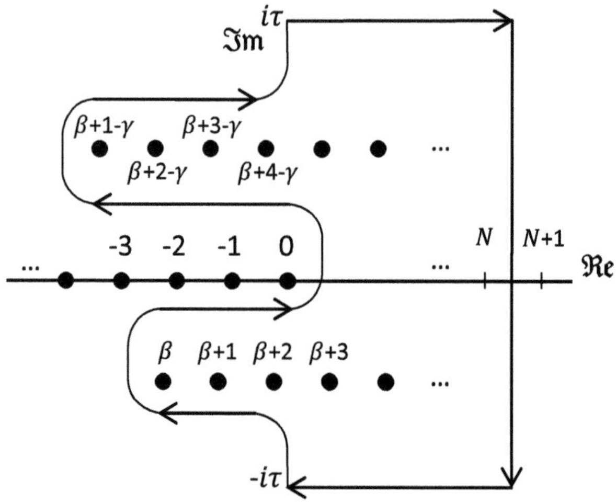

Fig. 16 Path of integration around the rectangle R', the dots represent the poles of the integrand of $\varphi(\beta, \gamma; z)$

$$\frac{1}{2\pi i}\int_{R'} I(s,z)\,ds \equiv \frac{1}{2\pi i}\left(\int_{N+\frac{1}{2}-i\tau}^{-i\tau} + \int_{-i\tau}^{i\tau} + \int_{i\tau}^{N+\frac{1}{2}+i\tau} + \int_{N+\frac{1}{2}+i\tau}^{N+\frac{1}{2}-i\tau}\right) I(s,z)\,ds$$
$$= -\sum_{n=0}^{M_1(N)} \operatorname*{Res}_{s=\beta+1-\gamma+n} I(s,z) - \sum_{n=0}^{M_2(N)} \operatorname*{Res}_{s=\beta+n} I(s,z)$$

where $M_1(N)$ and $M_2(N)$ are the number of poles $\beta+1-\gamma, \beta+2-\gamma, \ldots$ and $\beta, \beta+1, \ldots$ which lie inside the rectangle respectively. We examine these integrals under the limit $\tau \to \infty^+$ one-by-one, using the asymptotics (A.50) of the Gamma function:

1. By writing $s = x - i\tau$ in the first integral we obtain,

$$e^{\tau\left(\arg(z)-\frac{3\pi}{2}\right)} \int_{N+\frac{1}{2}}^{0} \mathcal{O}\left(|z|^x \tau^{\mathrm{Re}(2\beta-\gamma)-x-\frac{1}{2}}\right) dx,$$

which tends to zero as $\tau \to \infty^+$, thanks to $\arg(z) < \frac{3\pi}{2}$.

2. In the limit $\tau \to \infty^+$, the second integral becomes $\varphi(\beta, \gamma; z)$, by definition.

3. Similarly to the first integral, by writing $s = x + i\tau$ in the third integral, we obtain,

$$e^{-\tau\left(\arg(z)+\frac{3\pi}{2}\right)} \int_{i\tau}^{N+\frac{1}{2}+i\tau} \mathcal{O}\left(|z|^x \tau^{\mathrm{Re}(2\beta-\gamma)-x-\frac{1}{2}}\right) dx,$$

which also tends to zero as $\tau \to \infty^+$, thanks to $\arg(z) > -\frac{3\pi}{2}$.

4. We write $s = N + \frac{1}{2} + iy$ in the fourth integral, to obtain,

$$\begin{aligned}\int_{N+\frac{1}{2}+i\tau}^{N+\frac{1}{2}-i\tau} I(s,z)\,ds &= i\left(\int_0^{-\tau} + \int_\tau^0\right) I\left(N+\frac{1}{2}+iy, z\right) dy \\ &= iz^{N+\frac{1}{2}-\beta}\left(\int_0^\tau \mathcal{O}\left(|y|^{-N+\mathrm{Re}(2\beta-\gamma)-2} e^{-y\left(\frac{3\pi}{2}-\arg(z)\right)}\right) dy\right. \\ &\quad \left.-\int_0^\tau \mathcal{O}\left(|y|^{-N+\mathrm{Re}(2\beta-\gamma)-2} e^{-y\left(\arg(z)+\frac{3\pi}{2}\right)}\right) dy\right). \quad \text{(B.70)}\end{aligned}$$

Using the fact that $\lim_{\tau\to\infty^+}\int_0^\tau e^{-ky}\,dy$ for $k > 0$ exists, the limit as $\tau \to \infty^+$ of fourth integral exists, thanks to $|\arg(z)| < \frac{3\pi}{2}$. Moreover, for $|z|$ sufficiently small, this limit exists uniformly with respect to large N, due to the minus sign in the exponent of $|y|$. In particular, for $|z|$ sufficiently small,

$$\lim_{N\to\infty^+} \int_{N+\frac{1}{2}-i\tau}^{N+\frac{1}{2}-i\tau} I(s,z)\,ds = 0.$$

Summarising the above analysis, we have shown the following,

$$\varphi(\beta,\gamma;z) = -\sum_{n=0}^{M_1(N)} \operatorname*{Res}_{s=\beta+1-\gamma-n} I(s,z) - \sum_{n=0}^{M_2(N)} \operatorname*{Res}_{s=\beta-n} I(s,z) + \lim_{\tau\to\infty^+}\int_{N+\frac{1}{2}-i\tau}^{N+\frac{1}{2}+i\tau} I(s,z)\,ds,$$

where the convergence of the limit of this integral is uniform with respect to $N \to \infty^+$. As such, we may interchange the limits $\lim_{\tau\to\infty^+}$ and $\lim_{N\to\infty^+}$ as follows,

$$\begin{aligned}\varphi(\beta,\gamma;z) &= -\lim_{N\to\infty^+}\sum_{n=0}^{M_1(N)} \operatorname*{Res}_{s=\beta+1-\gamma+n} I(s,z) - \lim_{N\to\infty^+}\sum_{n=0}^{M_2(N)} \operatorname*{Res}_{s=\beta+n} I(s,z) \\ &\quad + \lim_{N\to\infty^+}\lim_{\tau\to\infty^+}\int_{N+\frac{1}{2}-i\tau}^{N+\frac{1}{2}+i\tau} I(s,z)\,ds, \\ &= -\sum_{n=0}^{\infty} \operatorname*{Res}_{s=\beta+1-\gamma+n} I(s,z) - \sum_{n=0}^{\infty} \operatorname*{Res}_{s=\beta+n} I(s,z) \\ &\quad + \lim_{\tau\to\infty^+}\lim_{N\to\infty^+}\int_{N+\frac{1}{2}-i\tau}^{N+\frac{1}{2}+i\tau} I(s,z)\,ds, \\ &= -\sum_{n=0}^{\infty} \operatorname*{Res}_{s=\beta+1-\gamma+n} I(s,z) - \sum_{n=0}^{\infty} \operatorname*{Res}_{s=\beta+n} I(s,z) + \lim_{\tau\to\infty^+} 0.\end{aligned}$$

We compute the residues to find,

$$\varphi(\beta,\gamma;z)=\frac{\Gamma(\gamma-1)}{\Gamma(\beta)}z^{1-\gamma}\sum_{n=0}^{\infty}\frac{(\beta+1-\gamma)_n z^n}{(2-\gamma)_n n!}+\frac{\Gamma(1-\gamma)}{\Gamma(\beta+1-\gamma)}\sum_{n=0}^{\infty}\frac{(\beta)_n z^n}{(\gamma)_n n!},$$

for $z\in\widetilde{\Sigma}_{-1}$ and $\widetilde{\Sigma}_0$ and the desired result is proved. □

Remark 11 Continuing with the issue raised in Remark 10, the fact is that integrating along the rectangle R to the left of the imaginary axis is only able to produce an asymptotic result because we do not have uniform convergence with respect to N in the integrals (B.68). This is to be expected, since we know $\varphi(\beta,\gamma;z)$ is analytic on sectors $\widetilde{\Sigma}_{-1}$ and $\widetilde{\Sigma}_0$, it certainly cannot be equal to a divergent ${}_2F_0$ series. However, when integrating along the rectangle R' to the right of the imaginary axis we produce an equality with a linear combination of convergent series, namely this is the analytic continuation of the solutions at $z=\infty$ to $z=0$. This is shown in (B.70), because the integrals here converge as $\tau\to\infty^+$ uniformly with respect to large N.

We conclude these computations by using Proposition B.14 to prove the formulae (3.12)–(3.13) of Lemma 3.3.

Proof of Lemma 3.3 Recall from the definitions (B.59) and (B.60) of solutions,

$$\tilde{y}_2^{(\infty,0)}(z)=-\varphi(\beta,\gamma;z)\quad\text{and}\quad \tilde{y}_1^{(\infty,0)}(z)=e^{i\pi(\beta-\gamma)}e^z\varphi\left(\gamma-\beta,\gamma;e^{-i\pi}z\right),\qquad z\in\widetilde{\Sigma}_0.$$

Let $\gamma_{\infty,0}$ be a curve as described at the beginning of this subsection. Proposition B.13 shows how to represent the solutions of Kummer equation (1.2) around $z=\infty$ using a Mellin-Barnes integral. Due to the analyticity of this integral, as shown in the first part of the proof of Proposition B.13, Proposition B.14 provides the formula for the analytic continuation of these solutions to $z=0$. That is to say,

$$\gamma_{\infty,0}\left[\tilde{y}_2^{(\infty,0)}\right](z)=-\frac{\Gamma(\gamma-1)}{\Gamma(\beta)}\tilde{y}_1^{(0)}(z)-\frac{\Gamma(1-\gamma)}{\Gamma(\beta+1-\gamma)}\tilde{y}_2^{(0)}(z).$$

By manipulating the parameters and variable as follows: $\beta\mapsto\gamma-\beta$, $\gamma\mapsto\gamma$, $z\mapsto e^{i\pi}z$, we also deduce,

$$\begin{aligned}\gamma_{\infty,0}\left[\tilde{y}_1^{(\infty,0)}\right](z)&=e^{i\pi(\beta-\gamma)}\frac{\Gamma(\gamma-1)}{\Gamma(\gamma-\beta)}e^{-i\pi(1-\gamma)}z^{1-\gamma}e^z{}_1F_1\left(\begin{matrix}1-\beta\\2-\gamma\end{matrix};-z\right)\\&\quad+e^{i\pi(\beta-\gamma)}\frac{\Gamma(1-\gamma)}{\Gamma(1-\beta)}e^z{}_1F_1\left(\begin{matrix}\gamma-\beta\\\gamma\end{matrix};-z\right).\end{aligned}$$

After applying Kummer transformation,

$$e^z {}_1F_1\left(\begin{matrix} a \\ c \end{matrix}; -z\right) \equiv {}_1F_1\left(\begin{matrix} c-a \\ c \end{matrix}; z\right),$$

we deduce the connection matrix as given in (3.13), namely,

$$\left(\gamma_{\infty,0}\left[\tilde{y}_1^{(\infty,0)}\right](z),\ \gamma_{\infty,0}\left[\tilde{y}_2^{(\infty,0)}\right](z)\right) = \left(\tilde{y}_1^{(0)}(z),\ \tilde{y}_2^{(0)}(z)\right)\widetilde{C}^{0\infty},$$

where,

$$\widetilde{C}^{0\infty} = \begin{pmatrix} e^{i\pi(\beta-1)}\frac{\Gamma(\gamma-1)}{\Gamma(\gamma-\beta)} & -\frac{\Gamma(\gamma-1)}{\Gamma(\beta)} \\ e^{i\pi(\beta-\gamma)}\frac{\Gamma(1-\gamma)}{\Gamma(1-\beta)} & -\frac{\Gamma(1-\gamma)}{\Gamma(\beta+1-\gamma)} \end{pmatrix}.$$

We now turn our attention to proving the formulae (3.12) for Stokes matrices. By Definition 3.2 of the Stokes matrices $\widetilde{S}_k$ and by the asymptotic behaviour (3.5) of the fundamental solutions $\widetilde{Y}^{(\infty,k)}(z)$, we have,

$$\begin{pmatrix} z^{\beta-\gamma}e^z & 0 \\ 0 & z^{1-\beta} \end{pmatrix}\widetilde{S}_k\begin{pmatrix} z^{\gamma-\beta}e^{-z} & 0 \\ 0 & z^{\beta-1} \end{pmatrix} \sim I, \quad \text{as } z\to\infty,\ \arg(z)-k\pi\in\left(\frac{\pi}{2},\frac{3\pi}{2}\right).$$

From this relation we easily deduce that $\widetilde{S}_{-1}$ is lower triangular and $\widetilde{S}_0$ is upper triangular, both with unit diagonals. Denote by $\tilde{s}_{-1}$ and $\tilde{s}_0$ the (2, 1) and (1, 2) elements of the matrices $\widetilde{S}_{-1}$ and $\widetilde{S}_0$ respectively. With the knowledge of the connection matrix $\widetilde{C}^{0\infty}$, we use the cyclic relation (3.15) as follows,

$$\begin{aligned}
&\widetilde{C}^{\infty 0}e^{2\pi i\widetilde{\Theta}_0}\widetilde{C}^{0\infty} = \left(\widetilde{S}_{-1}\right)^{-1}e^{-2\pi i\widetilde{\Theta}_\infty}\left(\widetilde{S}_0\right)^{-1},\\
\Leftrightarrow &\begin{pmatrix} e^{2\pi i(\beta-\gamma)} & \frac{-2\pi i e^{-i\pi\gamma}}{\Gamma(\beta)\Gamma(\beta+1-\gamma)} \\ \frac{-2\pi i e^{2\pi i(\beta-\gamma)}}{\Gamma(1-\beta)\Gamma(\gamma-\beta)} & 1-e^{2\pi i(\beta-\gamma)}+e^{2\pi i(1-\gamma)} \end{pmatrix}\\
&\qquad = \begin{pmatrix} 1 & 0 \\ -\tilde{s}_{-1} & 0 \end{pmatrix}\begin{pmatrix} e^{2\pi i(\beta-\gamma)} & 0 \\ 0 & e^{2\pi i(1-\beta)} \end{pmatrix}\begin{pmatrix} 1 & -\tilde{s}_0 \\ 0 & 1 \end{pmatrix},\\
\Leftrightarrow &\begin{cases} \tilde{s}_{-1} = \frac{2\pi i}{\Gamma(1-\beta)\Gamma(\gamma-\beta)}, \\ \tilde{s}_0 = \frac{2\pi i}{\Gamma(\beta)\Gamma(\beta+1-\gamma)}e^{i\pi(\gamma-2\beta)}, \end{cases}
\end{aligned} \tag{B.71}$$

which are indeed the Stokes multipliers found in the formulae (3.12) for the Stokes matrices.

Remark 12 If we had chosen to normalise the monodromy data of Kummer equation with respect to the fundamental solution $\widetilde{Y}^{(\infty,-1)}(z)$ then the signs of the exponents in $\widetilde{C}^{0\infty}$ would be inverted. Furthermore, the monodromy matrix around infinity would change as $\widetilde{M}_\infty \mapsto \widetilde{S}_0^{-1}\widetilde{M}_\infty\widetilde{S}_0$.

B.1 Gevrey Asymptotics and a Result of Ramis and Martinet

We close this subsection about Kummer confluent hypergeometric differential equation by examining Gevrey asymptotics and stating a result of Ramis and Martinet (1989). This also gives us the opportunity to show a contemporary approach to the theory of Stokes phenomenon, which we have learned from Balser (1994), Put et al. (2003). The contents of this additional subsection will not be necessary for our main theorems in Sect. 4, we include it for the curiosity of the reader.

We recall some definitions and facts regarding asymptotic theory. In the following, keep in mind that the role of the letter k will mirror the concept of a linear differential equation having a pole of Poincaré rank k, so that for Kummer equation we are specifically concerned with $k = 1$. Denote by $\mathbb{C}[[z^{-1}]]$ the field of formal series in z^{-1}.

Definition 4.1 Let f be a function analytic in a sector $\widetilde{\Sigma}$. We say that f has the series $\widehat{f} = \sum_{n=0}^{\infty} f_n z^{-n} \in \mathbb{C}[[z^{-1}]]$ as its Gevrey asymptotic expansion of order k^{-1} as $z \to \infty, z \in \widetilde{\Sigma}$, denoted $f \simeq_{\frac{1}{k}} \widehat{f}$, if for every closed subsector σ of $\widetilde{\Sigma}$, there exists a constant $K > 0$ such that, for all $N \in \mathbb{N}$ and $z \in \sigma$,

$$\left| z^N \left(f(z) - \sum_{n=0}^{N-1} f_n z^{-n} \right) \right| \leq K^N \Gamma \left(1 + \frac{N}{k} \right). \tag{B.72}$$

We denote by $\mathcal{A}_{\frac{1}{k}}(\widetilde{\Sigma})$ the set of analytic functions on $\widetilde{\Sigma}$ which have a Gevrey asymptotic expansion of order k^{-1}.

Gevrey asymptotics is a stronger definition than the usual one of Poincaré because it specifies how the right hand side of the inequality (B.72) depends on N. In Poincaré's definition of an asymptotic series the precise dependence on N is not relevant. If we denote by $\mathcal{A}(\widetilde{\Sigma})$ the set of analytic functions on a sector $\widetilde{\Sigma}$ which admit an asymptotic expansion then we have,

$$\mathcal{A}(\widetilde{\Sigma}) \supset \mathcal{A}_1(\widetilde{\Sigma}) \supset \mathcal{A}_{\frac{1}{2}}(\widetilde{\Sigma}) \supset \mathcal{A}_{\frac{1}{3}}(\widetilde{\Sigma}) \supset \dots, \tag{B.73}$$

since the asymptotic expansion (A.50) of the Gamma function implies:

$$\frac{\Gamma\left(1 + \frac{N}{k+1}\right)}{\Gamma\left(1 + \frac{N}{k}\right)} \to 0 \text{ as } N \to \infty.$$

We note that, if $f \in \mathcal{A}_{\frac{1}{k}}(\widetilde{\Sigma})$, with $f \simeq_{\frac{1}{k}} \sum_{n=0}^{\infty} f_n z^{-n}$, then these coefficients satisfy $|f_n| < K^n \Gamma\left(1 + \frac{n}{k}\right)$, for some positive constant K and $n \geq 1$. To see this, we add the following inequalities:

$$\left| f(z) - \sum_{n=0}^{N-1} f_n z^{-n} \right| \leq |z|^{-N} K^N \Gamma\left(1 + \frac{N}{k}\right),$$

$$\left| f(z) - \sum_{n=0}^{N} f_n z^{-n} \right| \leq |z|^{-N-1} K^{N+1} \Gamma\left(1 + \frac{N+1}{k}\right),$$

to obtain the following inequality for f_N,

$$|f_N| \leq K^N \Gamma\left(1 + \frac{N}{k}\right) + |z|^{-1} K^{N+1} \Gamma\left(1 + \frac{N+1}{k}\right),$$

from which we immediately find the claimed property by taking the limit $z \to \infty$. This motivates the following definition.

Definition 4.2 We call a series $\widehat{f} = \sum_{n=0}^{\infty} f_n z^{-n} \in \mathbb{C}[[z]]$ a Gevrey series of order k^{-1} if there exists a positive constant K such that, $|f_n| < K^n \Gamma\left(1 + \frac{n}{k}\right)$ for all $n \geq 1$. We denote by $\mathbb{C}[[z]]_{\frac{1}{k}}$ the set of all Gevrey series of order k^{-1}.

Consider the map $J : \mathcal{A}_{\frac{1}{k}}(\widetilde{\Sigma}) \to \mathbb{C}[[z]]_{\frac{1}{k}}$ which maps an analytic function f on the sector $\widetilde{\Sigma}$ to its Gevrey asymptotic expansion of order k^{-1}. We recall the following result, see for instance Balser (1994), Put et al. (2003).

Theorem B.15 *Assume $k > \frac{1}{2}$. The set $\mathcal{A}_{\frac{1}{k}}(\widetilde{\Sigma})$ is a differential algebra and the map J is a homomorphism. Moreover, if the sector $\widetilde{\Sigma}$ has an opening less than $\frac{\pi}{k}$, then J is surjective, otherwise, if $\widetilde{\Sigma}$ has an opening greater than $\frac{\pi}{k}$, then J is injective.*

This remarkable theorem draws the connection between Gevrey asymptotics and Stokes phenomenon. Given a formal Gevrey series of order k^{-1}, this theorem shows that there is a unique analytic function on a sector of opening greater than $\frac{\pi}{k}$ which has that series as its Gevrey asymptotic expansion of order k^{-1}. Observe that this is exactly parallel to the theory of Stokes phenomenon: given a differential equation with a pole of Poincaré rank k and a formal fundamental series solution at that point, there are unique analytic fundamental solutions on a sectors of openings greater than $\frac{\pi}{k}$ with the prescribed formal series as their asymptotic expansions.

Let $\varphi(\beta, \gamma; z)$ be defined as in (B.61). Ramis and Martinet prove the following result.

Theorem B.16 *The function $z^a \varphi(a, c; z)$ has ${}_2F_0\left(a, a+1-c; -z^{-1}\right)$ as its Gevrey asymptotic expansion of order one as $z \to \infty$ with $|arg(z)| < \frac{3\pi}{2}$. Similarly, $(-z)^{c-a}$ $\varphi(c-a, c; -z)$ has ${}_2F_0\left(c-a, 1-a; z^{-1}\right)$ as its Gevrey asymptotic expansion of order one with $|arg(-z)| < \frac{3\pi}{2}$.*

We have seen in the first part of the proof of Proposition B.13 that $\varphi(a, c; z)$ and $\varphi(c-a, c; -z)$ are analytic in the sectors $\widetilde{\Sigma}_{-1}$ and $\widetilde{\Sigma}_0$. In particular, since these sectors have openings greater than π, Theorem B.15 states that the map J :

$\mathcal{A}_1\left(\widetilde{\Sigma}_+\right) \to \mathbb{C}[[z]]_1$ is injective. In other words, there are unique analytic functions on these sectors which have the formal series solutions,

$$z^{-a}{}_2F_0\left(a, a+1-c; -z^{-1}\right) \quad \text{and} \quad (-z)^{a-c}e^z{}_2F_0\left(c-a, 1-a; z^{-1}\right), \tag{B.74}$$

as their Gevrey asymptotic expansions of order 1. Since we have seen that Gevrey asymptotics imply asymptotics in the usual sense, recall (B.73), this implies that such analytic functions on these sectors are in fact solutions to Kummer equation (1.2), by the uniqueness statement in Theorem 3.2. Since the formal series solutions (B.74) are clearly linearly independent, Ramis and Martinet's Theorem shows that the functions,

$$\varphi(a, c; z) \quad \text{and} \quad e^z\varphi(c-a, c; -z),$$

constitute a fundamental set of solutions of Kummer equation. Compared with our proof of this fact, stated as Proposition B.13, it is satisfying to deduce this from a different perspective.

References

Andrews, G.E., Askey, R., Roy, R.: Special Functions Encyclopedia of Mathematics and Its Applications, vol. 71. Cambridge University Press, Cambridge (1999)

Balser, W.: From Divergent Power Series to Analytic Functions. Theory and Application of Multisummable Power Series. Lecture Notes in Mathematics, vol. 1582. Springer, Berlin, Heidelberg, New York (1994)

Balser, W., Jurkat, W.B., Lutz, D.A.: Birkhoff invariants and stokes multipliers for meromorphic linear differential equations. J. Math. Anal. Appl. **71**(1), 48–94 (1979). Academic Press, Inc. Institute of Physics Publishing

Bateman, H., Erdélyi, A.: Higher Transcendental Functions Volumes 1 and 2. McGraw-Hill, New York (1953–1955)

Glutsyuk, A.: Stokes operators via limit monodromy of generic perturbation. J. Dyn. Control Syst. **5**(1), 101–135 (1999)

Horrobin, C., Matveeva, Mazzocco, M., Rubtsov V.: Stokes phenomenon arising in the confluence of two simple poles and corresponding Poisson structures, in progress, (2020)

Lambert, C., Rousseau, C.: The stokes phenomenon in the confluence of the hypergeometric equation using Riccati equation. J. Diff. Equ. **244**(10), 2641–2664 (2008)

van der Put, M., Singer, M.F.: Galois Theory of Linear Differential Equations. A Series of Comprehensive Studies in Mathematics, vol. 328. Springer, Berlin, Heidelberg, New York (2003)

Ramis, J.P., Martinet, J.: Computer Algebra and Differential Equations (Théorie de Galois Différentielle et Resommation, Chaptire 3: Les Fonctions Hypergeometriques Confluentes). Academic Press, New York, E. Tournier (1989)

Wasow, W.: Asymptotic Expansions for Ordinary Differential Equations. Interscience Publishers. A Division of John Wiley & Sons, Inc., New York, London, Sydney (1965)

Watanabe, H.: Sur la dégénérescence de quelques formules de connexion pour les fonctions hypergéométriques de Gauss. (French. English Summary) [Degeneration of some connection formulas for Gauss hypergeometric functions]. Aequationes Math. **73**(3), 201–221 (2007)

Whittaker, E.T., Watson, G.N.: A Course of Modern Analysis, 4th edn. AMS Press, New York (1979)

Periodic Trajectories of Ellipsoidal Billiards in the 3-Dimensional Minkowski Space

Vladimir Dragović and Milena Radnović

Abstract In this paper, we give detailed analysis and description of periodic trajectories of the billiard system within an ellipsoid in the 3-dimensional Minkowski space, taking into account all possibilities for the caustics. The conditions for periodicity are derived in algebro-geometric, analytic, and polynomial form.

Keywords Ellipsoidal billiards · Periodic trajectories · Poncelet theorem · Pseudo-Euclidean spaces · Hyper-elliptic curves · Pell's equation

1 Introduction

Discrete integrable systems occupy an important part of the scientific activity and legacy of Professor Nalini Joshi. There are several recent monographs related to discrete integrability as intensively developed field of pure and applied mathematics (see Duistermaat (2010), Bobenko and Suris (2008), Hietarinta et al. (2016), Joshi (2019)). Integrable billiards (see Kozlov and Treshchëv (1991), Dragović and Radnović (2011)) form an important class of discrete integrable systems. This paper is devoted to integrable billiards in the 3-dimensional Minkowski space, merging two lines of our previous studies.

We will derive here the periodicity conditions for such billiards in different forms: in algebro-geometric terms and in terms of polynomial functional equations. More

Dedicated to Professor Nalini Joshi on the occasion of her anniversary.

V. Dragović
Department of Mathematical Sciences, The University of Texas at Dallas, Richardson, USA
e-mail: vladimir.dragovic@utdallas.edu

M. Radnović (✉)
The University of Sydney, School of Mathematics and Statistics, Sydney, Australia
e-mail: milena.radnovic@sydney.edu.au

V. Dragović · M. Radnović
Mathematical Institute SANU, Belgrade, Serbia

© Springer Nature Switzerland AG 2020

F. Nijhoff et al. (eds.), *Asymptotic, Algebraic and Geometric Aspects of Integrable Systems*, Springer Proceedings in Mathematics & Statistics 338,
https://doi.org/10.1007/978-3-030-57000-2_8

about extremal polynomials and related Pell's equations one can find in Akhiezer (1990), Bogatyrev (2012) and references therein. Following Birkhoff and Morris (1962), Khesin and Tabachnikov (2009), we introduced notions of relativistic quadrics and applied them to billiards in the pseudo-Euclidean spaces in Dragović and Radnović (2012). In a more recent paper Dragović and Radnović (2019b), we established a fundamental relationship between periodic integrable billiards in the Euclidean spaces of arbitrary dimension and extremal polynomials and Pell's equations. We applied these ideas in more detail in the basic, planar cases in Dragović and Radnović (2019a) for the Euclidean metrics and in Adabrah et al. (2019) for the Minkowski metric. In this work, we deal with the three-dimensional Minkowski space, as a gateway to the study of billiards in higher-dimensional pseudo-Euclidean spaces. The results of this paper may provide an approach to the solution to a known open problem, Problem 5.2 from Genin et al. (2007), which is also Problem 7 from Tabachnikov (2015). See Remark 4 for more details.

The organization of the paper is as follows. Section 2 introduces basic notation. Section 3 is devoted to algebro-geometric formulation of periodicity conditions, while Sect. 4 derives the conditions of periodicity in terms of Pell's equations and related polynomial functional equations.

2 Confocal Families of Quadrics and Billiards

In this section, we recall necessary notions and properties related to confocal families of quadrics and billiards within ellipsoids in the Minkowski space. A more detailed account can be found in Genin et al. (2007), Khesin and Tabachnikov (2009), Dragović and Radnović (2012).

The Minkowski space $\mathbf{E}^{2,1}$ is $\mathbf{R}^3$ with *the Minkowski scalar product*: $\langle X, Y\rangle = X_1Y_1 + X_2Y_2 - X_3Y_3$.

The Minkowski distance between points X, Y is

$$\operatorname{dist}(X, Y) = \sqrt{\langle X - Y, X - Y\rangle}.$$

Since the scalar product can be negative, notice that the Minkowski distance can have imaginary values as well.

Let ℓ be a line in the Minkowski space, and v its vector. The line ℓ is called *space-like* if $\langle v, v\rangle > 0$; *time-like* if $\langle v, v\rangle < 0$; and *light-like* if $\langle v, v\rangle = 0$. Two vectors x, y are *orthogonal* in the Minkowski space if $\langle x, y\rangle = 0$. Note that a light-like vector is orthogonal to itself.

Confocal families. Denote by

$$\mathcal{E} : \frac{x_1^2}{a_1} + \frac{x_2^2}{a_2} + \frac{x_3^2}{a_3} = 1, \tag{1}$$

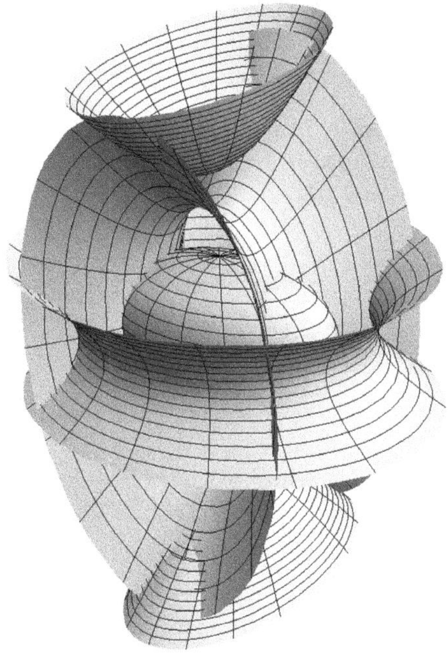

Fig. 1 Confocal quadrics in the three-dimensional Minkowski space

with $a_1 > a_2, a_3 > 0$, an ellipsoid. Let us remark that equation of any ellipsoid in the Minkowski space can be brought into the canonical form (1) using transformations that preserve the scalar product.

The family of quadrics confocal with $\mathcal{E}$ is:

$$\mathcal{Q}_\lambda\ :\ \frac{x_1^2}{a_1-\lambda}+\frac{x_2^2}{a_2-\lambda}+\frac{x_3^2}{a_3+\lambda}=1, \qquad \lambda\in\mathbf{R}. \tag{2}$$

The family (2) contains four geometrical types of quadrics:

- 1-sheeted hyperboloids oriented along x_3-axis, for $\lambda \in (-\infty, -a_3)$;
- ellipsoids, corresponding to $\lambda \in (-a_3, a_2)$;
- 1-sheeted hyperboloids oriented along x_2-axis, for $\lambda \in (a_2, a_1)$;
- 2-sheeted hyperboloids, for $\lambda \in (a, +\infty)$ – these hyperboloids are oriented along x_3-axis.

In Fig. 1, one non-degenerate quadric of each geometric type is shown. In addition, there are four degenerated quadrics: $\mathcal{Q}_{a_1}$, $\mathcal{Q}_{a_2}$, $\mathcal{Q}_{-a_3}$, $\mathcal{Q}_{\infty}$, that is planes $x_1 = 0$, $x_2 = 0$, $x_3 = 0$, and the plane at the infinity respectively.

The following theorem consists of a generalisation of the Chasles theorem to the Minkowski space and the additional conditions on the parameters of the quadrics touching a given line. The corresponding generalisation of the Chasles theorem was first obtained in Khesin and Tabachnikov (2009), while the classification of the types of confocal quadrics touching a given line for the 3-dimensional Minkowski space was considered in Genin et al. (2007), regarding the geodesics on an ellipsoid, and in Dragović and Radnović (2012) regarding billiards within ellipsoids in the pseudo-Euclidean space of arbitrary dimension.

Theorem 1 *In the Minkowski space* $\mathbf{E}^{2,1}$ *consider a line intersecting ellipsoid* $\mathcal{E}$ (1). *Then this line is touching two quadrics from* (2). *If we denote their parameters by* γ_1, γ_2 *and take:*

$$\{b_1, \ldots, b_p, c_1, \ldots, c_q\} = \{a_1, a_2, -a_3, \gamma_1, \gamma_2\},$$
$$c_q \leq \cdots \leq c_1 < 0 < b_1 \leq \cdots \leq b_p, \quad p + q = 5,$$

we will additionally have:

- *if the line is space-like, then* $p = 3$, $q = 2$, $a_1 = b_3$, $\gamma_1 \in \{b_1, b_2\}$ *for* $1 \leq i \leq k - 1$, *and* $\gamma_2 \in \{c_1, c_2\}$*;*
- *if the line is time-like, then* $p = 4$, $q = 1$, $c_1 = -a_3$, $\gamma_1 \in \{b_1, b_2\}$, $\gamma_2 \in \{b_3, b_4\}$*;*
- *if the line is light-like, then* $p = 4$, $q = 1$, $b_4 = \infty = \gamma_2$, $b_3 = a_1$, $\gamma_1 \in \{b_1, b_2\}$, *and* $c_1 = -a_3$.

Moreover, for each point on ℓ *inside* $\mathcal{E}$, *there are exactly* 3 *distinct quadrics from* (2) *containing it. More precisely, there is exactly one parameter of these quadrics in each of the intervals:*

$$[c_1, 0), \ (0, b_1], \ [b_2, b_3].$$

Remark 1 Since $[c_1, b_1] \subset [-a_3, a_2]$ and $[b_2, b_3] \subset [a_2, a_1]$, any given point within $\mathcal{E}$ lies at the intersection of two ellipsoids and one 1-sheeted hyperboloid oriented along x_2-axis.

For each quadric, its *tropic curves* are the sets of points where the induced metric on the tangent plane is degenerate. An ellipsoid is divided by its tropic curves into three connected components: two of them are "polar caps" mutually symmetric with respect to the x_1x_2-plane, while the third one is the "equatorial" annulus placed between them, see Fig. 2. Notice that the induced metric within each "polar cap" is Reimannian, while it is Lorentzian in the annulus between the tropic curves Genin et al. (2007).

Remark 2 From Theorem 1 and Remark 1, we have that a given point M within $\mathcal{E}$ lies at the intersection of three quadrics $\mathcal{Q}_{\lambda_1}$, $\mathcal{Q}_{\lambda_2}$, $\mathcal{Q}_{\lambda_3}$, $\lambda_1 < \lambda_2 < \lambda_3$. It is proved in Dragović and Radnović (2012) that M will belong to a "polar cap" of ellipsoid $\mathcal{Q}_{\lambda_1}$, and to the "equatorial belt" annulus of ellipsoid $\mathcal{Q}_{\lambda_2}$, with $-c \leq \lambda_1 < 0 < \lambda_2 \leq b$.

The triple $(\lambda_1, \lambda_2, \lambda_3)$ represents *generalised elliptic coordinates* of M.

Billiards in the Minkowski space. Let v be a vector and γ a hyper-plane in the Minkowski space. Decompose vector v into the sum $v = a + n_\gamma$ of a vector n_γ orthogonal to γ and a belonging to γ. Then vector $v' = a - n_\gamma$ is *the billiard reflection* of v on γ. It is easy to see that then v is also the billiard reflection of v' with respect to γ.

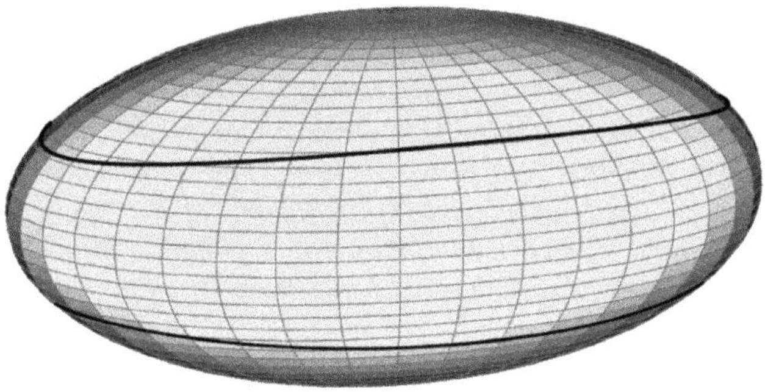

Fig. 2 Tropic curves on ellipsoid

Note that $v = v'$ if v is contained in γ and $v' = -v$ if it is orthogonal to γ. If n_γ is light-like, which means that it belongs to γ, then the reflection is not defined.

Line ℓ' is a billiard reflection of ℓ off a smooth surface $\mathcal{S}$ if their intersection point $\ell \cap \ell'$ belongs to $\mathcal{S}$ and the vectors of ℓ, ℓ' are reflections of each other with respect to the tangent plane of $\mathcal{S}$ at this point.

Remark 3 It can be seen directly from the definition of reflection that the type of line is preserved by the billiard reflection. Thus, the lines containing segments of a given billiard trajectory within $\mathcal{S}$ are all of the same type: they are all either space-like, time-like, or light-like.

If $\mathcal{S}$ is an ellipsoid, then it is possible to extend the reflection mapping to those points where the tangent planes contain the orthogonal vectors. At such points, a vector reflects into the opposite one, i.e. $v' = -v$ and $\ell' = \ell$. For the explanation, see Khesin and Tabachnikov (2009). As follows from the explanation given there, it is natural to consider each such reflection as two reflections: one reflection off the "polar cap" and one off the "equatorial belt".

The following version of the Chasles' theorem holds for billiards within ellipsoids in the Minkowski space:

Theorem 2 (Khesin and Tabachnikov (2009)) *In the Minkowski space* $\mathbf{E}^{2,1}$, *consider a billiard trajectory within ellipsoid* $\mathcal{E}$. *Then each segment of that trajectory is touching the same pair of quadrics confocal with* $\mathcal{E}$.

The two quadrics from Theorem 2 are called *the caustics* of the trajectory.

3 Periodic Trajectories

We will prove now the generalisation of the Poncelet theorem for the 3-dimensional Minkowski space. This proof is in the spirit of classical works of Jacobi and Darboux (see, for example Jacobi (1884), Darboux (1870)), and also resembles the proof of a

Poncelet theorem for light-like geodesics on a quadric in the Minkowski space from Genin et al. (2007).

Theorem 3 *In the Minkowski space $\mathbf{E}^{2,1}$, consider an n-periodic billiard trajectory within ellipsoid $\mathcal{E}$. Denote $n = m_1 + n_1$, where m_1 is the total number of reflections off the "polar caps", and n_1 the number of reflections off the "equatorial belt" of $\mathcal{E}$ along the trajectory. Then each billiard trajectory within $\mathcal{E}$ sharing the same pair of caustics is also n-periodic, with m_1 and n_1 reflections off the "polar caps" and "equatorial belt" respectively.*

Proof The differential equations in the elliptic coordinates of the lines touching two given quadrics $\mathcal{Q}_{\gamma_1}$ and $\mathcal{Q}_{\gamma_2}$ from (2) are:

$$\sum_{i=1}^{3} \frac{d\lambda_i}{\sqrt{\mathcal{P}(\lambda_i)}} = 0, \quad \sum_{i=1}^{3} \frac{\lambda_i d\lambda_i}{\sqrt{\mathcal{P}(\lambda_i)}} = 0,$$

with

$$\mathcal{P}(x) = \varepsilon(a_1 - x)(a_2 - x)(a_3 + x)(\gamma_1 - x)(\gamma_2 - x), \quad \varepsilon = \operatorname{sign}(\gamma_1\gamma_2). \tag{3}$$

Introduce constants $b_1, \ldots, b_p, c_1, \ldots, c_q$ as in Theorem 1.

Along billiard trajectory, each of the elliptic coordinates $\lambda_1, \lambda_2, \lambda_3$ takes values in segments $[c_1, 0]$, $[0, b_1]$, $[b_2, b_3]$ respectively, with local extrema being only the end-points of the segments.

The value $\lambda_1 = 0$ corresponds to the reflection off a "polar cap" of $\mathcal{E}$, and $\lambda_2 = 0$ to the reflection off the "equatorial belt". If $\lambda_1 = \lambda_2 = 0$, then that corresponds to hitting the tropic curve, which will be counted as two reflections – one off the "polar cap" and one off the "equatorial belt". Whenever one of the elliptic coordinates takes value $a_1, a_2, -a_3$, the particle is crossing the coordinate plane $x_1 = 0, x_2 = 0, x_3 = 0$ respectively. Values γ_1, γ_2 correspond to touching points with the caustics.

Similarly as in Darboux (1914), see also Dragović and Radnović (2004), integrating along the periodic trajectory gives:

$$m_1 \int_0^{c_1} \frac{\lambda_1^k d\lambda_1}{\sqrt{\mathcal{P}(\lambda_1)}} + n_1 \int_0^{b_1} \frac{\lambda_2^k d\lambda_2}{\sqrt{\mathcal{P}(\lambda_2)}} - n_2 \int_{b_2}^{b_3} \frac{\lambda_3^k d\lambda_3}{\sqrt{\mathcal{P}(\lambda_3)}} = 0, \quad k \in \{0, 1\}, \tag{4}$$

where n_2 is the number of times λ_3 traced the segment $[b_2, b_3]$ along the trajectory. Since these relations do not depend on the initial point, each trajectory with the same caustics will become closed after $\lambda_1, \lambda_2, \lambda_3$ traced the corresponding segments m_1, n_1, n_2 times respectively.

We will denote the underlying hyper-elliptic curve as:

$$\mathcal{C} \; : \; y^2 = \mathcal{P}(x), \tag{5}$$

with $\mathcal{P}(x)$ given by (3). The Weierstrass point on $\mathcal{C}$ corresponding to the value $x = \xi$, $\xi \in \{\gamma_1, \gamma_2, a_1, a_2, -a_3, \infty\}$ will be denoted by P_ξ. One of the points corresponding to $x = 0$ will be denoted by P_0.

We note that relation (4) implies the following equivalence on the Jacobian of the hyper-elliptic curve $\mathcal{C}$:

$$m_1(P_0 - P_{c_1}) + n_1(P_0 - P_{b_1}) + n_2(P_{b_2} - P_{b_3}) \sim 0. \tag{6}$$

In the following theorem, we present a detailed algebro-geometric characterisation of periodic trajectories whenever the curve $\mathcal{C}$ is non-singular.

Theorem 4 (Algebro-geometric conditions for periodicity) *Consider a billiard trajectory within ellipsoid $\mathcal{E}$ in the Minkowski space $\mathbf{E}^{2,1}$, with non-degenerate distinct caustics by $\mathcal{Q}_{\gamma_1}$ and $\mathcal{Q}_{\gamma_2}$. Then the trajectory is n-periodic if and only if one of the following is satisfied:*

– *The trajectory is space-like, and*

  (S1) both caustics are ellipsoids and either:

  - *n is even, and the divisor nP_0 equivalent to one of nP_∞, $(n-2)P_\infty + P_{\gamma_1} + P_{\gamma_2}$ on the Jacobian of the curve $\mathcal{C}$; or*
  - *n is odd, and the divisor nP_0 equivalent to one of $(n-1)P_\infty + P_{\gamma_1}$, $(n-1)P_\infty + P_{\gamma_2}$.*

  (S2) $\mathcal{Q}_{\gamma_1}$ is ellipsoid, $\mathcal{Q}_{\gamma_2}$ 1-sheeted hyperboloid along x_3-axis, and either:

  - *n is even and $nP_0 \sim nP_\infty$; or*
  - *n is odd and $nP_0 \sim (n-1)P_\infty + P_{\gamma_1}$.*

  (S3) one caustic is a 1-sheeted hyperboloid oriented along x_3-axis, the other a 1-sheeted hyperboloid oriented along x_2-axis, n is even, and $nP_0 \sim nP_\infty$.

  (S4) $\mathcal{Q}_{\gamma_1}$ is ellipsoid, $\mathcal{Q}_{\gamma_2}$ 1-sheeted hyperboloid oriented along x_2-axis, and either:

  - *n is even and $nP_0 \sim nP_\infty$; or*
  - *n is odd and $nP_0 \sim (n-1)P_\infty + P_{\gamma_2}$.*

– *The trajectory is time-like, and*

  (T1) $\mathcal{Q}_{\gamma_1}$ is ellipsoid, $\mathcal{Q}_{\gamma_2}$ 1-sheeted hyperboloid oriented along x_2-axis, and either:

  - *n is even and $nP_0 \sim nP_\infty$; or*
  - *n is odd and $nP_0 \sim (n-1)P_\infty + P_{\gamma_1}$.*

  (T2) $\mathcal{Q}_{\gamma_1}$ is ellipsoid, $\mathcal{Q}_{\gamma_2}$ 2-sheeted hyperboloid along x_3-axis, and either:

  - *n is even and $nP_0 \sim nP_\infty$; or*
  - *n is odd and $nP_0 \sim (n-1)P_\infty + P_{\gamma_1}$.*

(T3) both caustics are 1*-sheeted hyperboloids oriented along* x_2*-axis, n is even, and the divisor* nP_0 *is equivalent to either* nP_∞ *or* $(n-2)P_\infty + P_{\gamma_1} + P_{\gamma_2}$.
(T4) one caustic is 1*-sheeted hyperboloid oriented along* x_2*-axis, the other* 2*-sheeted hyperboloid oriented along* x_3*-axis, n is even, and* $nP_0 \sim nP_\infty$.

Proof For a space-like trajectory, according to Theorem 1, we have $\gamma_2 < 0 < \gamma_1 < a_1$, $\gamma_2 \in \{c_1, c_2\}$, $\gamma_1 \in \{b_1, b_2\}$. Thus, there are four possibilities of the types of the caustics.

Case S1 ($\gamma_2 = c_1$, $\gamma_1 = b_1$)*:* Both caustics are ellipsoids, $-a_3 < \gamma_2 < 0 < \gamma_1 < a_2 < a_1$.

The algebro-geometric condition (6) in this case is:

$$m_1(P_0 - P_{\gamma_1}) + n_1(P_0 - P_{\gamma_2}) + n_2(P_{a_2} - P_{a_1}) \sim 0.$$

n_2 is even, since it is the number of times the particle crossed the plane $x_2 = 0$ along the closed trajectory, so the condition is equivalent to:

$$nP_0 - m_1P_{\gamma_1} - n_1P_{\gamma_2} \sim 0.$$

From there:

$$nP_0 \sim \begin{cases} nP_\infty, & \text{if } m_1 \text{ and } n_1 \text{ are even;} \\ (n-2)P_\infty + P_{\gamma_1} + P_{\gamma_2}, & \text{if } m_1 \text{ and } n_1 \text{ are odd;} \\ (n-1)P_\infty + P_{\gamma_1}, & \text{if } m_1 \text{ is odd and } n_1 \text{ even;} \\ (n-1)P_\infty + P_{\gamma_2}, & \text{if } m_1 \text{ is even and } n_1 \text{ odd.} \end{cases}$$

Case S2 ($\gamma_2 = c_2$, $\gamma_1 = b_1$)*:* One caustic is an ellipsoid, and the other 1-sheeted hyperboloid along x_3-axis, $\gamma_2 < -a_3 < 0 < \gamma_1 < a_2 < a_1$.

The algebro geometric condition (6) for n-periodicity is:

$$m_1(P_0 - P_{-a_3}) + n_1(P_0 - P_{\gamma_1}) + n_2(P_{a_2} - P_{a_1}) \sim 0.$$

In this case, m_1 and n_2 must be even, so n and n_1 are of the same parity. Thus

$$nP_0 \sim \begin{cases} nP_\infty, & \text{if } n \text{ is even;} \\ (n-1)P_\infty + P_{\gamma_1}, & \text{if } n \text{ is odd.} \end{cases}$$

Case S3 ($\gamma_2 = c_2$, $\gamma_1 = b_2$)*:* One caustic is a 1-sheeted hyperboloid oriented along x_3-axis, and the other a 1-sheeted hyperboloid oriented along x_2-axis: $\gamma_2 < -a_3 < 0 < a_2 < \gamma_1 < a_1$.

The algebro geometric condition for n-periodicity is:

$$m_1(P_0 - P_{-a_3}) + n_1(P_0 - P_{a_2}) + n_2(P_{\gamma_1} - P_{a_1}) \sim 0,$$

where m_1, n_1, n_2 are all even, which implies $nP_0 \sim nP_\infty$.

Case S4 ($\gamma_2 = c_1$, $\gamma_1 = b_2$)*:* The caustics are a 1-sheeted hyperboloid oriented along x_2-axis and an ellipsoid: $-a_3 < \gamma_2 < 0 < a_2 < \gamma_1 < a_1$.

The algebro geometric condition for n-periodicity is:

$$m_1(P_0 - P_{\gamma_2}) + n_1(P_0 - P_{a_2}) + n_2(P_{\gamma_1} - P_{a_1}) \sim 0,$$

with even n_1, n_2, so n and m_1 are of the same parity. From there we get:

$$nP_0 \sim \begin{cases} nP_\infty, & \text{if } n \text{ is even;} \\ (n-1)P_\infty + P_{\gamma_2}, & \text{if } n \text{ is odd.} \end{cases}$$

For a time-like trajectory, Theorem 1 gives $0 < \gamma_1 < \gamma_2$, $\gamma_1 \in \{b_1, b_2\}$, $\gamma_2 \in \{b_3, b_4\}$. Again, there are four possibilities for the types of the caustics.

Case T1 ($\gamma_1 = b_1$, $\gamma_2 = b_3$)*:* One caustic is ellipsoid, the other is 1-sheeted hyperboloid oriented along x_2-axis, $-a_3 < 0 < \gamma_1 < a_2 < \gamma_2 < a_1$.

The algebro geometric condition for n-periodicity is:

$$m_1(P_0 - P_{-a_3}) + n_1(P_0 - P_{\gamma_1}) + n_2(P_{a_2} - P_{\gamma_2}) \sim 0,$$

where m_1 and n_2 must be even, so n and n_1 are of the same parity. Thus this is equivalent to $nP_0 - n_1P_{\gamma_1} - n_2P_{\gamma_2} \sim 0$, i.e.

$$nP_0 \sim \begin{cases} nP_\infty, & \text{if } n \text{ is even;} \\ (n-1)P_\infty + P_{\gamma_1}, & \text{if } n \text{ is odd.} \end{cases}$$

Case T2 ($\gamma_1 = b_1$, $\gamma_2 = b_4$)*:* The caustics are an ellipsoid and a 2-sheeted hyperboloid along x_3-axis, $-a_3 < 0 < \gamma_1 < a_2 < a_1 < \gamma_2$. This case is done identically as Case T1.

Case T3: Both caustics are 1-sheeted hyperboloids along y-axis: $\gamma_1 = b_2, \gamma_2 = b_3$. Here $-a_3 < 0 < a_2 < \gamma_1 < \gamma_2 < a_1$.

The algebro geometric condition for n-periodicity is:

$$m_1(P_0 - P_{-a_3}) + n_1(P_0 - P_{a_2}) + n_2(P_{\gamma_1} - P_{\gamma_2}) \sim 0,$$

where n_1, m_1 are both even, so n is also even. We get:

$$nP_0 \sim \begin{cases} nP_\infty, & \text{if } n_2 \text{ is even;} \\ (n-2)P_\infty + P_{\gamma_1} + P_{\gamma_2}, & \text{if } n_2 \text{ is odd.} \end{cases}$$

Case T4 ($\gamma_1 = b_2$, $\gamma_2 = b_4$)*:* The caustics are a 1-sheeted hyperboloid along x_2-axis and a 2-sheeted hyperboloid along x_3-axis, $-a_3 < 0 < a_2 < \gamma_1 < a_1 < \gamma_2$.

The algebro geometric condition for n-periodicity is:

$$m_1(P_0 - P_{-a_3}) + n_1(P_0 - P_{a_2}) + n_2(P_{\gamma_1} - P_{a_1}) \sim 0,$$

where m_1, n_1, n_2 are all even, which than gives $nP_0 \sim nP_\infty$.

The analytic Cayley-type conditions for periodic trajectories can be derived from Theorem 4 using the next Lemma.

Lemma 1 *Consider a non-singular curve* $\mathscr{C}$ (5). *Then:*

– $nP_0 \sim nP_\infty$ *for* n *even if and only if* $n \geq 6$ *and*

$$\operatorname{rank}\begin{pmatrix} A_4 & A_5 & \dots & A_{m+1} \\ A_5 & A_6 & \dots & A_{m+2} \\ \dots \\ A_{m+2} & A_{m+3} & \dots & A_{2m-1} \end{pmatrix} < m-2, \quad n = 2m,$$

with $\sqrt{\mathscr{P}(x)} = A_0 + A_1x + A_2x^2 + \dots$;
– $nP_0 \sim (n-2)P_\infty + P_{\gamma_1} + P_{\gamma_2}$ *for* n *even if and only if* $n \geq 4$ *and*

$$\operatorname{rank}\begin{pmatrix} B_2 & B_3 & \dots & B_m \\ B_3 & B_4 & \dots & B_{m+1} \\ \dots \\ B_{m+1} & B_{m+2} & \dots & B_{2m-1} \end{pmatrix} < m-1, \quad n = 2m,$$

with $\dfrac{\sqrt{\mathscr{P}(x)}}{(x-\gamma_1)(x-\gamma_2)} = B_0 + B_1x + B_2x^2 + \dots$;
– $nP_0 \sim (n-1)P_\infty + P_{\gamma_1}$ *for* n *odd if and only if* $n \geq 5$ *and*

$$\operatorname{rank}\begin{pmatrix} C_3 & C_4 & \dots & C_{m+1} \\ C_4 & C_5 & \dots & C_{m+2} \\ \dots \\ C_{m+2} & C_{m+3} & \dots & C_{2m} \end{pmatrix} < m-1, \quad n = 2m+1,$$

with $\dfrac{\sqrt{\mathscr{P}(x)}}{x-\gamma_1} = C_0 + C_1x + C_2x^2 + \dots$;
– $nP_0 \sim (n-1)P_\infty + P_{\gamma_2}$ *for* n *odd if and only if* $n \geq 5$ *and*

$$\operatorname{rank}\begin{pmatrix} D_3 & D_4 & \dots & D_{m+1} \\ D_4 & D_5 & \dots & D_{m+2} \\ \dots \\ D_{m+2} & D_{m+3} & \dots & D_{2m} \end{pmatrix} < m-1, \quad n = 2m+1,$$

with $\dfrac{\sqrt{\mathscr{P}(x)}}{x-\gamma_2} = D_0 + D_1x + D_2x^2 + \dots$.

Proof When $n = 2m$ is even, the basis for $\mathscr{L}(nP_\infty)$ is:

$$\{1, x, x^2, \dots, x^m, y, xy, \dots, x^{m-3}y\},$$

while $\mathscr{L}((n-2)P_\infty + P_{\gamma_1} + P_{\gamma_2})$ has basis

$$\left\{1, x, x^2, \dots, x^{m-1}, \frac{y}{(x-\gamma_1)(x-\gamma_2)}, \frac{xy}{(x-\gamma_1)(x-\gamma_2)}, \dots, \frac{x^{m-2}y}{(x-\gamma_1)(x-\gamma_2)}\right\}.$$

When $n = 2m+1$, the basis for $\mathscr{L}\left((n-1)P_\infty + P_{\gamma_1}\right)$ is:

$$\left\{1, x, x^2, \dots, x^m, \frac{y}{x-\gamma_1}, \frac{xy}{x-\gamma_1}, \dots, \frac{x^{m-2}y}{x-\gamma_1}\right\}.$$

In each case, the condition for the divisors equivalence is that there is a linear combination of the basis with a zero of order n at $x = 0$, which gives n linear equations for the coefficients. In order to get a non-trivial solutions, the rank of the system cannot be maximal, which gives the stated conditions, as it was done in Griffiths and Harris (1978), Dragović and Radnović (2011).

Next, we will consider the case when the two caustics coincide: $\gamma_1 = \gamma_2$. Then, the segments of a billiard trajectory within $\mathcal{E}$ are generatrices of the double caustic, which must be a 1-sheeted hyperboloid oriented along x_2-axis. Such a situation can be considered as a limit of the case T3 from Theorem 4, when $\gamma_2 \to \gamma_1$. The Cayley-type condition for periodicity is thus obtained by taking the limit of the corresponding analytic condition from Lemma 1.

Proposition 1 *A billiard trajectory within $\mathcal{E}$ with segments on 1-sheeted hyperboloid $\mathcal{Q}_{\gamma_1}$, which is oriented along x_2-axis, is n periodic if and only if n is even and either*

–

$$\operatorname{rank}\begin{pmatrix} A_4 & A_5 & \dots & A_{m+1} \\ A_5 & A_6 & \dots & A_{m+2} \\ \dots \\ A_{m+2} & A_{m+3} & \dots & A_{2m-1} \end{pmatrix} < m-2, \quad n = 2m \geq 6,$$

with $(\gamma_1 - x)\sqrt{(a_1-x)(a_2-x)(a_3+x)} = A_0 + A_1x + A_2x^2 + \dots$*; or*

–

$$\operatorname{rank}\begin{pmatrix} B_2 & B_3 & \dots & B_m \\ B_3 & B_4 & \dots & B_{m+1} \\ \dots \\ B_{m+1} & B_{m+2} & \dots & B_{2m-1} \end{pmatrix} < m-1, \quad n = 2m \geq 4,$$

with $\dfrac{\sqrt{(a_1-x)(a_2-x)(a_3+x)}}{\gamma_1 - x} = B_0 + B_1x + B_2x^2 + \dots.$

Finally, we will consider light-like trajectories. Such trajectories can be considered as a limit of Cases S2 and S3 from Theorem 4, when $\gamma_2 \to -\infty$, or a limit of Cases

T2 and T4, with $\gamma_2 \to +\infty$. The analytic conditions are obtained as the limit of the corresponding conditions from Lemma 1.

Proposition 2 *A light-like billiard trajectory within $\mathcal{E}$, with non-degenerate caustic $\mathcal{Q}_{\gamma_1}$, is n-periodic if and only if*

– *n is even, $n \geq 6$, and*

$$\operatorname{rank}\begin{pmatrix} A_4 & A_5 & \dots & A_{m+1} \\ A_5 & A_6 & \dots & A_{m+2} \\ \dots \\ A_{m+2} & A_{m+3} & \dots & A_{2m-1} \end{pmatrix} < m-2, \quad n = 2m,$$

with $\sqrt{(a_1 - x)(a_2 - x)(a_3 + x)(\gamma_1 - x)} = A_0 + A_1 x + A_2 x^2 + \dots$*; or*

– *$\mathcal{Q}_{\gamma_1}$ is an ellipsoid, n is odd, $n \geq 5$, and*

$$\operatorname{rank}\begin{pmatrix} B_3 & B_4 & \dots & B_{m+1} \\ B_4 & B_5 & \dots & B_{m+2} \\ \dots \\ B_{m+2} & B_{m+3} & \dots & B_{2m} \end{pmatrix} < m-1, \quad n = 2m+1,$$

with $\sqrt{\dfrac{(a_1 - x)(a_2 - x)(a_3 + x)}{\gamma_1 - x}} = B_0 + B_1 x + B_2 x^2 + \dots.$

4 Polynomial Equations

In this section, we express the periodicity conditions as polynomial functional equations.

Lemma 2 *Consider a non-singular curve $\mathcal{C}$* (5). *Then:*

– *$nP_0 \sim nP_\infty$ for $n = 2m$ if and only if $n \geq 6$ and there are real polynomials $p_m(s)$ and $q_{m-3}(s)$ of degrees m and $m-3$ respectively such that*

$$p_m^2(s) - s\left(s - \frac{1}{a_1}\right)\left(s - \frac{1}{a_2}\right)\left(s + \frac{1}{a_3}\right)\left(s - \frac{1}{\gamma_1}\right)\left(s - \frac{1}{\gamma_2}\right) q_{m-3}^2(s) = 1;$$

– *$nP_0 \sim (n-2)P_\infty + P_{\gamma_1} + P_{\gamma_2}$ for $n = 2m$ even if and only if $n \geq 4$ and there are real polynomials $p_{m-1}(s)$ and $q_{m-2}(s)$ of degrees $m-1$ and $m-2$ respectively such that*

$$\left(s - \frac{1}{\gamma_1}\right)\left(s - \frac{1}{\gamma_2}\right) p_{m-1}^2(s) - s\left(s - \frac{1}{a_1}\right)\left(s - \frac{1}{a_2}\right)\left(s + \frac{1}{a_3}\right) q_{m-2}^2(s) = \varepsilon,$$

with $\varepsilon = \operatorname{sign}(\gamma_1\gamma_2)$*;*

- $nP_0 \sim (n-1)P_\infty + P_{\gamma_1}$ *for* $n = 2m+1$ *odd and* $\gamma_1 > 0$ *if and only if* $n \geq 5$ *and there are real polynomials* $p_m(s)$ *and* $q_{m-2}(s)$ *of degrees* m *and* $m-2$ *respectively such that*

$$\left(s - \frac{1}{\gamma_1}\right) p_m^2(s) - s\left(s - \frac{1}{a_1}\right)\left(s - \frac{1}{a_2}\right)\left(s + \frac{1}{a_3}\right)\left(s - \frac{1}{\gamma_2}\right) q_{m-2}^2(s) = -1;$$

- $nP_0 \sim (n-1)P_\infty + P_{\gamma_2}$ *for* $n = 2m+1$ *odd and* $\gamma_2 < 0$ *if and only if* $n \geq 5$ *and there are real polynomials* $p_m(s)$ *and* $q_{m-2}(s)$ *of degrees* m *and* $m-2$ *respectively such that*

$$\left(s - \frac{1}{\gamma_2}\right) p_m^2(s) - s\left(s - \frac{1}{a_1}\right)\left(s - \frac{1}{a_2}\right)\left(s + \frac{1}{a_3}\right)\left(s - \frac{1}{\gamma_1}\right) q_{m-2}^2(s) = 1.$$

Proof It is clear from the proof of Lemma 1 that the relation $2mP_0 \sim 2mP_\infty$ is satisfied if and only if there are real polynomials $p_m^*(x)$ and $q_{m-3}^*(x)$ such that $p_m^*(x) + q_{m-3}^*(x)\sqrt{\mathscr{P}(x)}$ has a zero of multiplicity $2m$ at $x = 0$. Multiplying that expression by $p_m^*(x) - q_{m-3}^*(x)\sqrt{\mathscr{P}(x)}$, we get that the polynomial $(p_m^*(x))^2 - \mathscr{P}(x)(q_{m-3}^*(x))^2$, which is of degree $2m$, has a zero of order $2m$ at $x = 0$. Assuming that p_m^* is monic, we have:

$$(p_m^*(x))^2 - \mathscr{P}(x)(q_{m-3}^*(x))^2 = x^{2m}.$$

Dividing by x^{2m} and introducing $s = 1/x$, we get the needed relation.

The relation $2mP_0 \sim (2m-2)P_\infty + P_{\gamma_1} + P_{\gamma_2}$ is satisfied if and only if there are real polynomials $p_{m-1}^*(x)$ and $q_{m-2}^*(x)$ of degrees $m-1$ and $m-2$ such that

$$p_{m-1}^*(x) + q_{m-2}^*(x)\frac{\sqrt{\mathscr{P}(x)}}{(\gamma_1 - x)(\gamma_2 - x)}$$

has a zero of order $2m$ at $x = 0$. Multiplying by:

$$\varepsilon(\gamma_1 - x)(\gamma_2 - x)\left(p_{m-1}^*(x) - q_{m-2}^*(x)\frac{\sqrt{\mathscr{P}(x)}}{(\gamma_1 - x)(\gamma_2 - x)}\right),$$

we get that the polynomial:

$$\varepsilon(\gamma_1 - x)(\gamma_2 - x)(p_{m-1}^*(x))^2 - (a_1 - x)(a_2 - x)(a_3 + x)(q_{m-2}^*(x))^2,$$

which is of degree $2m$, has a zero of order $2m$ at $x = 0$. Thus, it equals εx^{2m}. Dividing by x^{2m} and introducing $s = 1/x$, we get the stated polynomial relation.

The relation $(2m+1)P_0 \sim 2mP_\infty + P_{\gamma_1}$ is satisfied if and only if there are real polynomials $p_m^*(x)$ and $q_{m-2}^*(x)$ of degrees m and $m-2$ respectively such that

$$p_m^*(x) + q_{m-2}^*(x)\frac{\sqrt{\mathscr{P}(x)}}{\gamma_1 - x}$$

has a zero of order $2m+1$ at $x=0$. Multiplying by:

$$(\gamma_1 - x)\left(p_m^*(x) - q_{m-2}^*(x)\frac{\sqrt{\mathcal{P}(x)}}{\gamma_1 - x}\right),$$

we get that the polynomial:

$$(\gamma_1 - x)(p_m^*(x))^2 - \varepsilon(a_1 - x)(a_2 - x)(a_3 + x)(\gamma_2 - x)(q_{m-2}^*(x))^2,$$

which is of degree $2m+1$, has a zero of order $2m+1$ at $x=0$. Assuming that $p_m^*(x)$ is monic, we get that the last expression equals $-x^{2m+1}$. Note that $\varepsilon = \operatorname{sign}(\gamma_1\gamma_2) = \operatorname{sign}(\gamma_2)$. Dividing by x^{2m+1}, introducing $s = 1/x$, we get the stated relation.

The relation $(2m+1)P_0 \sim 2mP_\infty + P_{\gamma_2}$ is satisfied if and only if there are real polynomials $p_m^*(x)$ and $q_{m-2}^*(x)$ of degrees m and $m-2$ respectively such that

$$p_m^*(x) + q_{m-2}^*(x)\frac{\sqrt{\mathcal{P}(x)}}{\gamma_2 - x}$$

has a zero of order $2m+1$ at $x=0$. Multiplying by:

$$(\gamma_2 - x)\left(p_m^*(x) - q_{m-2}^*(x)\frac{\sqrt{\mathcal{P}(x)}}{\gamma_2 - x}\right),$$

we get that the polynomial:

$$(\gamma_2 - x)(p_m^*(x))^2 - \varepsilon(a_1 - x)(a_2 - x)(a_3 + x)(\gamma_1 - x)(q_{m-2}^*(x))^2,$$

which is of degree $2m+1$, has a zero of order $2m+1$ at $x=0$. Assuming that $p_m^*(x)$ is monic, we get that the last expression equals $-x^{2m+1}$. Note that $\varepsilon = \operatorname{sign}(\gamma_1\gamma_2) = -\operatorname{sign}(\gamma_1)$. Dividing by $-x^{2m+1}$, introducing $s = 1/x$, we get the stated relation.

By taking the appropriate limits, we get the polynomial conditions for the case of a double caustic and the case of light-like trajectories:

Proposition 3 *(a) A billiard trajectory within $\mathcal{E}$ with segments on 1-sheeted hyperboloid $\mathcal{Q}_{\gamma_1}$, which is oriented along x_2-axis, is n periodic if and only if $n = 2m$ is even and either:*

- *$n \geq 6$ and there are real polynomials $p_m(s)$ and $q_{m-3}(s)$ of degrees m and $m-3$ respectively such that*

$$p_m^2(s) - s\left(s - \frac{1}{a_1}\right)\left(s - \frac{1}{a_2}\right)\left(s + \frac{1}{a_3}\right)\left(s - \frac{1}{\gamma_1}\right)^2 q_{m-3}^2(s) = 1;$$

- *$n \geq 4$ and there are real polynomials $p_{m-1}(s)$ and $q_{m-2}(s)$ of degrees $m-1$ and $m-2$ respectively such that*

$$\left(s - \frac{1}{\gamma_1}\right)^2 p_{m-1}^2(s) - s\left(s - \frac{1}{a_1}\right)\left(s - \frac{1}{a_2}\right)\left(s + \frac{1}{a_3}\right) q_{m-2}^2(s) = 1.$$

(b) A light-like billiard trajectory within $\mathcal{E}$, with non-degenerate caustic $\mathcal{Q}_{\gamma_1}$, is n-periodic if and only if

- *$n = 2m$ is even, $n \geq 6$, $\mathcal{Q}_{\gamma_1}$ is an ellipsoid or a 1-sheeted hyperboloid oriented along x_2-axis, and there are real polynomials $p_m(s)$ and $q_{m-3}(s)$ of degrees m and $m-3$ respectively such that*

$$p_m^2(s) - s^2\left(s - \frac{1}{a_1}\right)\left(s - \frac{1}{a_2}\right)\left(s + \frac{1}{a_3}\right)\left(s - \frac{1}{\gamma_1}\right) q_{m-3}^2(s) = 1;$$

- *$n = 2m+1$ is odd, $n \geq 5$, $\mathcal{Q}_{\gamma_1}$ is an ellipsoid, and there are real polynomials $p_m(s)$ and $q_{m-2}(s)$ of degrees m and $m-2$ respectively such that*

$$\left(s - \frac{1}{\gamma_1}\right) p_m^2(s) - s^2\left(s - \frac{1}{a_1}\right)\left(s - \frac{1}{a_2}\right)\left(s + \frac{1}{a_3}\right) q_{m-2}^2(s) = -1.$$

Corollary 1 *If the billiard trajectories within $\mathcal{E}$ with caustics $\mathcal{Q}_{\gamma_1}$ and $\mathcal{Q}_{\gamma_2}$ are n-periodic, then there exist real polynomials $\hat{p}_n$ and $\hat{q}_{n-3}$ of degrees n and $n-3$ respectively, which satisfy the Pell equation:*

$$\hat{p}_n^2(s) - s\left(s - \frac{1}{a_1}\right)\left(s - \frac{1}{a_2}\right)\left(s + \frac{1}{a_3}\right)\left(s - \frac{1}{\gamma_1}\right)\left(s - \frac{1}{\gamma_2}\right) \hat{q}_{n-3}^2(s) = 1.$$

Proof If $n = 2m$, we know that one of the first two cases of Lemma 2 is satisfied. In the first case, take $\hat{p}_n = 2p_m^2 - 1$ and $\hat{q}_{n-3} = 2p_m q_{m-3}$. In the second case, we set:

$$\hat{p}_n(s) = 2\left(s - \frac{1}{\gamma_1}\right)\left(s - \frac{1}{\gamma_2}\right) p_{m-1}^2(s) - \varepsilon, \quad \hat{q}_{n-3} = 2p_{m-1}q_{m-2}.$$

If $n = 2m+1$, one of the last two cases of Lemma 2 holds. In the third case, we set:

$$\hat{p}_n(s) = 2\left(s - \frac{1}{\gamma_1}\right) p_m^2(s) + 1, \quad \hat{q}_{n-3} = 2p_m q_{m-2},$$

and in the fourth one:

$$\hat{p}_n(s) = 2\left(s - \frac{1}{\gamma_2}\right) p_m^2(s) - 1, \quad \hat{q}_{n-3} = 2p_m q_{m-2}.$$

Remark 4 By considering light-like trajectories with an ellipsoid as caustic, and taking the limit when parameter of the caustic approaches zero, we get the light-like geodesics on the ellipsoid $\mathcal{E}$. Applying the appropriate limit to the analytic conditions for periodicity obtained in this work, may indicate the approach to the conditions for periodicity for the Poncelet-style closure theorem for light-like geodesics in the equatorial belt from Genin et al. (2007) can be obtained, thus solving the Problem 5.2 from that paper, see also Problem 7 from Tabachnikov (2015).

Acknowledgements The research of V. D. and M. R. was supported by the Discovery Project #DP190101838 *Billiards within confocal quadrics and beyond* from the Australian Research Council and Project #174020 *Geometry and Topology of Manifolds, Classical Mechanics and Integrable Systems* of the Serbian Ministry of Education, Technological Development and Science. The authors are grateful to the referee for careful reading and very useful comments and suggestions.

References

Adabrah, A.K., Dragović, V., Radnović, M.: Periodic billiards within conics in the Minkowski plane and Akhiezer polynomials. Regul. Chaotic Dyn. **24**, 464–501 (2019)

Akhiezer, N.I.: Elements of the theory of elliptic functions. Translations of Mathematical Monographs vol. 79, American Mathematical Society, Providence, RI (1990)

Birkhoff, G., Morris, R.: Confocal conics in space-time. Am. Math. Mon. **69**, 1–4 (1962)

Bobenko, A.I., Suris, Y.B.: Discrete differential geometry: integrable structure. Graduate Studies in Mathematics vol. 98, American Mathematical Society, Providence, RI (2008)

Bogatyrev, A.: Extremal Polynomials and Riemann Surfaces. Springer, Heidelberg (2012)

Darboux, G.: Sur les polygones inscrits et circonscrits à l'ellipsoïde. Bulletin de la Société philomathique **7**, 92–94 (1870)

Darboux, G.: Leçons sur la théorie générale des surfaces et les applications géométriques du calcul infinitesimal. Gauthier-Villars, Paris (1914)

Dragović, V., Radnović, M.: Cayley-type conditions for billiards within k quadrics in $\mathbf{R}^d$. J. Phys. A: Math. Gen. **37**, 1269–1276 (2004)

Dragović, V., Radnović, M.: Poncelet Porisms and Beyond. Springer Birkhauser, Basel (2011)

Dragović, V., Radnović, M.: Ellipsoidal billiards in pseudo-Euclidean spaces and relativistic quadrics. Adv. Math. **231**, 1173–1201 (2012)

Dragović, V., Radnović, M.: Caustics of Poncelet polygons and classical extremal polynomials. Regul. Chaotic Dyn. **24**, 1–35 (2019a)

Dragović, V., Radnović, M.: Periodic ellipsoidal billiard trajectories and extremal polynomials. Commun. Math. Phys. **372**, 183–211 (2019b)

Duistermaat, J.J.: Discrete Integrable Systems: QRT Maps and Elliptic Surfaces. Springer, New York (2010)

Genin, D., Khesin, B., Tabachnikov, S.: Geodesics on an ellipsoid in Minkowski space. L'Enseign. Math. **53**, 307–331 (2007)

Griffiths, P., Harris, J.: On Cayley's explicit solution to Poncelet's porism. Enseign. Math. **24**, 31–40 (1978)

Hietarinta, J., Joshi, N., Nijhoff, F.W.: Discrete Systems and Integrability. Cambridge University Press (2016)

Jacobi, C.: Vorlesungen über Dynamic. Supplementband. Berlin, Gesammelte Werke (1884)

Joshi, N.: Discrete Painlevé Equations. American Mathematical Society, Providence, RI (2019)

Khesin, B., Tabachnikov, S.: Pseudo-Riemannian geodesics and billiards. Adv. Math. **221**, 1364–1396 (2009)

Kozlov, V., Treshchëv, D.: Billiards. American Mathematical Society, Providence RI (1991)

Tabachnikov, S.: A baker's dozen of problems. Arnold Math. J. **1**, 59–67 (2015)

Analogues of Kahan's Method for Higher Order Equations of Higher Degree

A. N. W. Hone and G. R. W. Quispel

Abstract Kahan introduced an explicit method of discretization for systems of first order differential equations with nonlinearities of degree at most two (quadratic vector fields). Kahan's method has attracted much interest due to the fact that it preserves many of the geometrical properties of the original continuous system. In particular, a large number of Hamiltonian systems of quadratic vector fields are known for which their Kahan discretization is a discrete integrable system. In this note, we introduce a special class of explicit order-preserving discretization schemes that are appropriate for certain systems of ordinary differential equations of higher order and higher degree.

Keywords Discretization · Kahan's method · Symplectic integrator

1 Introduction

Kahan's method is a special discretization scheme that provides an explicit method for integrating quadratic vector fields, given by systems of first order ordinary differential equations (ODEs) of the form

$$\frac{\mathrm{d}x_i}{\mathrm{d}t} = f_i(x_1, \ldots, x_N), \qquad i = 1, \ldots, N, \tag{1}$$

A. N. W. Hone—Work begun on leave at School of Mathematics & Statistics, University of New South Wales, NSW 2052, Australia.

A. N. W. Hone (✉)
School of Mathematics, Statistics & Actuarial Science, University of Kent,
Canterbury CT2 7NF, UK
e-mail: A.N.W.Hone@kent.ac.uk

G. R. W. Quispel
Department of Mathematics and Statistics, La Trobe University, Bundoora, VIC 3086, Australia
e-mail: R.Quispel@latrobe.edu.au

© Springer Nature Switzerland AG 2020

F. Nijhoff et al. (eds.), *Asymptotic, Algebraic and Geometric Aspects of Integrable Systems*, Springer Proceedings in Mathematics & Statistics 338,
https://doi.org/10.1007/978-3-030-57000-2_9

where each function f_i is a polynomial of total degree two in the independent variables $x_1, \ldots, x_N$ (see Kahan (1993) or Kahan and Li (1997)). In order to specify Kahan's method, one should replace each derivative on the left-hand side of (1) by the forward difference, so that

$$\frac{\mathrm{d}x_i}{\mathrm{d}t} \to \Delta x_i := \frac{\overline{x}_i - x_i}{h},$$

while terms of degrees two, one and zero appearing in each f_i on the right-hand side are replaced according to the rules

$$x_j x_k \to \frac{1}{2}(\overline{x}_j x_k + x_j \overline{x}_k), \quad x_j \to \frac{1}{2}(x_j + \overline{x}_j), \quad c \to c, \tag{2}$$

where h is the time step and $\overline{x}_i$ denotes the approximation to $x_i(t+h)$.

It was noticed some time ago that Kahan's method provides an effective integration scheme for the classic two-species Lotka-Volterra model

$$\frac{\mathrm{d}x}{\mathrm{d}t} = \alpha x(1-y), \qquad \frac{\mathrm{d}y}{\mathrm{d}t} = y(x-1) \tag{3}$$

(with $\alpha > 0$ being an arbitrary parameter), retaining the qualitative features of the orbits of the continuous system, namely the stability of orbits around the elliptic fixed point at $(x, y) = (1, 1)$. This was subsequently explained by the fact that the Kahan discretization of (3), given by

$$\begin{aligned}(\overline{x} - x)/h &= \tfrac{\alpha}{2}\Big(x(1-\overline{y}) + \overline{x}(1-y)\Big), \\ (\overline{y} - y)/h &= \tfrac{1}{2}\Big(y(\overline{x}-1) + \overline{y}(x-1)\Big),\end{aligned}$$

preserves the same symplectic form

$$\omega = \frac{\mathrm{d}x \wedge \mathrm{d}y}{xy}$$

as the original Hamiltonian system (Sanz-Serna 1994). In the context of Lotka-Volterra models, a variant of Kahan's method with similar properties was discovered by Mickens (2003), who had previously considered various examples of nonstandard discretization methods (Mickens 1994), but a more rapid growth of interest in Kahan's method began when Hirota and Kimura independently proposed the rules (2) for the discretization of the Euler equations for rigid body motion, finding that the resulting discrete system is also completely integrable (Hirota and Kimura 2000), and this has led to the search for other discrete integrable systems arising in this way (Hone and Petrera 2009), with a survey of several results given in Petrera et al. (2011), and some more recent examples in Petrera and Zander (2017) and Petrera et al. (2019), for instance.

Many of the geometrical properties of Kahan's method for quadratic vector fields are based on the polarization identity for quadratic forms (Celledoni et al. 2013), and recently this has led to a generalization of Kahan's method that can cope with vector fields of degree three or more, by using higher degree analogues of polarization (Celledoni et al. 2015). One disadvantage of the latter method for higher degree vector fields is that, in common with multistep methods in numerical analysis, one must use extra grid points for the discretization, so the original ODE system does not provide enough initial values to start the iteration of the discrete version. However, if one is looking for a discretization scheme that preserves integrability or other geometric properties of ODEs, then it is desirable for the initial value space of the discrete system to have the same dimension as that of the continuous one. Here we would like to suggest a discretization scheme with the latter property, which is a natural generalization of Kahan's method to higher order and higher degree.

The idea is to consider a system of ODEs of order $n \geq 1$, with the right-hand sides being functions of the coordinates $x_1, \ldots, x_N$ only, of the form

$$\frac{\mathrm{d}^n x_i}{\mathrm{d}t^n} = f_i(x_1, \ldots, x_N), \qquad i = 1, \ldots, N, \tag{4}$$

where each function f_i is a polynomial of maximal degree $n + 1$. For $n = 1$ this is a quadratic vector field, which one can discretize using Kahan's method. In the next section, we present an explicit discretization scheme for systems of the form (4), valid for any $n \geq 1$, which reduces to Kahan's method when $n = 1$. The first new case is $n = 2$, corresponding to systems of Newton equations, which are relevant in many applications. We illustrate this in Sect. 3 by considering the discretization of the motion of a single particle moving in a quartic potential. The latter is one of the simplest examples of an integrable Hamiltonian system, and it turns out that the discrete version produced by the method is also integrable, with a conserved quantity and an invariant symplectic form. In Sect. 4, we consider a different example of fourth order, namely a nonlinear beam equation, and briefly compare the discretization obtained by the new method with another discretization obtained by applying an approach similar to Kahan's directly to the Lagrangian of the continuous system.

2 A Higher Order Version of Kahan's Method

For $n = 2$, (4) becomes a system of Newton equations, assumed to have polynomial forces of degree at most three, which can be conveniently written as

$$\frac{\mathrm{d}^2 x_i}{\mathrm{d}t^2} = \sum_{0 \leq j_1 \leq j_2 \leq j_3 \leq N} c_{i,j_1,j_2,j_3} x_{j_1} x_{j_2} x_{j_3}, \qquad i = 1, \ldots, N, \tag{5}$$

where $c_{ijk\ell}$ are arbitrary coefficients, and we have included an additional dummy variable $x_0 = 1$ to allow terms of degree less than three to be included within the same summation. Then to discretize (5) we propose the following:

$$\frac{\overline{x}_i - 2x_i + \underline{x}_i}{h^2} = \frac{1}{6}\sum_{\sigma\in S_3}\sum_{0\leq j_1\leq j_2\leq j_3\leq N} c_{i,j_1,j_2,j_3}\,\underline{x}_{j_{\sigma(1)}}x_{j_{\sigma(2)}}\overline{x}_{j_{\sigma(3)}}, \tag{6}$$

for $i = 1, \ldots, N$; the first summation is over permutations σ in S_3, the symmetric group on three symbols, and $\overline{x}_i = x^{(1)}$, $\underline{x}_i = x^{(-1)}$ are the approximations to $x_i(t \pm h)$, with time step h. For terms of degree three, with each variable x_j appearing at the three adjacent lattice points $\underline{x}_j = x_j^{(-1)}$, $x_j = x_j^{(0)}$, $\overline{x}_j = x_j^{(1)}$, the replacement rule is described explicitly by

$$x_jx_kx_\ell \to \frac{1}{6}\Big(\underline{x}_jx_k\overline{x}_\ell + \underline{x}_j\overline{x}_kx_\ell + x_j\underline{x}_k\overline{x}_\ell + x_j\overline{x}_k\underline{x}_\ell + \overline{x}_jx_k\underline{x}_\ell + \overline{x}_j\underline{x}_kx_\ell\Big), \tag{7}$$

while for terms of degree two the rule is obtained by setting $\ell = 0$, so that $x_\ell \to x_0 = 1$ in the above, and for the linear terms one can set $k = \ell = 0$, so that the rule for terms of degree less than three is

$$\begin{aligned} x_jx_k &\to \tfrac{1}{6}\Big(\underline{x}_jx_k + \underline{x}_j\overline{x}_k + x_j\underline{x}_k + x_j\overline{x}_k + \overline{x}_jx_k + \overline{x}_j\underline{x}_k\Big), \\ x_j &\to \tfrac{1}{3}\Big(\underline{x}_j + x_j + \overline{x}_j\Big), \qquad c \to c. \end{aligned} \tag{8}$$

Following the approach of Celledoni et al. (2015), a second order system of equations can be written in vector form as

$$\frac{\mathrm{d}^2\mathbf{x}}{\mathrm{d}t^2} = \mathbf{f}(\mathbf{x}), \tag{9}$$

where each component of the vector of functions $\mathbf{f} = (f_1, f_2, \ldots, f_N)^T$ is a polynomial of degree at most three, and then the replacement rules (7) and (8) are equivalent to the formula

$$\begin{aligned} \tfrac{1}{h^2}\Big(\overline{\mathbf{x}} - 2\mathbf{x} + \underline{\mathbf{x}}\Big) = {}& \tfrac{9}{2}\mathbf{f}\Big(\tfrac{\overline{\mathbf{x}}+\mathbf{x}+\underline{\mathbf{x}}}{3}\Big) - \tfrac{4}{3}\Big(\mathbf{f}\Big(\tfrac{\underline{\mathbf{x}}+\mathbf{x}}{2}\Big) + \mathbf{f}\Big(\tfrac{\mathbf{x}+\overline{\mathbf{x}}}{2}\Big) + \mathbf{f}\Big(\tfrac{\underline{\mathbf{x}}+\overline{\mathbf{x}}}{2}\Big)\Big) \\ & + \tfrac{1}{6}\Big(\mathbf{f}(\underline{\mathbf{x}}) + \mathbf{f}(\mathbf{x}) + \mathbf{f}(\overline{\mathbf{x}})\Big). \end{aligned} \tag{10}$$

Proposition 1 *The discretization (10) commutes with affine transformations*

$$\mathbf{y} \mapsto \mathbf{x} = A\mathbf{y} + \mathbf{b}, \tag{11}$$

where $A \in GL(N, \mathbb{R})$ is a constant matrix and $\mathbf{b} \in \mathbb{R}^N$ is a vector of constants.

Proof Under the transformation (11), $\mathbf{f}(\cdot)$ in (9) is replaced by $A^{-1}\mathbf{f}(A\cdot+\mathbf{b})$. Upon substituting (11) and its shifted versions into (10), it is not hard to check that the same occurs for each appearance of $\mathbf{f}$ on the right-hand side. □

The symmetric replacement rules above generalize to any order $n \geq 1$, so that for a system of nth order ODEs (4) with right-hand sides all of degree $n+1$ the discretization becomes

$$\Delta^n x_i = \frac{1}{(n+1)!}\sum_{\sigma\in S_{n+1}}\sum_{j_k\leq j_{k+1}} c_{i,j_1,\ldots,j_{n+1}}\, x_{j_{\sigma(1)}} x^{(1)}_{j_{\sigma(2)}} x^{(2)}_{j_{\sigma(3)}}\cdots x^{(n)}_{j_{\sigma(n+1)}}, \tag{12}$$

for $i = 1,\ldots,N$, with $x_i^{(1)} = \overline{x}_i$, $x_i^{(2)} = \overline{\overline{x}}_i,\ldots,x_j^{(n)}$ corresponding to shifts by steps of $h, 2h,\ldots,nh$, and the interior summation being for $0 \leq j_1 \leq \cdots \leq j_{n+1} \leq N$. On the left-hand side of (12) we have replaced the nth derivative by the nth power of the forward difference operator, and for convenience we have written everything on the right-hand side in terms of forward shifts of the variables x_j. The discretization (12) reduces to Kahan's method when $n = 1$, and to (6) when $n = 2$, modulo shifting the lattice points $-1, 0, 1$ in the latter up to $0, 1, 2$.

Clearly there are other choices of discrete nth derivative that one could take, and other affine combinations of terms with the same homogeneous degree could be chosen while preserving the continuum limit. We have taken the most symmetrical choice in (12), because it is manifestly linear in each of the highest shifts $x_1^{(n)}, x_2^{(n)}\ldots,x_N^{(n)}$, so it can be explicitly solved for each of these quantities to yield rational functions of all the lower shifts. It is also linear in each of the lowest shifts $x_i = x_i^{(0)}$ for $i = 1,\ldots,N$, so it can be explicitly solved for these as well. Thus (12) is an implicit way of writing an explicit birational map in dimension nN, corresponding to $\mathbf{X}\mapsto\bar{\mathbf{X}}$, where

$$\begin{aligned}\mathbf{X} &= (x_1^{(0)},\ldots,x_N^{(0)},x_1^{(1)},\ldots,x_N^{(1)},\ldots,x_1^{(n-1)},\ldots,x_N^{(n-1)}),\\ \bar{\mathbf{X}} &= (x_1^{(1)},\ldots,x_N^{(1)},x_1^{(2)},\ldots,x_N^{(2)},\ldots,x_1^{(n)},\ldots,x_N^{(n)}).\end{aligned}$$

3 Discretization of a Quartic Oscillator

To see why it might be worth investigating these higher Kahan-like schemes, we start by presenting the following example: $n = 2$ with a cubic force on a particle in one dimension, generated by a natural Hamiltonian with a quartic potential, that is

$$H = \frac{1}{2}p^2 + \frac{1}{4}ax^4 + \frac{1}{3}bx^3 + \frac{1}{2}cx^2 + dx,$$

which yields the Newton equation

$$\ddot{x} = -ax^3 - bx^2 - cx - d. \tag{13}$$

This is an integrable system par excellence, and the generic level sets $H =$ const are quartic curves of genus one in the (x, p) plane. The discretization (6) applied to (13) produces a difference equation of second order, given by

$$\overline{x} = \frac{(3-\gamma)x - \delta - (\beta x + \gamma)\underline{x}}{\beta x + \gamma + (\alpha x + \beta)\underline{x}}, \tag{14}$$

where

$$\alpha = ah^2, \quad \beta = \frac{bh^2}{3}, \quad \gamma = 1 + \frac{ch^2}{3}, \quad \delta = dh^2.$$

The map (14) is an example of a QRT map (Quispel et al. 1988), but let us suppose that we do not know the geometric properties of this map. To find these properties, such as the existence of a preserved measure, and first and second integrals of the map (14), we will look for preserved Darboux polynomials, as detailed in our recent work Celledoni et al. (2019) and Celledoni et al. (2020). To this end, we write the second order equation (14) as two first order ones, namely

$$\overline{x} = y, \qquad \overline{y} = \frac{(3-\gamma)y - \delta - (\beta y + \gamma)x}{\beta y + \gamma + (\alpha y + \beta)x}, \tag{15}$$

and look for polynomials P satisfying

$$P(\overline{x}, \overline{y}) = J(x, y) P(x, y), \tag{16}$$

where J is the Jacobian determinant of the map (15), i.e.

$$J(x, y) = \frac{(\beta y + \gamma)^2 + (\alpha y + \beta)\big((3-\gamma)y - \delta\big)}{(\alpha xy + \beta(x+y) + \gamma)^2}. \tag{17}$$

Substituting (17) into (16), and looking for polynomials up to total degree four in x and y, we find two linearly independent solutions, given by

$$P_1 = \alpha xy + \beta(x+y) + \gamma,$$

$$P_2 = (\alpha\gamma - \beta^2)x^2y^2 + \epsilon xy(x+y) + \zeta(x^2+y^2) - (3-\gamma)^2 xy + (3-\gamma)\delta(x+y) - \delta^2,$$

with

$$\epsilon = \alpha\delta + \beta(3-\gamma), \qquad \zeta = \beta\delta + \gamma(3-\gamma).$$

It follows that the map (14) is measure-preserving, with the invariant symplectic form

$$\frac{\mathrm{d}x \wedge \mathrm{d}y}{P_1} = \frac{\mathrm{d}x \wedge \mathrm{d}y}{\alpha xy + \beta(x+y) + \gamma}, \tag{18}$$

and the first integral

$$I = \frac{P_2}{P_1} \tag{19}$$

given by

$$\frac{(\alpha\gamma - \beta^2)x^2y^2 + \epsilon xy(x+y) + \zeta(x^2+y^2) - (3-\gamma)^2xy + (3-\gamma)\delta(x+y) - \delta^2}{\alpha xy + \beta(x+y) + \gamma}.$$

Hence the integrability is preserved by the discretization in this case, and we recover the standard property of a QRT map, that it preserves a pencil of biquadratic curves, here given by

$$\lambda P_1(x, y) + P_2(x, y) = 0.$$

Moreover, in the continuum limit $h \to 0$, the standard area form $\mathrm{d}x \wedge \mathrm{d}y$ and the Hamiltonian H are recovered from (18) and (19) respectively, since from $y = x + hp + O(h^2)$ we find

$$P_1 = 1 + O(h^2), \qquad P_2 = 4Hh^2 + O(h^3).$$

The Eq. (13) includes Duffing's equation, which is the case $b = d = 0$, and also the second order ODE for the Weierstrass $\wp$ function, which arises when $a = c = 0$. In Potts (1982), another replacement rule is used for the cubic and linear terms in Duffing's equation, somewhat less symmetrical than the one defined by (7), and it is shown that if the coefficients and denominator in the second difference operator are replaced by suitable functions of the parameters and the time step h then this alternative rule results in a discretization that is exact, in the sense that the iterates of the difference equation interpolate the solution of the original ODE. Similarly, in Potts (1987) an exact discretization is obtained for the case corresponding to the Weierstrass $\wp$ function, with only quadratic and constant terms on the right-hand side. However, in the latter case, the exact discretization (derived from the addition formula for the $\wp$ function) requires not only a different replacement rule for the quadratic terms compared with (8), but also extra cubic and linear terms that must be included, with a coefficient which is $O(h^2)$. When $a = c = 0$, the Eq. (13) can be rewritten as a quadratic vector field, namely

$$\frac{\mathrm{d}x}{\mathrm{d}t} = p, \qquad \frac{\mathrm{d}p}{\mathrm{d}t} = -bx^2 - d,$$

so that Kahan's method can be applied, as in Petrera et al. (2011), resulting in a first order discrete system which is equivalent to a second order difference equation for x, namely

$$\overline{x} + \underline{x} = \frac{4x - 2\delta}{3\beta x + 2} \tag{20}$$

(where we set $\beta = bh^2/3$, $\delta = dh^2$ as before). The Eq. (20) is a QRT map in additive form, clearly of a different type to (14), which becomes

$$\overline{x} = \frac{2x - \delta - (\beta x + 1)\underline{x}}{\beta(x + \underline{x}) + 1}$$

when $\alpha = 0$, $\gamma = 1$. To see that they are really different QRT maps, in the sense that they are not related to one another via so-called curve-dependent McMillan maps (Iatrou and Roberts 2001), observe that the pencil of invariant biquadratic curves corresponding to (20) is

$$\lambda - \beta^2 x^2 y^2 + \frac{4}{3}\beta xy(x + y) + \frac{4}{3}(x^2 + y^2) - \frac{2}{3}(4 + \beta\delta)xy + \frac{4}{3}\delta(x + y) = 0,$$

whereas when $\alpha = 0$, $\gamma = 1$ the pencil $\lambda P_1(x, y) + P_2(x, y) = 0$ for (14) reduces to one of a different type, namely

$$\lambda\Big(1 + \beta(x + y)\Big) - \beta^2 x^2 y^2 + 2\beta xy(x + y) + (\beta\delta + 2)(x^2 + y^2) - 4xy + 2\delta(x + y) - \delta^2 = 0.$$

4 Two Discretizations of a Nonlinear Beam Equation

Vibrating beams were considered by Leonardo da Vinci (1493), but the traditional theory of vibrations of a beam is usually attributed to Euler and Bernoulli (Han et al. 1999), being described by a partial differential equation (PDE) of fourth order, which in dimensionless form is given by

$$\frac{\partial^2 w}{\partial t^2} + \frac{\partial^4 w}{\partial x^4} = Q.$$

For the case of a static beam, the equation has the form

$$\frac{\mathrm{d}^4 w}{\mathrm{d}x^4} = Q, \tag{21}$$

where $w = w(x)$ is the vertical deflection of the beam, which lies horizontally along the x-axis. The standard beam model is linear, with the distributed load Q on the right-hand size being a constant (or more generally, a function of x, the independent variable). However, here we consider a more general nonlinear version of the model, derived from a second order Lagrangian of the form

$$L = \frac{1}{2}\left(\frac{\mathrm{d}^2 w}{\mathrm{d}x^2}\right)^2 - V(w), \tag{22}$$

which gives a nonlinear load function

$$Q(w) = \frac{\mathrm{d}V}{\mathrm{d}w}.$$

In the linear case, the model was considered recently from the viewpoint of a Hamilton-Jacobi approach to higher order implicit systems (Esen et al. 2020), while a coupled PDE system of beam equations with cubic nonlinearity was analysed in Shi and Xu (2020). From the second order Lagrangian (21), we can introduce the Ostrogradsky variables (see Błaszak (1998), for instance), given by

$$q_1 = w, \quad q_2 = w', \quad p_1 = \frac{\partial L}{\partial w'} - \frac{\mathrm{d}}{\mathrm{d}x}\left(\frac{\partial L}{\partial w''}\right) = -w''', \quad p_2 = \frac{\partial L}{\partial w''} = w'',$$

where the primes denote derivatives with respect to the independent variable x. Then (q_1, p_1), (q_2, p_2) provide two pairs of canonically conjugate positions and momenta, and the Euler-Lagrange equation

$$\frac{\mathrm{d}^2}{\mathrm{d}x^2}\left(\frac{\partial L}{\partial w''}\right) - \frac{\mathrm{d}}{\mathrm{d}x}\left(\frac{\partial L}{\partial w'}\right) + \frac{\partial L}{\partial w} = 0, \tag{23}$$

which for the Lagrangian (22) is given by (21) with $Q = \mathrm{d}V/\mathrm{d}w$, is equivalent to Hamilton's equations for the Hamiltonian function

$$H = \frac{1}{2}(p_2)^2 + q_2 p_1 + V(q_1).$$

For the sake of concreteness, we consider the case of an odd potential

$$V(w) = \frac{a}{5}w^5 + \frac{b}{3}w^3 + cw,$$

so that the nonlinear beam equation is given by

$$w'''' = aw^4 + bw^2 + c. \tag{24}$$

To begin with, we consider the result of applying the discretization rule (12) to the nonlinear beam equation (24), which produces a difference equation of fourth order, of the form

$$\Delta^4 w = F(w^{(0)}, w^{(1)}, w^{(2)}, w^{(3)}, w^{(4)}) \tag{25}$$

for a function F that is a sum of terms of total degree four, two and zero. This can be written more symmetrically by shifting down by two steps, to yield

$$\frac{w^{(-2)} - 4w^{(-1)} + 6w^{(0)} - 4w^{(1)} + w^{(2)}}{h^4} = F_4 + F_2 + c, \tag{26}$$

where the quartic terms are specified by

$$\begin{aligned} \tfrac{5}{a} F_4 = {} & w^{(-2)}w^{(-1)}w^{(0)}w^{(1)} + w^{(-2)}w^{(-1)}w^{(0)}w^{(2)} + w^{(-2)}w^{(-1)}w^{(1)}w^{(2)} \\ & + w^{(-2)}w^{(0)}w^{(1)}w^{(2)} + w^{(-1)}w^{(0)}w^{(1)}w^{(2)}, \end{aligned} \tag{27}$$

and the quadratic terms are given by

$$\begin{aligned} \tfrac{10}{b} F_2 = {} & w^{(-2)}w^{(-1)} + w^{(-2)}w^{(0)} + w^{(-2)}w^{(1)} + w^{(-2)}w^{(2)} + w^{(-1)}w^{(0)} \\ & + w^{(-1)}w^{(1)} + w^{(-1)}w^{(2)} + w^{(0)}w^{(1)} + w^{(0)}w^{(2)} + w^{(1)}w^{(2)}. \end{aligned} \tag{28}$$

It turns out that the birational map defined by (26) is measure-preserving. This is a consequence of the fact that the formula for the right-hand side of (25) is both linear and symmetric in its arguments, so that the derivatives with respect to the highest and lowest shifts, namely

$$\frac{\partial F}{\partial w^{(0)}} = G(w^{(1)}, w^{(2)}, w^{(3)}, w^{(4)}), \qquad \frac{\partial F}{\partial w^{(4)}} = H(w^{(0)}, w^{(1)}, w^{(2)}, w^{(3)}), \tag{29}$$

are very closely related to one another.

Proposition 2 *The discretization (26) preserves the volume form*

$$\Omega = \frac{1}{1 - h^2 H(w^{(-2)}, w^{(-1)}, w^{(0)}, w^{(1)})} \mathrm{d}w^{(-2)} \wedge \mathrm{d}w^{(-1)} \wedge \mathrm{d}w^{(0)} \wedge \mathrm{d}w^{(1)},$$

where H is defined by (29).

Proof Upon taking the differential of both sides of (26), we obtain the equation

$$\begin{aligned} & \Big(1 - h^2 G(w^{(-1)}, w^{(0)}, w^{(1)}, w^{(2)})\Big)\mathrm{d}w^{(-2)} \\ & + \Big(1 - h^2 H(w^{(-2)}, w^{(-1)}, w^{(0)}, w^{(1)})\Big)\mathrm{d}w^{(2)} + \cdots = 0, \end{aligned}$$

where the ellipsis denotes terms that are linear in $\mathrm{d}w^{(-1)}$, $\mathrm{d}w^{(0)}$ and $\mathrm{d}w^{(1)}$. The result then follows from taking the wedge product of the equation above with $\mathrm{d}w^{(-1)} \wedge \mathrm{d}w^{(0)} \wedge \mathrm{d}w^{(1)}$, and noting the identity

$$G(w^{(-1)}, w^{(0)}, w^{(1)}, w^{(2)}) = H(w^{(-1)}, w^{(0)}, w^{(1)}, w^{(2)}),$$

which follows from the symmetry of F. □

When a, b are not both zero, so that the nonlinear terms are present, the above discretization of (24) cannot be obtained from a second order discrete Lagrangian of the form

$$\mathcal{L} = \mathcal{L}(w^{(n)}, w^{(n+1)}, w^{(n+2)}),$$

since the discrete Euler-Lagrange equations

$$\sum_{i=0}^{2} \frac{\partial}{\partial w^{(n)}} \mathcal{L}(w^{(n-i)}, w^{(n+1-i)}, w^{(n+2-i)}) = 0 \tag{30}$$

do not generate terms containing products $w^{(j)}w^{k)}$ with $|j-k| > 2$. In order to obtain a discretization with a Lagrangian structure, we fix $n = 0$ and take a discrete Lagrangian of the form

$$\mathcal{L}(w^{(0)}, w^{(1)}, w^{(2)}) = \mathcal{T} - \mathcal{V},$$

where the discrete fourth derivative is generated by the term

$$\mathcal{T} = \frac{1}{2h^4}\Big(2(w^{(0)} - w^{(1)})^2 - (w^{(0)} - w^{(2)})^2 + 2(w^{(1)} - w^{(2)})^2\Big),$$

and the other terms are specified by

$$\mathcal{V} = \mathcal{V}_5 + \mathcal{V}_3 + \frac{c}{3}\Big(w^{(0)} + w^{(1)} + w^{(2)}\Big)$$

with

$$\begin{aligned}
\tfrac{5}{a}\mathcal{V}_5 = {} & \alpha_0 w^{(0)}(w^{(1)})^3 w^{(2)} + \tfrac{1}{2}\alpha_1\Big((w^{(0)})^2(w^{(1)})^3 + (w^{(1)})^2(w^{(2)})^3\Big) \\
& + \tfrac{1}{2}\alpha_2\Big((w^{(1)})^2(w^{(0)})^3 + (w^{(2)})^2(w^{(1)})^3\Big) \\
& + \tfrac{1}{2}\alpha_3\Big(w^{(0)}(w^{(1)})^4 + w^{(1)}(w^{(2)})^4\Big) \\
& + \tfrac{1}{2}\alpha_4\Big(w^{(1)}(w^{(0)})^4 + w^{(2)}(w^{(1)})^4\Big) \\
& + \tfrac{1}{3}\alpha_5\Big((w^{(0)})^5 + (w^{(1)})^5 + (w^{(2)})^5\Big),
\end{aligned} \tag{31}$$

and

$$\begin{aligned}
\tfrac{3}{b}\mathcal{V}_3 = {} & \beta_0 w^{(0)} w^{(1)} w^{(2)} + \tfrac{1}{2}\beta_1\Big(w^{(0)}(w^{(1)})^2 + w^{(1)}(w^{(2)})^2\Big) \\
& + \tfrac{1}{2}\beta_2\Big(w^{(1)}(w^{(0)})^2 + w^{(2)}(w^{(1)})^2\Big) \\
& + \tfrac{1}{3}\beta_3\Big((w^{(0)})^3 + (w^{(1)})^3 + (w^{(2)})^3\Big),
\end{aligned} \tag{32}$$

where, in (31) and (32) we have taken affine combinations, so that the coefficients are required to satisfy

$$\sum_{j=0}^{5} \alpha_j = 1 = \sum_{j=0}^{3} \beta_j$$

in order to ensure the correct continuum limit, and we have included all possible terms of degrees 5 and 3, respectively, except those whose discrete variational derivative produces expressions of degree greater than one in $w^{(-2)}$ or $w^{(2)}$ (we have also grouped together terms having the same variational derivative). Hence we arrive at a discretization of (24) which is explicit and birational, being given by

$$\frac{w^{(-2)} - 4w^{(-1)} + 6w^{(0)} - 4w^{(1)} + w^{(2)}}{h^4} = \hat{F}_4 + \hat{F}_2 + c, \tag{33}$$

where the quartic and quadratic terms are given by

$$\begin{aligned} \tfrac{5}{a}\hat{F}_4 = \alpha_0&\Big(w^{(-2)}(w^{(-1)})^3 + 3w^{(-1)}(w^{(0)})^2 w^{(1)} + (w^{(1)})^3 w^{(2)}\Big) \\ &+\alpha_1\Big(3(w^{(-1)})^2(w^{(0)})^2 + 2w^{(0)}(w^{(1)})^3\Big) \\ &+\alpha_2\Big(2(w^{(-1)})^3 w^{(0)} + 3(w^{(0)})^2(w^{(1)})^2\Big) \\ &+\alpha_3\Big(4w^{(-1)}(w^{(0)})^3 + (w^{(1)})^4\Big) + \alpha_4\Big((w^{(-1)})^4 + 4(w^{(0)})^3 w^{(1)}\Big) \\ &+5\alpha_5(w^{(0)})^4, \end{aligned} \tag{34}$$

$$\begin{aligned} \tfrac{3}{b}\hat{F}_2 = \beta_0&\Big(w^{(-2)}w^{(-1)} + w^{(-1)}w^{(1)} + w^{(1)}w^{(2)}\Big) \\ &+\beta_1\Big(2w^{(-1)}w^{(0)} + (w^{(1)})^2\Big) \\ &+\beta_2\Big((w^{(-1)})^2 + 2w^{(0)}w^{(1)}\Big) + 3\beta_3(w^{(0)})^2, \end{aligned} \tag{35}$$

respectively. A general approach to Lagrangian fourth-order difference equations and their continuum limits appears in the recent paper Gubbiotti (2020).

An advantage of using the Lagrangian discretization (33) is that it is symplectic; so it is a birational symplectic integrator. This can be seen from the discrete analogue of the Ostrogradsky transformation, introduced in Bruschi et al. (1991), which provides canonical variables q_1, p_1, q_2, p_2 via the formulae

$$\begin{aligned} q_1 &= w^{(0)}, \quad p_1 = \mathcal{L}_1(w^{(-1)}, w^{(0)}, w^{(1)}) + \mathcal{L}_2(w^{(-2)}, w^{(1)}, w^{(0)}), \\ q_2 &= w^{(1)}, \quad p_2 = \mathcal{L}_2(w^{(-1)}, w^{(0)}, w^{(1)}), \end{aligned} \tag{36}$$

where

$$\mathcal{L}_j = \frac{\partial \mathcal{L}}{\partial w^{(j)}}(w^{(0)}, w^{(1)}, w^{(2)}), \qquad j = 0, 1, 2.$$

In terms of these variables, the four-dimensional map defined by (33) preserves the canonical symplectic form

$$\omega = \mathrm{d}p_1 \wedge \mathrm{d}q_1 + \mathrm{d}p_2 \wedge \mathrm{d}q_2,$$

and this immediately implies that it preserves the volume form $\omega \wedge \omega$, so it is measure-preserving.

Qualitatively it appears that the approximate solutions of (24) provided by these two discretizations are somewhat similar. To see this, one can consider solutions in the neighbourhood of a fixed point. If $ab \neq 0$ then, by scaling w and x, the parameters can be taken as

$$a = 1, \qquad b = -2\epsilon, \qquad c = 1 - \delta,$$

with $\epsilon^2 = 1$ and δ arbitrary. Then (24) has fixed points at $w = \pm\sqrt{\epsilon \pm \sqrt{\delta}}$, so that $\delta \geq 0$ is a necessary condition for reality, and then generically there are either four, two or zero real fixed points depending on the choice of $\epsilon = \pm 1$ and the value of δ. In particular, let us take the case

$$\epsilon = 1, \qquad 0 < \delta < 1$$

when there are four real fixed points, one of which is at $w = w^*$, where

$$w^* = \sqrt{1 + \sqrt{\delta}}.$$

The eigenvalues of the linearization of (24) around this point consist of a real pair $\pm\gamma$ and an imaginary pair $\pm i\gamma$, for $\gamma = (4w^*\sqrt{\delta})^{1/4}$, corresponding to one stable direction, one unstable direction, and a two-dimensional centre manifold. The discretizations (26) and (33) both have the same fixed points as the original differential equation, and using the fact that (26) is reversible, and that (33) is symplectic (and also reversible), together with standard facts about linear stability of reversible/symplectic maps (see Howard and Mackay (1987) or Lahiri et al. (1995)), in each case the characteristic polynomial of the linearization around a fixed point is palindromic (equivalently, λ is a root if and only if λ^{-1} is). If we consider the linearization around w^*, then in both cases we find two real eigenvalues that are reciprocals of one another, corresponding to the stable and unstable directions, together with a complex conjugate pair of eigenvalues of modulus one, giving a two-dimensional centre manifold, just as for the differential equation; and similar considerations apply to the other fixed points. Thus, to a first approximation, the qualitative behaviour of the two discretizations is the same.

5 Conclusions

We have found that the higher order analogue of Kahan's method proposed here preserves integrability in the second order example of the quartic oscillator (13) that we have considered, while in the case of a nonlinear beam equation of fourth order the resulting discretization (26) is measure-preserving, and its qualitative behaviour looks similar to that of the Lagrangian discretization (33). In future work we would like to apply this discretization method to other ODE systems of higher order, as well as looking for first integrals of the particular fourth order maps (26) and (33) using the method of discrete Darboux polynomials as described in Celledoni et al. (2019) and Celledoni et al. (2020).

Acknowledgements ANWH is supported by Fellowship EP/M004333/1 from the Engineering & Physical Sciences Research Council, UK, and he thanks the School of Mathematics and Statistics, University of New South Wales, for hosting him as a Visiting Professorial Fellow with funding from the Distinguished Researcher Visitor Scheme. He is also grateful to John Roberts and Wolfgang Schief for providing additional support during his stay in Sydney, where the idea behind this work originated, and to his colleagues at La Trobe for hospitality during his visit to Melbourne in May 2019. Part of this work was carried out at the Isaac Newton Institute, during the GCS programme, supported by EPSRC grant EP/R014604/1. GRWQ is grateful to the Simons Foundation for a grant supporting this work.

References

Błaszak, M.: Multi-Hamiltonian Theory of Dynamical Systems. Springer (1998)

Bruschi, M., Ragnisco, O., Santini, P.M., Tu, G-Z.: Integrable symplectic maps. Physica D **49**, 273–294 (1991)

Celledoni, E., McLachlan, R.I., Owren, B., Quispel, G.R.W.: Geometric properties of Kahan's method. J. Phys. A **46**, 025201 (2013)

Celledoni, E., McLachlan, R.I., McLaren, D.I., Owren, B., Quispel, G.R.W.: Discretization of polynomial vector fields by polarization. Proc. R. Soc. A **471**, 20150390 (2015)

Celledoni, E., Evripidou, C., McLaren, D.I., Owren, B., Quispel, G.R.W., Tapley, B.K., van der Kamp, P.: Using discrete Darboux polynomials to detect and determine preserved measures and integrals of rational maps. J. Phys. A **52**, 31LT01 (2019)

Celledoni, E., Evripidou, C., McLaren, D.I., Owren, B., Quispel, G.R.W., Tapley, B.K.: Detecting and determining preserved measures and integrals of rational maps. arXiv:1902.04685v2

Da Vinci, L.: Codex Madrid, vol. I, 1493

Esen, O., de León, M., Sardón, C.: A Hamilton-Jacobi formalism for higher order implicit systems. J. Phys. **A53**, 075204 (2020)

Gubbiotti, G.: Lagrangians and integrability for additive fourth-order difference equations. arXiv:1910.11458

Han, S.M., Benaroya, H., Wei, T.: Dynamics of transversely vibrating beams using four engineering theories. J. Sound Vib. **225**, 935–988 (1999)

Hirota, R., Kimura, K.: Discretization of the Euler top. J. Phys. Soc. Jpn. **69**, 627–630 (2000)

Hone, A.N.W., Petrera, M.: Three-dimensional discrete systems of Hirota-Kimura type and deformed Lie-Poisson algebras. J. Geom. Mech. **1**, 55–85 (2009)

Howard, J.E., Mackay, R.S.: Linear stability of symplectic maps. J. Math. Phys. **28**, 1036–1051 (1987)

Iatrou, A., Roberts, J.A.G.: Integrable mappings of the plane preserving biquadratic invariant curves. J. Phys. A: Math. Gen. **34**, 6617–6636 (2001)

Kahan, W.: Unconventional numerical methods for trajectory calculations, unpublished lecture notes (1993)

Kahan, W., Li, R.-C.: Unconventional schemes for a class of ordinary differential equations - with applications to the Korteweg-de Vries equation. J. Comp. Phys. **134**, 316–331 (1997)

Lahiri, A., Bhowal, A., Roy, T.K.: Fourth order resonant collisions of multipliers in reversible maps: Period-4 orbits and invariant curves. Physica D **85**, 10–24 (1995)

Mickens, R.E.: Nonstandard finite difference models of differential equations. World Scientific (1994)

Mickens, R.E.: A nonstandard finite-difference scheme for the Lotka-Volterra system. Appl. Num. Math. **45**, 309–314 (2003)

Petrera, M., Pfadler, A., Suris, Yu.B.: On integrability of Hirota-Kimura type discretizations. Regul. Chaotic Dyn. **16**, 245–289 (2011)

Petrera, M., Zander, R.: New classes of quadratic vector fields admitting integral-preserving Kahan-Hirota-Kimura discretizations. J. Phys. A: Math. Theor. **50**, 205203 (2017)

Petrera, M., Smirin, J.: Suris, Yu.B: Geometry of the Kahan discretizations of planar quadratic Hamiltonian systems. Proc. R. Soc. A **475**, 20180761 (2019)

Potts, R.B.: Best difference equation approximation to Duffing's equation. J. Austral. Math. Soc. Ser. B **23**, 349–356 (1982)

Potts, R.B.: Weierstrass elliptic difference equations. J. Austral. Math. Soc. Ser. B **35**, 43–48 (1987)

Quispel, G.R.W., Roberts, J.A.G., Thompson, C.J.: Integrable mappings and soliton equations. Phys. Lett. A **126**, 419–421 (1988)

Sanz-Serna, J.M.: An unconventional symplectic integrator of W. Kahan. Appl. Num. Math. **16**, 245–250 (1994)

Shi, Y., Xu, J.: Quasi-periodic solutions for a class of coupled beam equation system. Discrete Cont. Dyn. B **25**, 31–53 (2020)

On Some Explicit Representations of the Elliptic Painlevé Equation

Masatoshi Noumi

Abstract We explain some details of the derivation of explicit representations for the elliptic Painlevé equation in the $\mathbb{P}^1 \times \mathbb{P}^1$ picture. We also discuss possibility of expressing the elliptic Painlevé equation in the factorized form. This paper is intended to be a supplement to the topical review *Geometric aspects of Painlevé equations* (Kajiwara et al. 2017).

Keywords Elliptic Painlevé equation · τ Function · Affine Weyl group

1 Introduction

This paper is intended to be a supplement to Sect. 5 of the topical review *Geometric aspects of Painlevé equations* (Kajiwara et al. 2017) by K. Kajiwara, Y. Yamada and the author. As explained in Kajiwara et al. (2017), the discrete Painlevé equations with affine Weyl group symmetries can be formulated in terms of point configurations in the product $\mathbb{P}^1 \times \mathbb{P}^1$ of two copies of the projective line. In this introduction, we briefly recall some basic facts concerning the elliptic Painlevé equation with affine Weyl group symmetry of type $E_8^{(1)}$ in the $\mathbb{P}^1 \times \mathbb{P}^1$ picture. We explain in the next section some details of the derivation of explicit representations as proposed in Kajiwara et al. (2017), for the elliptic Painlevé equation with respect to the translation by $\alpha_1 = H_x - H_y$. We also discuss possibility of expressing the elliptic Painlevé equation in the factorized form. For various approaches to the elliptic Painlevé equations, we refer the reader to Ohta-Ramani-Grammaticos (2001), Sakai (2001), Kajiwara et al. (2006), Noumi-Tsujimoto-Yamada (2013) and Noumi (2018).

Dedicated to Professor Nalini Joshi on her sixtieth birthday.

M. Noumi (✉)
Department of Mathematics, Kobe University, Rokko, Kobe 657-8501, Japan
e-mail: noumi@math.kobe-u.ac.jp

© Springer Nature Switzerland AG 2020
F. Nijhoff et al. (eds.), *Asymptotic, Algebraic and Geometric Aspects of Integrable Systems*, Springer Proceedings in Mathematics & Statistics 338,
https://doi.org/10.1007/978-3-030-57000-2_10

1.1 Affine Weyl Group $W(E_8^{(1)})$ and Kac Translations

We consider the free $\mathbb{Z}$-module of rank 10

$$L = \mathbb{Z}H_x \oplus \mathbb{Z}H_y \oplus \mathbb{Z}E_1 \oplus \cdots \oplus \mathbb{Z}E_8 \tag{1}$$

endowed with the scalar product (non-degenerate symmetric bilinear form) $(\ |\ ):$ $L \times L \to \mathbb{Z}$ such that

$$\begin{aligned} &(H_x \mid H_x) = (H_y \mid H_y) = 0, \quad (H_x \mid H_y) = (H_y \mid H_x) = -1, \\ &(E_i \mid E_j) = \delta_{i,j} \quad (i, j = 1, \ldots, 8), \\ &(H_x \mid E_j) = (E_j \mid H_x) = 0, \quad (H_y \mid E_j) = (E_j \mid H_y) = 0 \quad (j = 1, \ldots, 8). \end{aligned} \tag{2}$$

In geometric terms, L is the *Picard lattice* associated with the blowup of $\mathbb{P}^1 \times \mathbb{P}^1$ at generic eight points $p_1, p_2, \ldots, p_8$, and $(\ |\ )$ denotes the intersection number of divisor classes multiplied by -1. In the inhomogeneous coordinates (x, y) of $\mathbb{P}^1 \times \mathbb{P}^1$, H_x and H_y represent the classes of lines $x =$ const. and $y =$ const. respectively, and each E_j corresponds to the exceptional divisor attached to p_j for $j = 1, \ldots, 8$. In this paper we use the notation H_x, H_y instead of H_1, H_2 in Kajiwara et al. (2017). We denote $\mathfrak{h} = L \otimes_{\mathbb{Z}} \mathbb{C}$ the complexification of L, and naturally extend the scalar product $(\ |\ )$ on L to a symmetric $\mathbb{C}$-bilinear form on $\mathfrak{h}$. We realize the *simple roots* $\alpha_0, \alpha_1, \ldots, \alpha_8$ of type $E_8^{(1)}$ in the Picard lattice L as

$$\begin{aligned} &\alpha_0 = E_1 - E_2, \quad \alpha_1 = H_x - H_y, \quad \alpha_2 = H_y - E_1 - E_2, \\ &\alpha_j = E_{j-1} - E_j \quad (j = 3, 4, \ldots, 8). \end{aligned} \tag{3}$$

They satisfy

$$\begin{aligned} &(\alpha_i \mid \alpha_i) = 2 \qquad (0 \le i \le 8), \\ &(\alpha_0 \mid \alpha_3) = -1, \quad (\alpha_0 \mid \alpha_j) = 0 \quad (1 \le j \le 8;\ j \ne 3), \\ &(\alpha_i \mid \alpha_{i+1}) = -1 \quad (1 \le i \le 7), \\ &(\alpha_i \mid \alpha_j) = 0 \quad (1 \le i, j \le 8;\ |j - i| \ge 2). \end{aligned} \tag{4}$$

In terms of the Dynkin diagram

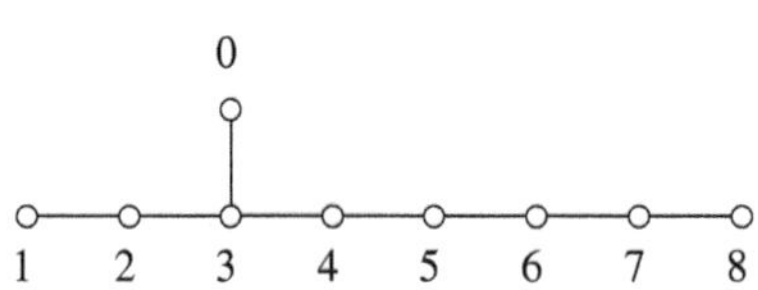

(5)

$(\alpha_i \mid \alpha_i) = 2$ for each index i, and $(\alpha_i \mid \alpha_j) = -1$ or 0 for distinct indices i, j, according as the corresponding nodes are connected by an edge or not. We denote by

$$Q = Q(E_8^{(1)}) = \mathbb{Z}\alpha_0 \oplus \mathbb{Z}\alpha_1 \oplus \cdots \oplus \mathbb{Z}\alpha_8 \subset L \tag{6}$$

the *root lattice* of type $E_8^{(1)}$. This Q contains a special element

$$\begin{aligned} \delta &= 2H_x + 2H_y - E_1 - \cdots - E_8 \\ &= 3\alpha_0 + 2\alpha_1 + 4\alpha_2 + 6\alpha_3 + 5\alpha_4 + 4\alpha_5 + 3\alpha_6 + 2\alpha_7 + \alpha_8, \end{aligned} \tag{7}$$

called the *null root*. It satisfies $(\delta \,|\, \alpha_j) = 0$ $(j = 0, 1, \ldots, 8)$, and hence $(\delta \,|\, \delta) = 0$. We remark that $\alpha_0, \alpha_1, \ldots, \alpha_7$ are the simple roots of the E_8 root system. Note also that

$$Q(E_8^{(1)}) = Q(E_8) \oplus \mathbb{Z}\delta, \quad Q(E_8) = \mathbb{Z}\alpha_0 \oplus \mathbb{Z}\alpha_1 \oplus \cdots \oplus \mathbb{Z}\alpha_7. \tag{8}$$

We denote by $W = W(E_8^{(1)}) = \langle s_0, s_1, \ldots, s_8 \rangle$ the affine Weyl group of type $E_8^{(1)}$. By definition this group is generated by the *simple reflections* s_j $(j = 0, 1, \ldots, 8)$ subject to the fundamental relations

$$\begin{aligned} &s_i^2 = 1 \quad (0 \le i \le 8), \\ &s_0 s_3 s_0 = s_3 s_0 s_3, \quad s_0 s_j = s_j s_0 \quad (1 \le j \le 8;\; j \ne 3), \\ &s_i s_{i+1} s_i = s_{i+1} s_i s_{i+1} \quad (1 \le i \le 7), \\ &s_i s_j = s_j s_i \quad (1 \le i, j \le 8;\; |j - i| \ge 2). \end{aligned} \tag{9}$$

Namely, $s_i^2 = 1$ for each index i; for distinct indices i, j, s_i and s_j satisfy the braid relation $s_i s_j s_i = s_i s_j s_i$ if the two nodes are connected by an edge, and they commute otherwise. For each $\alpha \in \mathfrak{h}$ with $(\alpha \,|\, \alpha) \ne 0$, we define the *reflection* $r_\alpha : \mathfrak{h} \to \mathfrak{h}$ with respect to α by

$$r_\alpha(h) = h - \left(h \,|\, \alpha^\vee\right)\alpha \quad (h \in \mathfrak{h}), \tag{10}$$

where $\alpha^\vee = 2\alpha/(\alpha \,|\, \alpha)$. Note that r_α is a $\mathbb{C}$-linear isometry: $\left(r_\alpha(h) \,|\, r_\alpha(h')\right) = \left(h \,|\, h'\right)$ for any $h, h' \in \mathfrak{h}$. In this setting, the reflections $s_j = r_{\alpha_j} : \mathfrak{h} \to \mathfrak{h}$ $(j = 0, 1, \ldots, 8)$ by the simple roots satisfy the fundamental relations (9). In fact it is known that the Weyl group $W = \langle s_0, s_1, \ldots, s_8 \rangle$ is isomorphic to the subgroup of $\mathrm{GL}(\mathfrak{h})$ generated by $s_j = r_{\alpha_j}$ $(j = 0, 1, \ldots, 8)$. We remark that each $w \in W$ defines an isometry $w : \mathfrak{h} \xrightarrow{\sim} \mathfrak{h}$; in other words, the scalar product $(\ |\ ) : \mathfrak{h} \times \mathfrak{h} \to \mathbb{C}$ is W-invariant. This action of W on $\mathfrak{h}$ stabilizes the Picard lattice L, as well as the root lattice $Q = Q(E_8^{(1)})$. Since $(\delta \,|\, \alpha_j) = 0$ $(j = 0, 1, \ldots, 8)$, the null root δ is invariant under the action of W. The simple reflections s_i $(i = 0, 1, \ldots, 8)$ act on H_x, H_y and E_j $(j = 1, \ldots, 8)$ as follows:

$$\begin{aligned}
&s_0(H_x)=H_x,\ \ s_0(H_y)=H_y,\ \ s_0(E_j)=E_{(12)j},\ \ (j=1,2,\ldots,8),\\
&s_1(H_x)=H_y,\ \ s_1(H_y)=H_x,\ \ s_1(E_j)=E_j\ \ (j=1,2,\ldots,8),\\
&s_2(H_x)=H_x+H_y-E_1-E_2,\ \ s_2(H_y)=H_y,\\
&s_2(E_1)=H_y-E_2,\ \ s_2(E_2)=H_y-E_1,\ \ s_2(E_j)=E_j\ \ (j=3,4,\ldots,8),
\end{aligned} \tag{11}$$

and for $i=3,4,\ldots,8$,

$$s_i(H_x)=H_x,\ \ s_i(H_y)=H_y,\ \ s_i(E_j)=E_{(i-1,i)j},\ \ (j=1,2,\ldots,8), \tag{12}$$

where (i,j) denotes the transposition of i and j. For example,

$$\begin{aligned}
&\alpha_8=E_7-E_8\xrightarrow{s_8s_7}\alpha_7=E_6-E_7\xrightarrow{s_7s_6}\cdots\xrightarrow{s_4s_3}\alpha_3=E_2-E_3\xrightarrow{s_0s_3}\alpha_0=E_1-E_2,\\
&\alpha_3=E_2-E_3\xrightarrow{s_3s_2}\alpha_2=H_y-E_1-E_2\xrightarrow{s_2s_1}\alpha_1=H_x-H_y.
\end{aligned} \tag{13}$$

We denote by $\Delta^{\mathrm{re}}=W.\alpha_8=W.\{\alpha_0,\alpha_1,\ldots,\alpha_8\}\subset Q$ the set of *real roots* of type $E_8^{(1)}$, which is the W-orbit of simple roots in Q. Since $(\alpha_i\,|\,\alpha_i)=2$ for all $i=0,1,\ldots,8$, all the real roots $\alpha\in\Delta^{\mathrm{re}}$ satisfy $(\alpha\,|\,\alpha)=2$ in this root system. We list here some typical real roots:

$$\begin{aligned}
&H_x-H_y,\ \ E_i-E_j\ \ (1\le i<j\le 8),\\
&H_x-E_i-E_j,\ \ H_y-E_i-E_j\ \ (1\le i<j\le 8),\\
&H_x+H_y-E_{j_1}-E_{j_2}-E_{j_3}-E_{j_4}\ \ (1\le j_1<j_2<j_3<j_4\le 8).
\end{aligned} \tag{14}$$

For instance, for $2\le j\le 8$

$$H_y-E_1-E_j=w(H_y-E_1-E_2),\ \ H_x-E_1-E_j=s_1w(H_y-E_1-E_2) \tag{15}$$

with $w=s_js_{j-1}\cdots s_3$, and for $2\le i<j\le 8$ we have

$$H_y-E_i-E_j=w(H_y-E_1-E_2),\ \ H_x-E_i-E_j=s_1w(H_y-E_1-E_2) \tag{16}$$

with $w=s_is_{i-1}\cdots s_3s_0s_js_{j-1}\cdots s_3$. If $\alpha,\beta\in\Delta^{\mathrm{re}}$ and $w(\alpha)=\beta$ for some $w\in W$, then $wr_\alpha w^{-1}=r_\beta$. This implies that the reflections r_α for all real roots are contained in $W=\langle s_0,s_1,s_2,\ldots,s_8\rangle$. We remark that the affine Weyl group $W=W(E_8^{(1)})=\langle s_0,s_1,\ldots,s_8\rangle$ contains the symmetric group $\mathfrak{S}_8=\langle s_0,s_3,s_4,\ldots,s_8\rangle$ of degree 8 which permutes E_j $(j=1,\ldots,8)$. It also contains the finite Weyl group $W(E_8)=\langle s_0,s_1,s_2,\ldots,s_7\rangle$ of type E_8. The set of E_8 roots is denoted by $\Delta(E_8)=W(E_8).\alpha_7=W(E_8)\{\alpha_0,\alpha_1,\ldots,\alpha_7\}\subset Q(E_8)$. The element

$$\phi=\delta-\alpha_8=3\alpha_0+2\alpha_1+4\alpha_2+6\alpha_3+5\alpha_4+4\alpha_5+3\alpha_6+2\alpha_7\in\Delta(E_8) \tag{17}$$

is called the *highest root* of type E_8.

Setting $\mathfrak{h}_0 = \{h \in \mathfrak{h} \mid (\delta \mid h) = 0\}$, for each $\alpha \in \mathfrak{h}$ with $(\delta \mid \alpha) = 0$, we define the *Kac translation* $T_\alpha : \mathfrak{h} \to \mathfrak{h}$ by

$$T_\alpha(h) = h + (\delta \mid h)\,\alpha - \big(\tfrac{1}{2}\,(\alpha \mid \alpha)\,(\delta \mid h) + (\alpha \mid h)\big)\,\delta \quad (h \in \mathfrak{h}). \tag{18}$$

These linear transformations $T_\alpha \in \mathrm{GL}(\mathfrak{h})$ $(\alpha \in \mathfrak{h}_0)$ satisfy

$$\begin{aligned}
&(1)\quad \big(T_\alpha(h) \,|\, T_\alpha(h')\big) = \big(h \,|\, h'\big) \quad (\alpha \in \mathfrak{h}_0;\;\; h, h' \in \mathfrak{h}),\\
&(2)\quad T_\alpha T_\beta = T_\beta T_\alpha = T_{\alpha+\beta} \quad (\alpha, \beta \in \mathfrak{h}_0); \quad T_{k\delta} = 1 \quad (k \in \mathbb{Z}),\\
&(3)\quad w T_\alpha w^{-1} = T_{w(\alpha)} \quad (\alpha \in \mathfrak{h}_0;\, w \in W),\\
&(4)\quad \alpha \in \mathfrak{h}_0,\, (\alpha \,|\, \alpha) \neq 0 \quad \Longrightarrow \quad T_\alpha = r_{\delta-\alpha^\vee} r_{\alpha^\vee}.
\end{aligned} \tag{19}$$

Applying (4) to $\alpha = \phi$ and $\delta - \alpha = \alpha_8$, we have $T_\phi = r_{\delta-\phi} r_\phi \in W$. This implies $T_\alpha \in W$ for all $\alpha \in \Delta^{\mathrm{re}}$, and hence for all $\alpha \in Q$. Noting that $T_{k\delta} = 1$ $(k \in \mathbb{Z})$, we see that $W = W(E_8^{(1)})$ contains the commutative subgroup $T(Q(E_8)) = \{T_\alpha \mid \alpha \in Q(E_8)\}$ of Kac translations. Furthermore, it decomposes into the semi-direct product $W(E_8^{(1)}) = T(Q(E_8)) \rtimes W(E_8)$; this means that any $w \in W(E_8^{(1)})$ is uniquely expressed in the form $w = T_\alpha v$, where $\alpha \in Q(E_8)$ and $v \in W(E_8)$.

The $\mathbb{C}$-vector space $\mathfrak{h} = L \otimes_{\mathbb{Z}} \mathbb{C}$ can regarded as the Cartan subalgebra of the affine Lie algebra of type $E_8^{(1)}$. We also consider the dual space

$$\mathfrak{h}^* = \mathrm{Hom}_{\mathbb{C}}(\mathfrak{h}, \mathbb{C}) = \mathbb{C}h_x \oplus \mathbb{C}h_y \oplus \mathbb{C}e_1 \oplus \cdots \oplus \mathbb{C}e_8 \tag{20}$$

of $\mathfrak{h}$ with the basis defined by the linear functions $h_x = (H_x \mid \cdot)$, $h_y = \big(H_y \mid \cdot\big)$ and $e_j = \big(E_j \mid \cdot\big)$ $(j = 1, \ldots, 8)$. We remark that the $\mathbb{C}$-isomorphism

$$\iota :\; \mathfrak{h} \xrightarrow{\sim} \mathfrak{h}^*, \quad \iota(h)(h') = \big(h \,|\, h'\big) \quad (h, h' \in \mathfrak{h}) \tag{21}$$

induced by the scalar product is in fact a W-isomorphism; namely, W acts on the linear functions h_x, h_y, e_j $(j = 1, \ldots, 8)$ in the same way as it does on H_x, H_y, E_j $(j = 1, \ldots, 8)$. As to the null root, we use the same symbol $\delta = 2h_x + 2h_y - e_1 - \cdots - e_8$, regarding it as a constant. This δ plays the role of the unit length of difference equations.

In the following, we regard $(h_x, h_y, e_1, \ldots, e_8)$ as a coordinate system of the 10-dimensional $\mathbb{C}$-vector space $\mathfrak{h}$. We denote by $\mathcal{M}(\mathfrak{h})$ the field of meromorphic functions on $\mathfrak{h}$. Then the affine Weyl group $W = \langle s_0, s_1, \ldots, s_8\rangle$ acts naturally on $\mathcal{M}(\mathfrak{h})$; for each $\varphi \in \mathcal{M}(\mathfrak{h})$, $w(\varphi)(h) = \varphi(w.h)$ for generic $h \in \mathfrak{h}$.

1.2 *Reference Curve of Bidegree* (2, 2) *in* $\mathbb{P}^1 \times \mathbb{P}^1$

We fix an additive subgroup $\Omega = \mathbb{Z}\omega_1 \oplus \mathbb{Z}\omega_2 \subset \mathbb{C}$ of rank 2, generated by nonzero complex numbers $\omega_1, \omega_2 \in \mathbb{C}$ such that $\mathrm{Im}(\omega_2/\omega_1) > 0$. In the following, we denote by

$$\sigma(u) = \sigma(u|\Omega) = u \prod_{\omega\in\Omega;\ \omega\neq 0} \left(1 - \frac{u}{\omega}\right) e^{u^2/2\omega^2 + u/\omega} \qquad (u \in \mathbb{C}) \tag{22}$$

the *Weierstrass sigma function* associated with the period lattice Ω. It is an odd entire function and has a simple zero at each point $\omega \in \Omega$; $\sigma(u)$ is normalized by the condition $\sigma(u) = u + O(u^5)$ as $u \to 0$. With the notation $e(u) = e^{2\pi\sqrt{-1}u}$, $\sigma(u)$ satisfies quasi-periodicity of the form

$$\sigma(u + \omega) = \epsilon_\omega e\big(\lambda_\omega(u + \tfrac{\omega}{2})\big)\, \sigma(u) \qquad (\omega \in \Omega), \tag{23}$$

where $\epsilon_\omega = +1$ or -1 according as $\omega \in 2\Omega$ or $\notin 2\Omega$, and $\lambda_\omega \in \mathbb{C}$ are complex constants depending additively on $\omega \in \Omega$. Note also that $\lambda_{\omega_1}\omega_2 - \lambda_{\omega_2}\omega_1 = 1$ (Legendre relation).

For $n \in \mathbb{Z}_{\geq 0}$ and $\mu \in \mathbb{C}$, we denote by $\mathcal{O}(\mathbb{C}; n, \mu)$ the $\mathbb{C}$-vector space of all holomorphic functions $f(u)$ satisfying the quasi-periodicity *of type* (n, μ) in the sense that

$$f(u + \omega) = (\epsilon_\omega)^n e\left(\lambda_\omega(n(u + \tfrac{\omega}{2}) - \mu)\right) f(u) \quad (\omega \in \Omega). \tag{24}$$

A typical quasi-periodic function of type (n, μ) is given by

$$f(u) = \text{const.}\, \sigma(u - a_1)\sigma(u - a_2) \cdots \sigma(u - a_n), \quad a_1 + \cdots + a_n = \mu. \tag{25}$$

We remark that $\dim_{\mathbb{C}} \mathcal{O}(\mathbb{C}; n, \mu) = n$ for $n > 0$. In fact, if we take an $n$tuple of generic points $b_1, \ldots, b_n \in \mathbb{C}$, then any $f(u) \in \mathcal{O}(\mathbb{C}; n, \mu)$ is expressed as

$$f(u) = \sum_{i=1}^{n} f(b_i) \frac{\sigma(u - b_i + \nu - \mu)}{\sigma(\nu - \mu)} \prod_{j\neq i} \frac{\sigma(u - b_j)}{\sigma(b_i - b_j)}, \tag{26}$$

where $\nu = b_1 + \cdots + b_n$. This Lagrange interpolation formula at $u = b_1, \ldots, b_n$ is equivalent to the partial fraction decomposition

$$\frac{f(u)}{\prod_{j=1}^{n} \sigma(u - b_j)} = \sum_{i=1}^{n} \frac{\sigma(u - b_i + \nu - \mu)}{\sigma(u - b_i)\sigma(\nu - \mu)} \frac{f(b_i)}{\prod_{j\neq i} \sigma(b_i - b_j)}. \tag{27}$$

Note also that $\dim_{\mathbb{C}} \mathcal{O}(\mathbb{C}; 0, \mu) = 1$ or 0, according as $\mu \in \Omega$ or $\notin \Omega$; if $\mu = m_1\omega_1 + m_2\omega_2 \in \Omega$, then $\mathcal{O}(\mathbb{C}; 0, \mu) = \mathbb{C}\, e(-\alpha u)$ where $\alpha = m_1\lambda_{\omega_1} + m_2\lambda_{\omega_2}$.

In the following, we denote by (ξ, η) homogeneous coordinates of $\mathbb{P}^1 \times \mathbb{P}^1$, where $\xi = (\xi_1 : \xi_2)$ and $\eta = (\eta_1 : \eta_2)$. Among the coordinates $h_x, h_y, e_1, e_2, \ldots, e_8$ of $\mathfrak{h}$, we use four parameters h_x, h_y, e_1, e_2 to define a curve $C_0 \subset \mathbb{P}^1 \times \mathbb{P}^1$ of bidegree $(2, 2)$, which we call the *reference curve*. By the parameters $h_x, h_y \in \mathbb{C}$, we define two functions

$$\varphi(t,u)=\sigma(t-u)\sigma(h_x-t-u),\quad \psi(t,u)=\sigma(t-u)\sigma(h_y-t-u) \tag{28}$$

in $(t,u)\in\mathbb{C}\times\mathbb{C}$. Regarded as functions in u, $\varphi(t,u)$ and $\psi(t,u)$ are quasi-periodic functions of type $(2,h_x)$ and of type $(2,h_y)$, respectively. Also, note that $\varphi(u,t)=-\varphi(t,u)$, $\varphi(t,t)=0$ and $\psi(u,t)=-\psi(t,u)$, $\psi(t,t)=0$. A characteristic property of $\varphi(t,u)$, as well as $\psi(t,u)$, is the three-term relation (of Hirota type)

$$\varphi(a,u)\varphi(b,c)+\varphi(b,u)\varphi(c,a)+\varphi(c,u)\varphi(a,b)=0, \tag{29}$$

or equivalently,

$$\varphi(c;u)=\frac{1}{\varphi(a,b)}\left(\varphi(b,u)\varphi(a,c)-\varphi(b,c)\varphi(a,u)\right). \tag{30}$$

Since $\dim_{\mathbb{C}}\mathcal{O}(\mathbb{C};2,h_x)=2$, three functions $\varphi(a,u),\varphi(b,u),\varphi(c,u)$ in this vector space should satisfy a nontrivial linear relation. Formula (30) is a special case of (26) where $f(u)=\varphi(c,u)$ and $b_1=a$, $b_2=b$. When we specialize t to e_i, we write

$$\varphi_i(u)=\varphi(e_i,u),\quad \psi_i(u)=\psi(e_i,u)\quad (i=1,\dots,8) \tag{31}$$

for short. With the parameters $e_1,e_2\in\mathbb{C}$, we consider the holomorphic mapping $p:\mathbb{C}\to\mathbb{P}^1\times\mathbb{P}^1$ specified as

$$p(u)=(\varphi(u),\psi(u)),\quad \varphi(u)=(\varphi_1(u):\varphi_2(u)),\quad \psi(u)=(\psi_1(u):\psi_2(u)), \tag{32}$$

by the substitution $\xi_i=\varphi_i(u),\eta_i=\psi_i(u)$ $(i=1,2)$, and set $C_0=p(\mathbb{C})\subset\mathbb{P}^1\times\mathbb{P}^1$. In terms of the inhomogeneous coordinates (x,y) of $\mathbb{P}^1\times\mathbb{P}^1$ such that $x=\xi_2/\xi_1$, $y=\eta_2/\eta_1$, we are considering the reference curve $C_0\subset\mathbb{P}^1\times\mathbb{P}^1$ defined by the parametrization

$$C_0:\left\{\begin{aligned} x&=\frac{\varphi_2(u)}{\varphi_1(u)}=\frac{\sigma(e_2-u)\sigma(h_x-e_2-u)}{\sigma(e_1-u)\sigma(h_x-e_1-u)},\\ y&=\frac{\psi_2(u)}{\psi_1(u)}=\frac{\sigma(e_2-u)\sigma(h_y-e_2-u)}{\sigma(e_1-u)\sigma(h_y-e_1-u)}.\end{aligned}\right. \tag{33}$$

Note that this curve passes through $p_1=p(e_1)=(\infty,\infty)$ and $p_2=p(e_2)=(0,0)$. For generic $h_x,h_y,e_1,e_2\in\mathbb{C}$, one can verify that the mapping $p:\ \mathbb{C}\to\mathbb{P}^1\times\mathbb{P}^1$ induces the isomorphism $E_\Omega=\mathbb{C}/\Omega\xrightarrow{\sim}C_0$ of elliptic curves.

This reference curve $C_0\subset\mathbb{P}^1\times\mathbb{P}^1$ is expressed as the zero locus of a homogenous polynomial $P(\xi,\eta)$ in ξ and η of bidegree (2, 2). Under the substitution $\xi=\varphi(u)$, $\eta=\psi(u)$, as functions in u, the monomials

$$\xi_1^{2-i}\xi_2^i\eta_1^{2-j}\eta_2^j=\varphi_1(u)^{2-i}\varphi_2(u)^i\psi_1(u)^{2-j}\psi_2(u)^j\quad (i,j=0,1,2) \tag{34}$$

of bidegree $(2, 2)$ are quasi-periodic of type $(8, 2h_x + 2h_y)$. Since $\dim_{\mathbb{C}} \mathcal{O}(\mathbb{C}; 8, 2h_x + 2h_y) = 8$, the nine functions in (34) are subject to a nontrivial linear relation of the form

$$P(\xi, \eta) = \sum_{i,j=0}^{2} c_{i,j}\, \xi_1^{2-i} \xi_2^{i} \eta_1^{2-j} \eta_2^{j} = 0. \tag{35}$$

For generic h_x, h_y, e_1, e_2, the coefficients $(c_{i,j})_{i,j=0}^{2}$ are determined as follows uniquely up to a multiplicative constant:

$$\begin{aligned} c_{0,0} &= 0, \quad c_{0,1} = -\frac{\sigma(h_x - 2e_2)}{\sigma(h_x - e_1 - e_2)}, \quad c_{0,2} = \frac{\sigma(h_x - h_y + e_1 - e_2)}{\sigma(h_x - h_y)}, \\ c_{1,0} &= \frac{\sigma(h_y - 2e_2)}{\sigma(h_y - e_1 - e_2)}, \quad c_{1,2} = -\frac{\sigma(h_y - 2e_1)}{\sigma(h_y - e_1 - e_2)}, \\ c_{2,0} &= -\frac{\sigma(h_x - h_y - e_1 + e_2)}{\sigma(h_x - h_y)}, \quad c_{2,1} = \frac{\sigma(h_x - 2e_1)}{\sigma(h_x - e_1 - e_2)}, \quad c_{2,2} = 0. \end{aligned} \tag{36}$$

and $c_{1,1}$ is expressed as

$$c_{1,1} = -\sum_{(i,j)\neq(1,1)} c_{ij}\varphi_1(t)^{1-i}\varphi_2(t)^{i-1}\psi_1(t)^{1-j}\psi_2(t)^{j-1}, \tag{37}$$

for a generic $t \in \mathbb{C}$; the right-hand side of (37) does not depend on the choice of t. From the condition that C_0 passes through $p_1 = (\infty, \infty)$ and $p_2 = (0, 0)$, it follows that $c_{2,2} = 0$ and $c_{0,0} = 0$. In terms of the inhomogeneous coordinates (x, y), Eq. (35) is then written as

$$F(x, y) = c_{0,1}x^{-1} + c_{0,2}x^{-1}y + c_{1,0}y^{-1} + c_{1,1} + c_{1,2}y + c_{2,0}xy^{-1} + c_{2,1}x = 0. \tag{38}$$

If we set $G(x, y) = F(x, y) - c_{1,1}$, $G(x, y)$ should be constant after the substitution $x = \varphi(u) = \varphi_2(u)/\varphi_1(u)$, $y = \psi(u) = \psi_2(u)/\psi_1(u)$. By the condition for the residues of $G(\varphi(u), \psi(u))$ at the six points $u = e_i, h_x - e_i, h_y - e_i$ $(i = 1, 2)$ to be zero, the constants $c_{i,j}$ except $c_{1,1}$ are determined as (36). Then $G(x, y)$ is given explicitly by

$$\begin{aligned} &G(x, y) \\ &= -\frac{\sigma(h_x - 2e_2)}{\sigma(h_x - e_1 - e_2)}x^{-1} + \frac{\sigma(h_x - h_y + e_1 - e_2)}{\sigma(h_x - h_y)}x^{-1}y - \frac{\sigma(h_y - 2e_1)}{\sigma(h_y - e_1 - e_2)}y \\ &\quad + \frac{\sigma(h_y - 2e_2)}{\sigma(h_y - e_1 - e_2)}y^{-1} - \frac{\sigma(h_x - h_y - e_1 + e_2)}{\sigma(h_x - h_y)}xy^{-1} + \frac{\sigma(h_x - 2e_1)}{\sigma(h_x - e_1 - e_2)}x, \end{aligned} \tag{39}$$

and the defining equation of the reference curve C_0 is expressed as $G(x, y) = \text{const}$. We remark that, setting

$$x = \frac{\varphi_2(e_3)}{\varphi_1(e_3)} f, \quad y = \frac{\psi_2(e_3)}{\psi_1(e_3)} g, \tag{40}$$

one can pass to the original inhomogeneous coordinates (f, g) of Kajiwara et al. (2017) in which $p_1 = (\infty, \infty)$, $p_2 = (0, 0)$ and $p_3 = p(e_3) = (1, 1)$.

1.3 Homogeneous Coordinates and τ-Functions

In the following we use the parameters $(h_x, h_y, e_1, \ldots, e_8)$ as the coordinate system of the $\mathbb{C}$-vector space $\mathfrak{h}$. Keeping the notations

$$\begin{aligned} \varphi(t, u) &= \sigma(t - u)\sigma(h_x - t - u) \in \mathcal{O}(\mathbb{C}; 2, h_x), \\ \psi(t, u) &= \sigma(t - u)\sigma(h_y - t - u) \in \mathcal{O}(\mathbb{C}; 2, h_y), \end{aligned} \tag{41}$$

we set

$$\varphi_i(u) = \varphi(e_i, u), \quad \psi_i(u) = \psi(e_i, u) \qquad (i = 1, 2, \ldots, 8). \tag{42}$$

Also, we define the *reference curve* $p : E_\Omega = \mathbb{C}/\Omega \to \mathbb{P}^1 \times \mathbb{P}^1$ by

$$C_0: \ \xi_i = \varphi_i(u), \quad \eta_i = \psi_i(u) \quad (u \in \mathbb{C}; i = 1, 2). \tag{43}$$

We simply express this parametrization as $\xi = \varphi(u)$, $\eta = \psi(u)$. Also, for a homogeneous polynomial $P = P(\xi, \eta)$ in ξ and η, we denote this substitution by

$$P\big|_{C_0} = P(\xi, \eta)\big|_{\xi=\varphi(u), \eta=\psi(u)} = P(\varphi(u), \psi(u)) \qquad (u \in \mathbb{C}). \tag{44}$$

We define eight points $p_1, \ldots, p_8 \in C_0$ by $p_j = p(e_j) = (\varphi(e_j), \psi(e_j))$ $(j = 1, \ldots, 8)$. We denote by

$$\mathbb{K} = \mathbb{C}(\sigma(\alpha); \alpha \in \Delta^{\mathrm{re}}) \subset \mathcal{M}(\mathfrak{h}) \tag{45}$$

the field of meromorphic functions generated by $\sigma(\alpha) = \sigma((\alpha \,|\, \cdot))$ $(\alpha \in \Delta^{\mathrm{re}})$. Note that the affine Weyl group $W = \langle s_0, s_1, \ldots, s_8 \rangle$ acts naturally on $\mathbb{K}$ as a group of $\mathbb{C}$-automorphism such that $w(\sigma(\alpha)) = \sigma(w.\alpha)$.

Let $\mathcal{K} = \mathbb{K}(\xi_1, \xi_2, \eta_1, \eta_2; \tau_1, \ldots, \tau_8)$ be the field of rational functions in the homogeneous coordinates ξ_i, η_i $(i = 1, 2)$ and τ variables τ_j $(j = 1, 2, \ldots, 8)$ with coefficients in $\mathbb{K}$. We extend the action of s_i $(i = 0, 1, \ldots, 8)$ on $\mathbb{K}$ to $\mathcal{K} = \mathbb{K}(\xi, \eta; \tau)$ as follows:

$$\begin{aligned}
&s_0(\xi_1)=\xi_2,\quad s_0(\xi_2)=\xi_1,\quad s_0(\eta_1)=\eta_2,\quad s_0(\eta_2)=\eta_1,\quad s_0(\tau_j)=\tau_{(1,2)j},\\
&s_1(\xi_1)=\eta_1,\quad s_1(\xi_2)=\eta_2,\quad s_1(\eta_1)=\xi_2,\quad s_1(\eta_2)=\xi_1,\quad s_1(\tau_j)=\tau_j,\\
&s_2(\xi_1)=\frac{\xi_1\eta_2}{\tau_1\tau_2},\quad s_2(\xi_2)=\frac{\xi_2\eta_1}{\tau_1\tau_2},\quad s_2(\eta_1)=\eta_2,\quad s_2(\eta_2)=\eta_1,\\
&s_2(\tau_1)=\frac{\eta_2}{\tau_2},\quad s_2(\tau_2)=\frac{\eta_1}{\tau_1},\quad s_2(\tau_j)=\tau_j\quad(j=3,4,\ldots,8),\\
&s_3(\xi_1)=\xi_1,\quad s_3(\xi_2)=\xi_3,\quad s_3(\eta_1)=\eta_1,\quad s_3(\eta_2)=\eta_2,\quad s_3(\tau_j)=\tau_{(2,3)j},
\end{aligned}\tag{46}$$

and for $i=4,5,\ldots,8$,

$$s_i(\xi_1)=\xi_1,\quad s_i(\xi_2)=\xi_2,\quad s_i(\eta_1)=\eta_2,\quad s_i(\eta_2)=\eta_2,\quad s_i(\tau_j)=\tau_{(i-1,i)j}.\tag{47}$$

In the definition of s_3, we have used the notation of linear functions

$$\xi_i=\frac{\varphi(e_1,e_i)\xi_2-\varphi(e_2,e_i)\xi_1}{\varphi(e_1,e_2)},\quad \eta_i=\frac{\psi(e_1,e_i)\eta_2-\psi(e_2,e_i)\eta_1}{\psi(e_1,e_2)}\tag{48}$$

for $i=1,2,\ldots,8$. Note that, when restricted to the reference curve C_0, ξ_i and η_i give

$$\begin{aligned}
\xi_i\big|_{C_0}&=\frac{\varphi(e_1,e_i)\varphi(e_2,u)-\varphi(e_2,e_i)\varphi(e_1,u)}{\varphi(e_1,e_2)}=\varphi(e_i,u)=\varphi_i(u),\\
\eta_i\big|_{C_0}&=\frac{\psi(e_1,e_i)\psi(e_2,u)-\psi(e_2,e_i)\psi(e_1,u)}{\psi(e_1,e_2)}=\psi(e_i,u)=\psi_i(u).
\end{aligned}\tag{49}$$

In particular, the linear functions ξ_i and η_i vanish at p_i.

Theorem 1 *The automorphisms $s_i:\mathcal{K}\to\mathcal{K}$ $(i=0,1,\ldots,8)$ defined as above satisfy the fundamental relations* (9) *for the simple reflections of the affine Weyl group $W=W(E_8^{(1)})$. Furthermore, this W action on $\mathcal{K}=\mathbb{K}(\xi,\eta;\tau)$ is compatible with the restriction to the reference curve C_0 defined by*

$$\xi_i=\varphi_i(u),\quad \eta_i=\psi_i(u)\quad(i=1,2);\quad \tau_j=\sigma(e_j-u)\quad(j=1,\ldots,8).\tag{50}$$

By the "compatibility with the restriction to C_0", we mean: For any rational function $R(\xi,\eta,\tau)\in\mathbb{K}(\xi,\eta;\tau)$, and for any $w\in W$, we have

$$(w.R(\xi,\eta;\tau))\big|_{\xi=\varphi(u),\eta=\psi(u),\tau=\sigma(e-u)}=w.R(\varphi(u),\psi(u);\sigma(e-u)),\tag{51}$$

where $\sigma(e-u)=(\sigma(e_1-u),\ldots,\sigma(e_8-u))$. Here we understand that the parameter u is invariant under the action of W.

Besides the linear functions ξ_i,η_i, it is convenient to introduce linear functions $\xi(t),\eta(t)$ depending on a parameter $t\in\mathbb{C}$ as follows:

$$\xi(t)=\frac{\varphi(e_1,t)\xi_2-\varphi(e_2,t)\xi_1}{\varphi(e_1,e_2)},\quad \eta(t)=\frac{\psi(e_1,t)\eta_2-\psi(e_2,t)\eta_1}{\psi(e_1,e_2)}\tag{52}$$

so that $\xi(e_i) = \xi_i$, $\eta(e_i) = \eta_i$ $(i = 1, 2, \ldots, 8)$. On the reference curve C_0, they are expressed as

$$\xi(t)|_{C_0} = \varphi(t, u), \quad \eta(t)|_{C_0} = \psi(t, u), \tag{53}$$

and hence, $\xi(t)$ and $\eta(t)$ vanish at $p(t) = (\varphi(t), \psi(t))$. These functions also satisfy the three-term relations

$$\begin{aligned} &\xi(a)\varphi(b, c) + \xi(b)\varphi(c, a) + \xi(c)\varphi(a, b) = 0, \\ &\eta(a)\varphi(b, c) + \eta(b)\varphi(c, a) + \eta(c)\varphi(a, b) = 0. \end{aligned} \tag{54}$$

In particular we have

$$\begin{aligned} &\xi_i\, \varphi(e_j, e_k) + \xi_j\, \varphi(e_k, e_i) + \xi_k\, \varphi(e_i, e_j) = 0, \\ &\eta_i\, \psi(e_j, e_k) + \eta_j\, \psi(e_k, e_i) + \eta_k\, \psi(e_i, e_j) = 0 \end{aligned} \tag{55}$$

for $i, j, k = 1, \ldots, 8$. Note that the three-term relations in (54) reduce to those of $\varphi(t, u)$ and $\psi(t, u)$ when restricted to the reference curve C_0.

For each element

$$\Lambda = d_x H_x + d_y H_y - m_1 E_1 - \cdots - m_8 E_8 \in L \quad (d_x, d_y \in \mathbb{Z}, m_j \in \mathbb{Z}) \tag{56}$$

of the Picard lattice, we denote by $L(\Lambda)$ the $\mathbb{K}$-vector space of all polynomials $P(\xi, \eta) \in \mathbb{K}[\xi, \eta]$, homogenous in ξ and η of bidegree (d_x, d_y), satisfying the vanishing condition $\mathrm{ord}_{p_j} P(\xi, \eta) \geq m_j$ at $p_j = (\varphi(e_j), \psi(e_j))$ $(j = 1, 2, \ldots, 8)$. For a homogeneous polynomial $P(\xi, \eta)$ of bidegree (d_x, d_y), and a point $p = (a, b) \in \mathbb{P}^1 \times \mathbb{P}^1$, $a = (a_1 : a_2)$, $b = (b_1 : b_2)$, the condition $\mathrm{ord}_p P(\xi, \eta) \geq m$ means that, after the substitution $\xi_i = a_i + u_i$, $\eta_i = b_i + v_i$ $(i = 1, 2,)$, $P(a + u, b + v)$ is expressed as a sum of homogeneous polynomials in $(u, v) = (u_1, u_2, v_1, v_2)$ of total degree $> m$. We also set

$$\mathcal{L}(\Lambda) = L(\Lambda)\tau^{\Lambda}, \quad \tau^{\Lambda} = \tau_1^{-m_1} \cdots \tau_8^{-m_8}. \tag{57}$$

Here we understand the superscript notation τ^{Λ} by $\tau^{H_x} = \tau^{H_y} = 1$ and $\tau^{E_i} = \tau_i$. Note that

$$\begin{aligned} &\mathcal{L}(H_x) = \mathbb{K}\xi_1 \oplus \mathbb{K}\xi_2, \quad \mathcal{L}(H_y) = \mathbb{K}\eta_1 \oplus \mathbb{K}\eta_2, \\ &\mathcal{L}(E_j) = \mathbb{K}\tau_j, \quad (j = 1, 2, \ldots, 8). \\ &\mathcal{L}(H_x - E_j) = \mathbb{K}\,\xi_j\tau_j^{-1}, \quad \mathcal{L}(H_y - E_j) = \mathbb{K}\,\eta_j\tau_j^{-1} \quad (j = 1, 2, \ldots, 8). \end{aligned} \tag{58}$$

Theorem 2 *Each element $w \in W = W(E_8^{(1)})$ induces $\mathbb{C}$-isomorphisms $w.\,\mathcal{L}(\Lambda) \xrightarrow{\sim} \mathcal{L}(w.\Lambda)$ for all $\Lambda \in L$. Furthermore,* $\dim_{\mathbb{K}} L(\Lambda) = \dim_{\mathbb{K}} L(w.\Lambda)$ *for all $\Lambda \in L$.*

This theorem is the key for computing the action of the affine Weyl group W on $\mathcal{K} = \mathbb{K}(\xi, \eta; \tau)$; it can be proved by the argument of Kajiwara et al. (2017, Sect. 5.1).

We denote by

$$M = W\{E_1, \dots, E_8\} = W E_8 \subset L \tag{59}$$

the W-orbit of $E_1, \dots, E_8$ in L. This set can also be characterized as

$$M = \{\Lambda \in L \mid (\delta \mid \Lambda) = -1,\ (\Lambda \mid \Lambda) = 1\}. \tag{60}$$

Since $E_8 \in L$ is $W(E_8)$-invariant, the correspondence $\alpha \to T_\alpha.E_8$ gives a bijection $Q(E_8) \overset{\sim}{\to} M$. For each $\Lambda \in M$, take an element $w \in W$ such that $w.E_8 = \Lambda$. Noting that $\mathcal{L}(E_8) = \mathbb{K}\,\tau_8$, and $w : \mathcal{L}(E_8) \overset{\sim}{\to} \mathcal{L}(w.E_8) = \mathcal{L}(\Lambda)$, we set $\tau(\Lambda) = w.\tau_8$. From the fact that the stabilizer of E_8 in W coincides with $W(E_8)$, and that τ_8 is $W(E_8)$-invariant, it follows that this definition of $\tau(\Lambda) \in \mathcal{L}(\Lambda)$ does not depend on the choice of $w \in W$ such that $w.E_8 = \Lambda$. In this way, we obtain a family of τ-functions $\{\tau(\Lambda) \mid \Lambda \in M\}$ such that

$$\tau(E_j) = \tau_j \quad (j = 1, \dots, 8), \quad w(\tau(\Lambda)) = \tau(w.\Lambda) \quad (w \in W, \Lambda \in M). \tag{61}$$

For each $\Lambda = d_x H_x + d_y H_y - m_1 E_1 - \cdots - m_8 E_8 \in M$, $\tau(\Lambda) \in \mathcal{L}(\Lambda)$, it is expressed as

$$\tau(\Lambda) = \phi_\Lambda(\xi, \eta)\tau_1^{-m_1} \cdots \tau_8^{-m_8}, \quad \phi_\Lambda(\xi, \eta) \in L(\Lambda), \tag{62}$$

where $\phi_\Lambda(\xi, \eta)$ is a homogeneous polynomial in ξ and η of bidegree (d_x, d_y) satisfying the condition $\mathrm{ord}_{p_j}\phi_\Lambda(\xi, \eta) \geq m_j$ for $j = 1, \dots, 8$. Since $\dim_{\mathbb{K}} L(\Lambda) = 1$, such a polynomial is determined up to a constant multiple. However, this ambiguity can be fixed by the compatibility of the W action and the restriction to the reference curve C_0. In fact

$$\tau(\Lambda)\big|_{C_0} = w(\tau_8\big|C_0) = w(\sigma(e_8 - u)) = \sigma(\lambda - u), \tag{63}$$

where $\lambda = w(e_8) = d_x h_x + d_y h_y - m_1 e_1 - \cdots - m_8 e_8$. On the other hand,

$$\phi_\Lambda(\xi, \eta)\tau_1^{-m_1} \cdots \tau_8^{-m_8}\big|_{C_0} = \phi_\Lambda(\varphi(u), \psi(u))\sigma(e_1 - u)^{-m_1} \cdots \sigma(e_8 - u)^{-m_8}. \tag{64}$$

Hence we have

$$\phi_\Lambda(\varphi(u), \psi(u)) = \sigma(\lambda - u)\sigma(e_1 - u)^{m_1} \cdots \sigma(e_8 - u)^{m_8}. \tag{65}$$

If this formula is valid, $\phi_\Lambda(\xi, \eta)$ manifestly satisfies the vanishing condition. To summarize,

Theorem 3 *For each $\Lambda \in M = W.E_8$ as in* (56)*, the τ function $\tau(\Lambda)$ is expressed as*

$$\tau(\Lambda) = \phi_\Lambda(\xi, \eta)\tau_1^{-m_1} \cdots \tau_8^{-m_8}, \tag{66}$$

where $\phi_\Lambda(\xi,\eta)$ is characterized as the unique homogeneous polynomial in ξ and η of bidegree (d_x, d_y) satisfying the restriction formula (65) *on the reference curve C_0.*

In terms of the τ functions $\tau(\Lambda)$ $(\Lambda \in M)$, the homogenous coordinates ξ_i, η_i and τ variables τ_i are expressed as

$$\xi_i = \tau(E_i)\tau(H_x - E_i), \quad \eta_i = \tau(E_i)\tau(H_y - E_i), \quad \tau_i = \tau(E_i) \tag{67}$$

for $i = 1, \ldots, 8$. From the three-term relation of (55), we obtain

$$\begin{aligned}
&\varphi(e_j, e_k)\tau(E_i)\tau(H_x - E_i) + \varphi(e_k, e_i)\tau(E_j)\tau(H_x - E_j)\\
&\qquad + \varphi(e_i, e_j)\tau(E_k)\tau(H_x - E_k) = 0,\\
&\psi(e_j, e_k)\tau(E_i)\tau(H_y - E_i) + \psi(e_k, e_i)\tau(E_j)\tau(H_y - E_j)\\
&\qquad + \psi(e_i, e_j)\tau(E_k)\tau(H_y - E_k) = 0.
\end{aligned} \tag{68}$$

These are the fundamental bilinear equations for the τ functions of the elliptic Painlevé equation.

Applying $w \in W$ to (67), we obtain

$$\begin{aligned}
w(\xi_i) &= w.\tau(E_i)w.\tau(H_x - E_i) = \tau(w.E_i)\tau(w.(H_x - E_i))\\
&= \phi_{w.E_i}(\xi,\eta)\tau^{w.E_i}\phi_{w.(H_x-E_i)}(\xi,\eta)\tau^{w.(H_x-E_i)}.
\end{aligned} \tag{69}$$

Hence we have

$$\begin{aligned}
w(\xi_i) &= \phi_{w.E_i}(\xi,\eta)\,\phi_{w.(H_x-E_i)}(\xi,\eta)\,\tau^{w.H_x},\\
w(\eta_i) &= \phi_{w.E_i}(\zeta,\eta)\,\phi_{w.(H_y-E_i)}(\xi,\eta)\,\tau^{w.H_y}
\end{aligned} \tag{70}$$

for $i = 1, \ldots, 8$. In particular, $w \in W$ acts on the inhomogeneous coordinates $x = \xi_2/\xi_1$, $y = \eta_2/\eta_1$ by the formulas

$$\begin{aligned}
w(x) &= \frac{\phi_{w.E_2}(x,y)\,\phi_{w.(H_x-E_2)}(x,y)}{\phi_{w.E_1}(x,y)\,\phi_{w.(H_x-E_1)}(x,y)},\\
w(y) &= \frac{\phi_{w.E_2}(x,y)\,\phi_{w.(H_y-E_2)}(x,y)}{\phi_{w.E_1}(x,y)\,\phi_{w.(H_y-E_1)}(x,y)}.
\end{aligned} \tag{71}$$

Here we used the notation $\phi_\Lambda(x,y)$ to refer to the polynomial in the inhomogeneous coordinates (x,y) obtained from $\phi_\Lambda(\xi,\eta)$ by the substitution $\xi = (1:x)$, $\eta = (1:y)$. Note that, since $w.\xi_1, w.\xi_2 \in \mathcal{L}(w.H_x)$, the two polynomials $\phi_{w.E_1}(\xi,\eta)\phi_{w.(H_x-E_1)}(\xi,\eta)$ and $\phi_{w.E_2}(\xi,\eta)\phi_{w.(H_x-E_2)}(\xi,\eta)$ have the same bidegree. In particular, for each $\alpha \in Q(E_8)$, we have

$$\begin{aligned}
T_\alpha(x) &= \frac{\phi_{T_\alpha.E_2}(x,y)\,\phi_{T_\alpha.(H_x-E_2)}(x,y)}{\phi_{T_\alpha.E_1}(x,y)\,\phi_{T_\alpha.(H_x-E_1)}(x,y)},\\
T_\alpha(y) &= \frac{\phi_{T_\alpha.E_2}(x,y)\,\phi_{T_\alpha.(H_y-E_2)}(x,y)}{\phi_{T_\alpha.E_1}(x,y)\,\phi_{T_\alpha.(H_y-E_1)}(x,y)}.
\end{aligned} \tag{72}$$

This is the elliptic Painlevé equation with respect to the translation by $\alpha \in Q(E_8)$ in the inhomogeneous coordinates.

2 Explicit Representations of the Elliptic Painlevé Equation

One can apply previous discussions for determining explicit expression for the Kac translations T_α $(\alpha \in Q(E_8))$. We consider exclusively the case of the translation by the simple root $\alpha_1 = H_x - H_y$, which can be regarded as the non-autonomous version of the QRT mapping.

2.1 Action of the Kac Translation T_{α_1}

Our goal is to find an explicit representation for the action of the translation T_{α_1} with respect to the simple root $\alpha_1 = H_x - H_y$. This translation is decomposed into the product $T_{\alpha_1} = w^y w^x$ of two involutions w^x and w^y, called the *horizontal* and *vertical* flips, where

$$\begin{aligned} w^x &= w^x_{12} w^x_{34} w^x_{56} w^x_{78}, \qquad w^x_{ij} = r_{E_i - E_j} r_{H_x - E_i - E_j} \quad (i \neq j), \\ w^y &= w^y_{12} w^y_{34} w^y_{56} w^y_{78}, \qquad w^y_{ij} = r_{E_i - E_j} r_{H_y - E_i - E_j} \quad (i \neq j). \end{aligned} \tag{73}$$

These w^x, w^y act on the Picard lattice by

$$\begin{aligned} &w^x.H_x = H_x, \quad w^x.H_y = 4H_x + H_y - E_1 - \cdots - E_8, \\ &w^x.E_j = H_x - E_j \quad (j = 1, \ldots, 8), \\ &w^y.H_x = H_x + 4H_y - E_1 - \cdots - E_8, \quad w^y.H_y = H_y, \\ &w^y.E_j = H_y - E_j \quad (j = 1, \ldots, 8). \end{aligned} \tag{74}$$

Since

$$\xi_i = \tau(E_i)\tau(H_x - E_i), \quad \eta_i = \tau(E_i)\tau(H_y - E_i), \tag{75}$$

we have

$$w^x(\xi_i) = \xi_i, \quad w^y(\eta_i) = \eta_i \quad (i = 1, \ldots, 8). \tag{76}$$

This implies

$$T_{\alpha_1}(\xi_i) = w^y(\xi_i), \quad T^{-1}_{\alpha_1}(\eta_i) = w^x(\eta_i) \quad (i = 1, \ldots, 8). \tag{77}$$

We first investigate $w^y(\xi_i)$. By (70),

$$w^y(\xi_i) = \phi_{w^y.E_i}(\xi, \eta)\phi_{w^y.(H_x-E_i)}(\xi, \eta)\,\tau^{w^y.H_x}. \tag{78}$$

Note that $w^y.E_i = H_y - E_i$ and $\tau(H_y - E_i) = \eta_i \tau_i^{-1}$, and hence, $\phi_{w.E_i}(\xi, \eta) = \eta_i$. On the other hand,

$$\begin{aligned} w^y.(H_x) &= H_x + 4H_y - \textstyle\sum_{1\le j\le 8} E_j = \delta - H_x + 2H_y, \\ w^y.(H_x - E_i) &= H_x + 3H_y - \textstyle\sum_{1\le j\le 8;\, j\ne i} E_j = \delta - H_x + H_y + E_i. \end{aligned} \tag{79}$$

Hence $\phi_{w^y.(H_x-E_i)}(\xi, \eta)$ is characterized as the unique homogeneous polynomial of bidegree $(1, 3)$ such that

$$\phi_{w^y.(H_x-E_i)}(\xi, \eta)\big|_{C_0} = \sigma(\delta - h_x + h_y + e_i - u) \prod_{1\le j\le 8;\, j\ne i} \sigma(e_j - u). \tag{80}$$

Similarly, we have $\phi_{w^x.E_i}(\xi, \eta) = \xi_i$, and $\phi_{w^x.(H_y-E_i)}(\xi, \eta)$ is characterized as the unique homogeneous polynomial of bidegree (3, 1) such that

$$\phi_{w^x.(H_y-E_i)}(\xi, \eta)\big|_{C_0} = \sigma(\delta + h_x - h_y + e_i - u) \prod_{1\le j\le 8;\, j\ne i} \sigma(e_j - u). \tag{81}$$

Putting $P_i(\xi, \eta) = \phi_{w^y.(H_x-E_i)}(\xi, \eta)$ and $Q_i(\xi, \eta) = \phi_{w^x.(H_y-E_i)}(\xi, \eta)$, we obtain the following description of the Kac translation by α_1.

Lemma 1 *For each $i = 1, \ldots, 8$, there exist unique polynomial $P_i(\xi, \eta)$ and $Q_i(\xi, \eta)$, homogeneous in ξ and η of bidegree $(1, 3)$ and of bidegree $(3, 1)$, respectively, such that*

$$\begin{aligned} P_i(\varphi(u), \psi(u)) &= \sigma(\delta - h_x + h_y + e_i - u)\textstyle\prod_{1\le j\le 8;\, j\ne i}\sigma(e_j - u), \\ Q_i(\varphi(u), \psi(u)) &= \sigma(\delta + h_x - h_y + e_i - u)\textstyle\prod_{1\le j\le 8;\, j\ne i}\sigma(e_j - u). \end{aligned} \tag{82}$$

The Kac translation with respect to the simple root $\alpha_1 = H_x - H_y$ is given by

$$T_{\alpha_1}(\tau_i) = P_i(\xi, \eta)\tau_i(\tau_1 \cdots \tau_8)^{-1}, \quad T_{\alpha_1}^{-1}(\tau_i) = Q_i(\xi, \eta)\tau_i(\tau_1^{-1} \cdots \tau_8)^{-1} \tag{83}$$

and

$$T_{\alpha_1}(\xi_i) = \eta_i P_i(\xi, \eta)(\tau_1 \cdots \tau_8)^{-1}, \quad T_{\alpha_1}^{-1}(\eta_i) = \xi_i Q_i(\xi, \eta)(\tau_1 \cdots \tau_8)^{-1}. \tag{84}$$

In terms of the inhomogeneous coordinates (x, y) of $\mathbb{P}^1 \times \mathbb{P}^1$ such that $x = \xi_2/\xi_1$ and $y = \eta_2/\eta_1$,

$$T_{\alpha_1}(x) = y\,\frac{P_2(x, y)}{P_1(x, y)}, \quad T_{\alpha_1}^{-1}(y) = x\,\frac{Q_2(x, y)}{Q_1(x, y)}. \tag{85}$$

for $i = 1, \ldots, 8$.

We remark that this type of representation of the elliptic Painlevé equation was first proposed by Murata (2004).[1] A geometric interpretation to this expression is also given by Carstea-Dzhamay-Takenawa (2017).

In order to write down $P_i(\xi, \eta)$ and $Q_i(\xi, \eta)$, it is convenient to divide the indexing set $\{1, \ldots, 8\}$ into two parts $\{1, 2, 3, 4\}$ and $\{5, 6, 7, 8\}$. Then, for $i = 1, 2, 3, 4$, we have

$$\begin{aligned} P_i(\xi, \eta) &= \sum_{k=5}^{8} \frac{\sigma(h_x - e_i - e_k - \delta)}{\sigma(h_x - h_y)} \frac{\prod_{1\le j\le 4;\, j\ne i} \sigma(h_y - e_k - e_j)}{\prod_{5\le j\le 8;\, j\ne k} \sigma(e_k - e_j)} \xi_k \prod_{5\le j\le 8;\, j\ne k} \eta_j, \\ Q_i(\xi, \eta) &= \sum_{k=5}^{8} \frac{\sigma(h_y - e_i - e_k - \delta)}{\sigma(h_y - h_x)} \frac{\prod_{1\le j\le 4;\, j\ne i} \sigma(h_x - e_k - e_j)}{\prod_{5\le j\le 8;\, j\ne k} \sigma(e_k - e_j)} \eta_k \prod_{5\le j\le 8;\, j\ne k} \xi_j. \end{aligned} \tag{86}$$

We explain below how to derive these explicit formulas, taking $P_1(\xi, \eta)$ as an example. In this case, $P_1(\xi, \eta)$ is of bidegree $(1, 3)$ and the curve $P_1(\xi, \eta) = 0$ passes through $p_2, p_3, \ldots, p_8$. Note that $\mathbb{K}$-vector space of homogeneous polynomials in η of degree 3 has dimension 4. Choosing three out of the four homogeneous linear functions $\eta_5, \eta_6, \eta_7, \eta_8$, we can take

$$\eta_6\eta_7\eta_8, \quad \eta_5\eta_7\eta_8, \quad \eta_5\eta_6\eta_8, \quad \eta_5\eta_6\eta_7 \tag{87}$$

for a basis of this vector space. The polynomial $P_1(\xi, \eta)$ of bidegree $(1, 3)$ is then expressed as

$$P_1(\xi, \eta) = a_5(\xi)\eta_6\eta_7\eta_8 + a_6(\xi)\eta_5\eta_7\eta_8 + a_7(\xi)\eta_5\eta_6\eta_8 + a_8(\xi)\eta_5\eta_6\eta_7, \tag{88}$$

where $a_j(\xi)$ $(j = 5, 6, 7, 8)$ are homogeneous of degree 1. By the condition $P_1(\xi, \eta) = 0$ passes through p_5, p_6, p_7, p_8, we see that $a_j(\xi) = c_j\xi_j$, $c_j \in \mathbb{K}$ for $j = 5, 6, 7, 8$. Hence we have

$$\begin{aligned} P_1(\xi, \eta) &= c_5\xi_5\eta_6\eta_7\eta_8 + c_6\xi_6\eta_5\eta_7\eta_8 + c_7\xi_7\eta_5\eta_6\eta_8 + c_8\xi_8\eta_5\eta_6\eta_7 \\ &= \left(c_5\frac{\xi_5}{\eta_5} + c_6\frac{\xi_6}{\eta_6} + c_7\frac{\xi_7}{\eta_7} + c_8\frac{\xi_8}{\eta_8} \right) \eta_5\eta_6\eta_7\eta_8. \end{aligned} \tag{89}$$

We fix the constants c_5, c_6, c_7, c_8 so that the restriction formula

$$P_1(\varphi(u), \psi(u)) = \sigma(\delta - h_x + h_y + e_1 - u) \prod_{j=2}^{8} \sigma(e_j - u) \tag{90}$$

holds. Note that (89) implies

[1]The author is grateful to the referee for directing his attention to this work.

$$\frac{P_1(\xi,\eta)}{\eta_5\eta_6\eta_7\eta_8} = \sum_{k=5}^{8} c_k \frac{\xi_k}{\eta_k}. \tag{91}$$

By the restriction to C_0 we obtain

$$\frac{\sigma(\delta - h_x + h_y + e_1 - u)\prod_{j=2}^{8}\sigma(e_j - u)}{\prod_{j=5}^{8}\psi_j(u)} = \sum_{k=5}^{8} c_k \frac{\varphi_k(u)}{\psi_k(u)}, \tag{92}$$

and hence

$$\frac{\sigma(\delta - h_x + h_y + e_1 - u)\prod_{j=2}^{4}\sigma(e_j - u)}{\prod_{j=5}^{8}\sigma(h_y - e_j - u)} = \sum_{k=5}^{8} c_k \frac{\sigma(h_x - e_k - u)}{\sigma(h_y - e_k - u)}. \tag{93}$$

This is an identity of quasi-periodic meromorphic functions of type $(0, h_x - h_y)$ with simple poles at $u = h_y - e_j$ $(j = 5, 6, 7, 8)$. Comparing the residues at these poles, we have

$$c_k = -\frac{\sigma(\delta - h_x + e_1 + e_k)}{\sigma(h_x - h_y)} \frac{\prod_{j=2}^{4}\sigma(h_y - e_k - e_j)}{\prod_{5\le j\le 8;\, j\ne k}\sigma(e_k - e_j)} \tag{94}$$

for $i = 5, 6, 7, 8$. Finally we obtain

$$\begin{aligned} P_1(\xi,\eta) &= \eta_5\eta_6\eta_7\eta_8 \sum_{k=5}^{8} \frac{\sigma(h_x - e_1 - e_k - \delta)}{\sigma(h_x - h_y)} \frac{\prod_{j=2}^{4}\sigma(h_y - e_k - e_j)}{\prod_{5\le j\le 8;\, j\ne k}\sigma(e_k - e_j)} \frac{\xi_k}{\eta_k} \\ &= \sum_{k=5}^{8} \frac{\sigma(h_x - e_1 - e_k - \delta)}{\sigma(h_x - h_y)} \frac{\prod_{j=2}^{4}\sigma(h_y - e_k - e_j)}{\prod_{5\le j\le 8;\, j\ne k}\sigma(e_k - e_j)} \xi_k \prod_{5\le j\le 8;\, j\ne k} \eta_j. \end{aligned} \tag{95}$$

2.2 Coordinates Depending on a Parameter

Fixing a generic (W-invariant) constant $t \in \mathbb{C}$, we consider the action of T_{α_1} on

$$\xi(t) = \frac{\varphi(e_1,t)\xi_2 - \varphi(e_2,t)\xi_1}{\varphi(e_1,e_2)}, \quad \eta(t) = \frac{\psi(e_1,t)\eta_2 - \psi(e_2,t)\eta_1}{\psi(e_1,e_2)}. \tag{96}$$

Since $w^x\varphi(e_i,t) = \varphi(e_i,t)$ and $w^x\varphi(e_i,e_j) = \varphi(e_i,e_j)$, we have $w^x\xi(t) = \xi(t)$, and $T_{\alpha_1}\xi(t) = w^y\xi(t)$. Since w^y induces a $\mathbb{C}$-isomorphism

$$w^y : \mathcal{L}(H_x) \xrightarrow{\sim} \mathcal{L}(H_x + 4H_y - E_1 - E_2 - \ldots - E_8), \tag{97}$$

$w^y(\xi(t))$ is expressed as

$$w^y(\xi(t)) = \Phi(\xi, \eta; t)(\tau_1 \cdots \tau_8)^{-1}, \tag{98}$$

with a homogeneous polynomial $\Phi(\xi, \eta; t) \in \mathbb{K}[\xi, \eta]$ of bidegree $(1, 4)$. By the restriction to C_0, from

$$\xi(t)|_{C_0} = \varphi(t, u) = \sigma(t - u)\sigma(h_x - t - u) \tag{99}$$

we obtain

$$\begin{aligned} w^y(\xi(t))|_{C_0} &= \sigma(t - u)\sigma(h_x + 4h_y - e_1 - \cdots - e_8 - t - u) \\ &= \sigma(\delta - h_x + 2h_y - t - u). \end{aligned} \tag{100}$$

Hence we have

$$\sigma(t - u)\sigma(\delta - h_x + 2h_y - t - u) = \Phi(\varphi(u), \psi(u); t) \prod_{j=1}^{8} \sigma(e_j - u)^{-1}, \tag{101}$$

in other words,

$$\Phi(\varphi(u), \psi(u); t) = \sigma(\delta - h_x + 2h_y - t - u)\sigma(t - u) \prod_{j=1}^{8} \sigma(e_j - u). \tag{102}$$

This means that the curve $\Phi(\xi, \eta) = 0$ of bidegree $(1, 4)$ passes through the 9 points $p_j = p(e_j)$ $(j = 1, \ldots, 8)$ together with $p(t)$; from now on, we set

$$e_9 = t, \quad p_9 = p(t), \quad \xi_9 = \xi(t), \quad \eta_9 = \eta(t). \tag{103}$$

Noting that the vector space of homogeneous polynomials in η of degree 4 has 5 dimension, we take the 5 homogeneous linear functions

$$\eta_5 \cdots \widehat{\eta_k} \cdots \eta_9 \quad (k = 5, 6, 7, 8, 9). \tag{104}$$

Then, we can express $\Phi(\xi, \eta; t)$ in the form

$$\Phi(\xi, \eta; t) = \sum_{k=5}^{9} a_k(\xi)\eta_5 \cdots \widehat{\eta_k} \cdots \eta_9 = \sum_{k=5}^{9} c_k \xi_k \eta_5 \cdots \widehat{\eta_k} \cdots \eta_9 \tag{105}$$

since $\Phi(\xi, \eta; t) = 0$ passes through $p_5, p_6, \ldots, p_9$. As before, rewriting this in the form

$$\frac{\Phi(\xi, \eta; t)}{\prod_{j=5}^{9} \eta_j} = \sum_{k=5}^{9} c_k \frac{\xi_k}{\eta_k}, \tag{106}$$

by the restriction to C_0 we obtain

$$\frac{\sigma(\delta - h_x + 2h_y - e_9 - u)\prod_{j=1}^{4}\sigma(e_j - u)}{\prod_{j=5}^{9}\sigma(h_y - e_j - u)} = \sum_{k=5}^{9} c_k \frac{\sigma(h_x - e_k - u)}{\sigma(h_y - e_k - u)}. \tag{107}$$

Taking the residues at $u = h_y - e_k$ ($k = 5, 6, \ldots, 9$), we can determine the constants c_k as

$$c_k = \frac{\sigma(\delta - h_x + h_y - e_9 + e_k)\prod_{j=1}^{4}\sigma(h_y - e_k - e_j)}{\sigma(h_x - h_y)\prod_{5\le j\le 9;\, j\ne k}\sigma(e_k - e_j)}. \tag{108}$$

Hence we have

$$\Phi(\xi, \eta; t) = \sum_{k=5}^{9} \frac{\sigma(\delta - h_x + h_y - e_9 + e_k)\prod_{j=1}^{4}\sigma(h_y - e_k - e_j)}{\sigma(h_x - h_y)\prod_{5\le j\le 9;\, j\ne k}\sigma(e_k - e_j)}\xi_k\eta_5\cdots\widehat{\eta_k}\cdots\eta_9. \tag{109}$$

Replacing e_9 by t, we obtain the following expression for the Kac translation of $\xi(t)$ and $\eta(t)$ with respect to $\alpha_1 = H_x - H_y$.

Lemma 2 *As to the Kac translation $T_{\alpha_1} = w^y w^x$ of*

$$\xi(t) = \frac{\varphi(e_1, t)\xi_2 - \varphi(e_2, t)\xi_1}{\varphi(e_1, e_2)}, \quad \eta(t) = \frac{\psi(e_1, t)\eta_2 - \psi(e_2, t)\eta_1}{\psi(e_1, e_2)} \tag{110}$$

by $\alpha_1 = H_x - H_y$, we have

$$\begin{aligned} T_{\alpha_1}(\xi(t)) &= w^y(\xi(t)) = \Phi(\xi, \eta; t)(\tau_1\cdots\tau_8)^{-1}, \\ T_{\alpha_1}^{-1}(\eta(t)) &= w^x(\eta(t)) = \Psi(\xi, \eta; t)(\tau_1\cdots\tau_8)^{-1} \end{aligned} \tag{111}$$

where

$$\begin{aligned} &\Phi(\xi, \eta; t) \\ &= \frac{\sigma(\delta - h_x + h_y)\prod_{j=1}^{4}\sigma(h_y - e_j - t)}{\sigma(h_x - h_y)\prod_{j=5}^{8}\sigma(e_j - t)}\xi(t)\eta_5\eta_6\eta_7\eta_8 \\ &\quad + \sum_{k=5}^{8}\frac{\sigma(\delta - h_x + h_y + e_k - t)\prod_{j=1}^{4}\sigma(h_y - e_k - e_j)}{\sigma(h_x - h_y)\sigma(e_k - t)\prod_{5\le j\le 8;\, j\ne k}\sigma(e_k - e_j)}\xi_k\eta_5\cdots\widehat{\eta_k}\cdots\eta_8\eta(t), \\ &\Psi(\xi, \eta; t) \\ &= \frac{\sigma(\delta - h_y + h_x)\prod_{j=1}^{4}\sigma(h_x - e_j - t)}{\sigma(h_y - h_x)\prod_{j=5}^{8}\sigma(e_j - t)}\xi_5\xi_6\xi_7\xi_8\eta(t) \\ &\quad + \sum_{k=5}^{8}\frac{\sigma(\delta - h_y + h_x + e_k - t)\prod_{j=1}^{4}\sigma(h_x - e_k - e_j)}{\sigma(h_y - h_x)\sigma(e_k - t)\prod_{5\le j\le 8;\, j\ne k}\sigma(e_k - e_j)}\xi_5\cdots\widehat{\xi_k}\cdots\xi_8\xi(t)\eta_k. \end{aligned} \tag{112}$$

From this lemma, we have

$$T_{\alpha_1}\left(\frac{\xi(t_2)}{\xi(t_1)}\right)=\frac{\Phi(\xi,\eta;t_2)}{\Phi(\xi,\eta;t_1)},\qquad T_{\alpha_1}^{-1}\left(\frac{\eta(t_2)}{\eta(t_1)}\right)=\frac{\Psi(\xi,\eta;t_2)}{\Psi(\xi,\eta;t_1)} \tag{113}$$

for generic W-invariant parameters $t_1, t_2 \in \mathbb{C}$. In the following we use the notation $T_{\alpha_1}(f)=\overline{f}$ and $T_{\alpha_1}^{-1}(f)=\underline{f}$ for simplicity, so that

$$\frac{\overline{\xi}(t_2)}{\overline{\xi}(t_1)}=\frac{\Phi(\xi,\eta;t_2)}{\Phi(\xi,\eta;t_1)},\qquad \frac{\underline{\eta}(t_2)}{\underline{\eta}(t_1)}=\frac{\Psi(\xi,\eta;t_2)}{\Psi(\xi,\eta;t_1)}, \tag{114}$$

where

$$\begin{aligned}
\overline{\xi}(t)&=\frac{\overline{\varphi}_1(t)\overline{\xi}_2-\overline{\varphi}_2(t)\overline{\xi}_1}{\overline{\varphi(e_1,e_2)}},\quad \underline{\eta}(t)=\frac{\underline{\psi}_1(t)\underline{\eta}_2-\underline{\psi}_2(t)\underline{\eta}_1}{\underline{\psi(e_1,e_2)}},\\
\overline{\varphi}_i(t)&=\sigma(\overline{e}_i-t)\sigma(\overline{h}_x-\overline{e}_i-t)=\sigma(e_i-h_x+h_y+\delta-t)\sigma(h_y-e_i-t),\\
\underline{\psi}_i(t)&=\sigma(\underline{e}_i-t)\sigma(\underline{h}_y-\underline{e}_i-t)=\sigma(e_i+h_x-h_y+\delta-t)\sigma(h_x-e_i-t).
\end{aligned} \tag{115}$$

Note that $\eta(t)=(\psi_1(t)\eta_2-\psi_2(t)\eta_1)/\psi(e_1,e_2)$ vanishes when we specialize the η coordinate by $\eta=\psi(t)$, namely $\eta_i=\psi_i(t)$ $(i=1,2)$. Hence, by the substitution $\eta=\psi(t)$, $\Phi(\xi,\eta;t)$ simplifies to a single term:

$$\begin{aligned}
\Phi(\xi,\psi(t);t)&=\frac{\sigma(\delta-h_x+h_y)\prod_{j=1}^{4}\sigma(h_y-e_j-t)}{\sigma(h_x-h_y)\prod_{j=5}^{8}\sigma(e_j-t)}\xi(t)\prod_{j=5}^{8}\psi_j(t)\\
&=\frac{\sigma(\delta-h_x+h_y)\prod_{j=1}^{8}\sigma(h_y-e_j-t)}{\sigma(h_x-h_y)}\xi(t).
\end{aligned} \tag{116}$$

We use a new parameter $v\in\mathbb{C}$ for this specialization of η assuming that $w^y(v)=v$. Since $\psi_i(v)=\psi_i(h_y-v)$, we have

$$\begin{aligned}
\Phi(\xi,\psi(v);v)&=\frac{\sigma(\delta-h_x+h_y)\prod_{j=1}^{8}\sigma(h_y-e_j-v)}{\sigma(h_x-h_y)}\xi(v),\\
\Phi(\xi,\psi(v);h_y-v)&=\frac{\sigma(\delta-h_x+h_y)\prod_{j=1}^{8}\sigma(e_j-v)}{\sigma(h_x-h_y)}\xi(h_y-v),
\end{aligned} \tag{117}$$

and hence

$$\frac{\Phi(\xi,\psi(v);h_y-v)}{\Phi(\xi,\psi(v);v)}=\frac{\xi(h_y-v)}{\xi(v)}\prod_{j=1}^{8}\frac{\sigma(e_j-v)}{\sigma(h_y-e_j-v)}. \tag{118}$$

This implies

$$\frac{\overline{\xi}(h_y-v)}{\overline{\xi}(v)}=\frac{\xi(h_y-v)}{\xi(v)}\prod_{j=1}^{8}\frac{\sigma(e_j-v)}{\sigma(h_y-e_j-v)} \tag{119}$$

with the parameter v such that $\eta = \psi(v)$. Similarly, we use a parameter $u \in \mathbb{C}$ for the specialization $\xi = \varphi(u)$ assuming that $w^x(u) = u$. Then we have

$$\frac{\underline{\eta}(h_x - u)}{\underline{\eta}(u)} = \frac{\eta(h_x - u)}{\eta(u)} \prod_{j=1}^{8} \frac{\sigma(e_j - u)}{\sigma(h_x - e_j - u)} \tag{120}$$

with the parameter u such that $\xi = \varphi(u)$. This system of equations (119), (120) is the explicit representation of the elliptic Painlevé equation proposed in Noumi-Tsujimoto-Yamada (2013) and Kajiwara-Noumi-Yamada (2017, Sect. 5.5, (5.82), (5.83)).

Finally, we comment on specializing $\xi = \varphi(u)$ and $\eta = \psi(v)$ with two parameters u, v, simultaneously. In geometric terms, this is equivalent to considering the covering

$$\pi: \ E_\Omega \times E_\Omega \to \mathbb{P}^1 \times \mathbb{P}^1 : \ \pi(u, v) = (\varphi(u), \psi(v)) \quad ((u, v) \in E_\Omega \times E_\Omega). \tag{121}$$

In the inhomogeneous coordinates $(x, y) \in \mathbb{P}^1 \times \mathbb{P}^1$, this covering is defined by

$$x = \frac{\sigma(e_2 - u)\sigma(h_x - e_2 - u)}{\sigma(e_1 - u)\sigma(h_x - e_1 - u)}, \quad y = \frac{\sigma(e_2 - v)\sigma(h_y - e_2 - v)}{\sigma(e_1 - v)\sigma(h_y - e_1 - v)}. \tag{122}$$

Under this parametrization, $\xi(v)$ and $\eta(u)$ are expressed as

$$\begin{aligned} \xi(v) &= \frac{\varphi_1(v)\xi_2 - \varphi_2(v)\xi_1}{\varphi_1(e_2)} = \frac{\varphi_1(v)\varphi_2(u) - \varphi_2(v)\varphi_1(u)}{\varphi_1(e_2)} = -\varphi(u, v), \\ \eta(u) &= \frac{\psi_1(u)\eta_2 - \psi_2(u)\eta_1}{\psi_1(e_2)} = \frac{\psi_1(u)\psi_2(v) - \psi_2(u)\psi_1(v)}{\psi_1(e_2)} = \psi(u, v). \end{aligned} \tag{123}$$

Hence, Eqs. (119) and (120) lifted to $E_\Omega \times E_\Omega$ should be written as

$$\begin{aligned} \frac{\overline{\varphi}(\overline{u}, h_y - v)}{\overline{\varphi}(\overline{u}, v)} &= \frac{\varphi(u, h_y - v)}{\varphi(u, v)} \prod_{j=1}^{8} \frac{\sigma(e_j - v)}{\sigma(h_y - e_j - v)}, \\ \frac{\underline{\psi}(h_x - u, \underline{v})}{\underline{\psi}(u, \underline{v})} &= \frac{\psi(h_x - u, v)}{\psi(u, v)} \prod_{j=1}^{8} \frac{\sigma(e_j - u)}{\sigma(h_x - e_j - u)} \end{aligned} \tag{124}$$

in an implicit factorized form. This system of equations is still somewhat ambiguous in the point how one should understand the evolutions $\overline{\varphi}$ and $\underline{\psi}$. A possible idea would be to set

$$\begin{aligned} \overline{\varphi}(\overline{u}, v) &= \overline{\sigma}(\overline{u} - v)\overline{\sigma}(\overline{h}_x - \overline{u} - v), \quad \overline{h}_x = \delta - h_x + 2h_y, \\ \underline{\psi}(u, \underline{v}) &= \underline{\sigma}(u - \underline{v})\underline{\sigma}(\underline{h}_y - u - \underline{v}), \quad \underline{h}_y = \delta + 2h_x - h_y, \end{aligned} \tag{125}$$

using sigma functions $\overline{\sigma}(u) = \sigma(u|\overline{\Omega})$, $\underline{\sigma}(u) = \sigma(u|\underline{\Omega})$ associated with appropriately evolved period lattices $\overline{\Omega}$, $\underline{\Omega}$. It would be an intriguing problem to pursuit the idea of *modulation* of the elliptic nome in the elliptic Painlevé equation. We expect that such an approach could also provide a method for constructing generic solutions to this equation in the geometric terms of plane curves.

Acknowledgements This work was supported by JSPS Kakenhi Grant (B)15H03626.

References

Carstea, A.S., Dzhamay, A., Takenawa, T.: Fiber-dependent deautonomization of integrable 2D mappings and discrete Painlevé equations. J. Phys. A **50**(40), 405202, 41pp (2017)

Kajiwara, K., Masuda, T., Noumi, M., Ohta, Y., Yamada, Y.: Point configurations, Cremona transformations and the elliptic Painlevé equations, Séminaires et Congrès **14**, 169–198 (2006)

Kajiwara, K., Noumi, M., Yamada, Y.: Geometric aspects of Painlevé equations. J. Phys. A **50**(7), 073001, 164pp (2017)

Murata, M.: New expressions for discrete Painlevé equations. Funkcial. Ekvac. **47**, 291–305 (2004)

Noumi, M.: Remarks on τ-functions for the difference Painlevé equations of type E_8. Advanced Studies in Pure Mathematics, vol. 76 (2018), Representation Theory, Special Functions and Painlevé equations — RIMS 2015, pp. 1–65

Noumi, M., Tsujimoto, S., Yamada, Y.: Padé interpolation for elliptic Painlevé equation. Symmetries, Integrable Systems and Representations (Kenji Iohara, Sophie Morier-Genoud, Bertrand Rémy Eds.), pp. 463–482, Springer Proceedings in Mathematics and Statistics, vol. 40. Springer (2013)

Ohta, Y., Ramani, A., Grammaticos, B.: An affine Weyl group approach to the eight parameter discrete Painlevé equations. J. Phys. A: Math. Gen. **34**, 10523–10532 (2001)

Sakai, H.: Rational surfaces associated with affine root systems and geometry of the Painlevé equations. Commun. Math. Phys. **220**, 165–229 (2001)

On the Lattice Potential KP Equation

Cewen Cao, Xiaoxue Xu, and Da-jun Zhang

Abstract The paper presents an approach to derive finite genus solutions to the lattice potential Kadomtsev-Petviashvili (lpKP) equation introduced by F.W. Nijhoff, et al. This equation is rederived from compatible conditions of three replicas of the discrete ZS-AKNS spectral problem, which is a Darboux transformation of the continuous ZS-AKNS spectral problem. With the help of these links and by means of the so called nonlinearization technique and Liouville platform, finite genus solutions of the lpKP equation are derived. Semi-discrete potential KP equations with one and two discrete arguments, respectively, are also discussed.

Keywords Lattice potential KP equation · Finite genus solutions · Nonlinearization · ZS-AKNS spectral problem · Liouville platform

1 Introduction

Discrete integrable systems and the problem of integrable discretization of given soliton equations have attracted more and more attention in recent years (Grammaticos et al. 2004; Hietarinta et al. 2016; Suris 2003). The main purpose of this paper is to investigate the lattice potential Kadomtsev-Petviashvili (lpKP) equation

C. Cao · X. Xu
School of Mathematics and Statistics, Zhengzhou University, Zhengzhou Henan 450001, People's Republic of China
e-mail: cwcao@zzu.edu.cn

X. Xu
e-mail: xiaoxuexu@zzu.edu.cn

D. Zhang (✉)
Department of Mathematics, Shanghai University, Shanghai 200444, People's Republic of China
e-mail: djzhang@staff.shu.edu.cn

© Springer Nature Switzerland AG 2020

F. Nijhoff et al. (eds.), *Asymptotic, Algebraic and Geometric Aspects of Integrable Systems*, Springer Proceedings in Mathematics & Statistics 338,
https://doi.org/10.1007/978-3-030-57000-2_11

$$\Xi^{\mathrm{lpKP}}_{(\beta_1,\beta_2,\beta_3)} \equiv (\beta_1 - \tilde{W})(\beta_2 - \beta_3 + \hat{\tilde{W}} - \bar{\tilde{W}}) + (\beta_2 - \bar{W})(\beta_3 - \beta_1 + \tilde{\bar{W}} - \hat{\bar{W}}) \\ +(\beta_3 - \hat{W})(\beta_1 - \beta_2 + \bar{\hat{W}} - \tilde{\hat{W}}) = 0, \tag{1}$$

and present an approach to construct finite genus solutions to 3D integrable lattice equations. This equation is first discovered by Nijhoff, Capel, Wiersma and Quispel by using the Bäcklund transformation approach, and later derived through an analysis of the Cauchy matrix (Nijhoff et al. 1984; Hietarinta et al. 2016).

To build the integrability of the lpKP equation (1) and calculate its finite genus solutions, we will introduce Lax triads from the ZS-AKNS spectral problem. Compatibility of these triads, respectively, give rise to the lattice potential KP equations with 3, 2 and 1 discrete arguments, as (see Sect. 2 for the derivation)

$$\Xi^{(0,3)} \equiv \frac{1}{2}[(\tilde{W} - \bar{W})\bar{\tilde{W}} + (\bar{W} - \hat{W})\hat{\bar{W}} + (\hat{W} - \tilde{W})\tilde{\hat{W}}] \\ +\gamma_1(\bar{\tilde{W}} - \hat{\tilde{W}} - \bar{W} + \hat{W}) + \gamma_2(\hat{\bar{W}} - \tilde{\bar{W}} - \hat{W} + \tilde{W}) \\ +\gamma_3(\tilde{\hat{W}} - \bar{\hat{W}} - \tilde{W} + \bar{W}) = 0, \tag{2}$$

$$\Xi^{(1,2)} \equiv (\tilde{W} - \bar{W})_x - \left[\frac{1}{2}(\tilde{W} - \bar{W}) + \gamma_1 - \gamma_2\right](\bar{\tilde{W}} - \tilde{W} - \bar{W} + W) = 0, \tag{3}$$

$$\Xi^{(2,1)} \equiv (\tilde{W} - W)_y - \left[(\tilde{W} + W)_x + 2\gamma_1(\tilde{W} - W) + \frac{1}{2}(\tilde{W} - W)^2\right]_x = 0. \tag{4}$$

Note that (2) is equivalent to the lpKP equation (1) with $\beta_k = -2\gamma_k,\ k = 1, 2, 3$, as $\Xi^{\mathrm{lpKP}}_{(-2\gamma_1,-2\gamma_2,-2\gamma_3)} = 2\Xi^{(0,3)}$. It also turns out that all these equations have the same potential KP (pKP) equation,

$$\Xi^{(3,0)} \equiv W_{xt} - \frac{1}{4}(W_{xxx} + 3W_x^2)_x - \frac{3}{4}W_{yy} = 0, \tag{5}$$

as their continuum limits (see Proposition 4).

The method of finite-gap integration originated in solving the periodic initial problem of the Korteweg-de Vries (KdV) equation (cf. Matveev (2008) and the references therein). Recently, an approach to deriving finite genus solutions for 2D discrete integrable systems, the lattice potential KdV equation (Cao and Xu 2012) and the lattice nonlinear Schrödinger (lNLS) model (Cao and Zhang 2012) were developed. In this paper, just as in the 2D case, explicit analytic solutions of the lattice pKP equations (2, 3, 4), together with the pKP equation (5), will be calculated by means of the finite-dimensional integrable flows of continuous and discrete types, i.e. Hamiltonian phase flows and integrable symplectic maps. These flows are constructed through nonlinearization of the continuous and discrete spectral problems (see Sects. 3, 4). It is surprising that they share same Liouville integrals, same Lax matrix $L(\lambda; p, q)$ and same algebraic curve $\mathcal{R}$. Thus the calculations can be done on the same Liouville platform. The Abel-Jacobi variable $\vec{\phi}$ in the Jacobian variety $J(\mathcal{R})$ straightens out both the H_j- and the S_{γ_k}-flow with the velocities $\vec{\Omega}_j$ and $\vec{\Omega}_{\gamma_k}$, respectively. As a result, we have a clear evolution picture for the lattice pKP equations as well as for the pKP equation, as the following,

$$\begin{aligned}
\varXI^{(0,3)}: &\quad \vec{\phi} \equiv \vec{\phi}_0 + m_1\vec{\Omega}_{\gamma_1} + m_2\vec{\Omega}_{\gamma_2} + m_3\vec{\Omega}_{\gamma_3}, \quad (\mathrm{mod}\,\mathscr{T}),\\
\varXI^{(1,2)}: &\quad \vec{\phi} \equiv \vec{\phi}_0 + x\vec{\Omega}_1 + m_1\vec{\Omega}_{\gamma_1} + m_2\vec{\Omega}_{\gamma_2}, \quad (\mathrm{mod}\,\mathscr{T}),\\
\varXI^{(2,1)}: &\quad \vec{\phi} \equiv \vec{\phi}_0 + x\vec{\Omega}_1 + y\vec{\Omega}_2 + m_1\vec{\Omega}_{\gamma_1}, \quad (\mathrm{mod}\,\mathscr{T}),\\
\varXI^{(3,0)}: &\quad \vec{\phi} \equiv \vec{\phi}_0 + x\vec{\Omega}_1 + y\vec{\Omega}_2 + t\vec{\Omega}_3, \quad (\mathrm{mod}\,\mathscr{T}),
\end{aligned}$$

which will provide the basic part of the explicit analytic solutions (see Sect. 5, 6).

The paper is organized as follows. Section 2 shows that how the lattice pKP equations (2), (3) and (4) arise from their Lax triads. Continuum limits of these lattice pKP equations give rise to the same pKP equation. In Sect. 3, a finite-dimensional integrable Hamiltonian system related to the ZS-AKNS spectral problem is introduced to provide integrals, spectral curve and Abel-Jacobi variables. In Sect. 4 we construct an integrable symplectic map S_γ in tilde direction, develop an algebro-difference analogue of the Burchnall-Chaundy's theory on commuting differential operators by which we express the potential functions in terms of theta function. This allows us to derive finite genus solutions for the lpKP equation in Sect. 5 and for other two semi-discrete and one continuous pKP equations in Sect. 6. Finally, concluding remarks are given in the last Section.

2 The Discretized pKP Equations

2.1 The KP Equation

In order to find the suitable discrete spectral problems for (1), let us first recall the usual continuous KP equation,

$$w_t = \frac{1}{4}(w_{xx} + 3w^2)_x + \frac{3}{4}\partial_x^{-1} w_{yy}. \tag{6}$$

It is well-known that the KP equation has a close relation with the ZS-AKNS spectral problem (U_1) (Cao et al. 1999; Konopelchenko et al. 1991),

$$\partial_x \chi = U_1 \chi = \begin{pmatrix} \lambda/2 & u \\ v & -\lambda/2 \end{pmatrix} \chi. \tag{7}$$

In fact, there is a hierarchy of isospectral equations (X_k) related to (7),

$$\partial_{\tau_k}(u, v) = X_k, \quad (k = 2, 3, \ldots), \tag{8}$$

in which the first two nonlinear members, $(y = \tau_2, t = \tau_3)$, the NLS equation and the modified KdV (mKdV) equation, respectively, are

$$\partial_y(u, v) = X_2 = (u_{xx} - 2u^2 v, \ -v_{xx} + 2uv^2), \tag{9a}$$

$$\partial_t(u, v) = X_3 = (u_{xxx} - 6uvu_x, \ v_{xxx} - 6uvv_x). \tag{9b}$$

Corresponding to the hierarchy (8), there exist a series of linear spectral problems (U_k),

$$\partial_{\tau_k}\chi = U_k\chi, \quad (k = 1, 2, \ldots), \tag{10}$$

where, apart from Eq. (7)$|_{x=\tau_1}$, we also have (with $y = \tau_2, \ t = \tau_3$)

$$\partial_y\chi = U_2\chi = \begin{pmatrix} \lambda^2/2 - uv & \lambda u + u_x \\ \lambda v - v_x & -\lambda^2/2 + uv \end{pmatrix}\chi, \tag{11a}$$

$$\partial_t\chi = U_3\chi = \begin{pmatrix} \lambda^3/2 - \lambda uv - u_x v + uv_x & \lambda^2 u + \lambda u_x + u_{xx} - 2u^2 v \\ \lambda^2 v - \lambda v_x + v_{xx} - 2uv^2 & -\lambda^3/2 + \lambda uv + u_x v - uv_x \end{pmatrix}\chi. \tag{11b}$$

The Lax pair for (X_k) is composed of (U_1) and (U_k). It is found that if (u, v) is a compatible solution of (X_2) and (X_3), then $w = -2uv$ solves the KP equation (6) (Cao et al. 1999; Konopelchenko et al. 1991). Thus the compatible conditions of (U_1), (U_2) and (U_3) implies the KP equation. In other words, (U_1, U_2, U_3) is the Lax triad for the KP equation and hence for the pKP equation via $w = W_x$.

2.2 *The Discrete pKP Equations*

The above facts of the KP equation lead us to consider discretization of the ZS-AKNS problem (7), by which we hope to find the Lax representation for the lpKP equation (1). One discretization of (7) is known as the Ablowitz-Ladik spectral problem (Ablowitz and Ladik 1975, 1976), which leads to a spatially discretized NLS equation. In this paper, we employ the following linear problem, $(D^{(\gamma)})$, adopted in Cao and Zhang (2012),

$$\tilde{\chi} = D^{(\gamma)}\chi, \quad D^{(\gamma)}(\lambda, a, b) = \begin{pmatrix} \lambda - \gamma + ab & a \\ b & 1 \end{pmatrix}, \tag{12}$$

which provides a second discretization for (7) (Merola et al. 1994) but is different from Ablowitz-Ladik's spectral problem (cf. Chen et al. (2017)). Note that (12) is also known as a Darboux transformation of the ZS-AKNS spectral problems (10) (Adler and Yamilov 1994). Here, for the above notation, let T_1 be shift operator along the m_1 direction, defined for any function $f : \mathbb{Z}^3 \to \mathbb{R}$ as

$$(T_1 f)(m_1, m_2, m_3) = \tilde{f}(m_1, m_2, m_3) = f(m_1 + 1, m_2, m_3).$$

Similarly, $T_2 f = \bar{f}$, $T_3 f = \hat{f}$ are shifts along the m_2 and m_3 direction, respectively.

Two basic relations,

$$(a, b) = (u, \tilde{v}), \quad (u\tilde{v})_x = \tilde{u}\tilde{v} - uv, \tag{13}$$

are derived from the compatibility condition of equations (7) and (12) (see Cao and Zhang (2012)). The former bridges their potential functions, while the latter suggests the setting of difference relation $\tilde{W} - W = -2u\tilde{v}$ as $W_x = w = -2uv$. These facts lead to the consideration of three replicas of Eq. (12) with distinct non-zero parameters $\gamma = \gamma_1, \gamma_2, \gamma_3$,

$$T_1\chi \equiv \tilde{\chi} = D^{(\gamma_1)}(\lambda, u, \tilde{v})\chi, \tag{14a}$$
$$T_2\chi \equiv \bar{\chi} = D^{(\gamma_2)}(\lambda, u, \bar{v})\chi, \tag{14b}$$
$$T_3\chi \equiv \hat{\chi} = D^{(\gamma_3)}(\lambda, u, \hat{v})\chi, \tag{14c}$$

which are denoted by $(D^{(\gamma_k)})$, $k = 1, 2, 3$, respectively, for short. Besides, auxiliary equations will be assigned for each special occasion from the following list,

$$\tilde{W} - W = -2u\tilde{v}, \tag{15a}$$
$$\bar{W} - W = -2u\bar{v}, \tag{15b}$$
$$\hat{W} - W = -2u\hat{v}, \tag{15c}$$
$$\partial_x W = -2uv. \tag{15d}$$

At first glance, these equations seem fairly hard to deal with. Here we remark that on the platform of Liouville integrability, a pair of functions (u, v) of discrete arguments m_1, m_2, m_3 can be constructed, which are finite genus potential for each of the discrete spectral problems (14a, 14b, 14c); and W can be solved with the help of the theta function and meromorphic differentials on the associated Riemann surface. This will lead to explicit analytic solutions to the lpKP equation (see Sect. 5). The approach can also be extended to the semi-discrete and purely continuous cases (see Sect. 6).

To derive the discrete pKP equations (2), (3) and (4), we replace (U_j) in the Lax triad (U_1, U_2, U_3) successively by $(D^{(\gamma_k)})$, and then we come to the following new Lax triads,

$$(D^{(\gamma_1)}, D^{(\gamma_2)}, D^{(\gamma_3)}),\ (U_1, D^{(\gamma_1)}, D^{(\gamma_2)}),\ (U_1, U_2, D^{(\gamma_1)}). \tag{16}$$

With the auxiliary relations (14), the compatibility of these triads lead to the discrete pKP equations (2), (3) and (4). We present the procedure of derivation via the following lemmas and propositions.

Lemma 1 *Let $(u, v) : \mathbb{Z}^2 \to \mathbb{R}^2$ be a pair of functions such that (i) Eqs. (14a, 14b) have compatible solution χ for one value of the spectral parameter λ; (ii) the system (15a, 15b) has a solution W. Then (u, v) solve the lNLS equation (Cao and Zhang 2012; Konopelchenko 1982)*

$$\Xi_1^{(0,2)} \equiv (\tilde{u} - \bar{u})(u\bar{\tilde{v}} + 1) + (\gamma_1 - \gamma_2)u = 0, \tag{17a}$$

$$\Xi_2^{(0,2)} \equiv (\tilde{v} - \bar{v})(u\bar{\tilde{v}} + 1) - (\gamma_1 - \gamma_2)\bar{\tilde{v}} = 0, \tag{17b}$$

and $W,\ u,\ v$ satisfy the relation

$$\begin{aligned} Y(\gamma_1, \gamma_2) &\equiv 2\Big(\tilde{v}\, \Xi_1^{(0,2)}(\gamma_1, \gamma_2) + \bar{u}\, \Xi_2^{(0,2)}(\gamma_1, \gamma_2)\Big) \\ &= \left[\frac{1}{2}(\tilde{W} - \bar{W}) + \gamma_1 - \gamma_2\right](\bar{\tilde{W}} - \tilde{W} - \bar{W} + W) + 2(\tilde{u}\tilde{v} - \bar{u}\bar{v}), \end{aligned} \tag{18}$$

which is equal to zero due to Eqs. (17).

Proof By direct calculations we see that the cross action $(T_1T_2 - T_2T_1)\chi$ is equal to

$$(\tilde{D}^{(\gamma_2)}D^{(\gamma_1)} - \bar{D}^{(\gamma_1)}D^{(\gamma_2)})\chi = \begin{pmatrix} \Upsilon_{11} & \Xi_1^{(0,2)} \\ \Xi_2^{(0,2)} & 0 \end{pmatrix}\begin{pmatrix} \chi^{(1)} \\ \chi^{(2)} \end{pmatrix}, \tag{19}$$

where

$$\begin{aligned} \Upsilon_{11} =& \frac{\lambda}{\gamma_1 - \gamma_2}[(\tilde{v} - \bar{v})\Xi_1^{(0,2)} - (\tilde{u} - \bar{u})\Xi_2^{(0,2)}] \\ &+ \frac{1}{\gamma_1 - \gamma_2}[(\gamma_1\bar{v} - \gamma_2\tilde{v})\Xi_1^{(0,2)} - (\gamma_2\bar{u} - \gamma_1\tilde{u})\Xi_2^{(0,2)}]. \end{aligned}$$

With (14a, 14b) one can rewrite $(T_1T_2 - T_2T_1)\chi$ in the form

$$\frac{1}{\gamma_1 - \gamma_2}\begin{pmatrix} (\gamma_1 - \lambda)\bar{\chi}^{(2)} - (\gamma_2 - \lambda)\tilde{\chi}^{(2)} & [(\gamma_1 - \lambda)\tilde{u} - (\gamma_2 - \lambda)\bar{u}]\chi^{(1)} \\ 0 & (\gamma_1 - \gamma_2)\chi^{(1)} \end{pmatrix}\begin{pmatrix} \Xi_1^{(0,2)} \\ \Xi_2^{(0,2)} \end{pmatrix}. \tag{20}$$

Usually the coefficient determinant is not zero. Since $(T_1T_2 - T_2T_1)\chi = 0$, we have $\Xi^{(0,i)} = 0$. Further, in light of Eqs. (15a, 15b), the left-hand side of Eq. (17) can be written as

$$\Xi_1^{(0,2)} = (\tilde{u} - \bar{u}) + \left[\frac{1}{2}(\tilde{W} - \bar{W}) + \gamma_1 - \gamma_2\right] u,$$

$$\Xi_2^{(0,2)} = (\tilde{v} - \bar{v}) - \left[\frac{1}{2}(\tilde{W} - \bar{W}) + \gamma_1 - \gamma_2\right] \tilde{\bar{v}},$$

which imply Eq. (18) by direct calculations.

Proposition 1 *Let $(u, v) : \mathbb{Z}^3 \to \mathbb{R}^2$ be a pair of functions such that (i) Eqs. (14a, 14b, 14c) have compatible solution χ for one value of λ; (ii) the system (15a, 15b, 15c) has a solution W. Then W solves Eq. (2), i.e. $\Xi^{(0,3)} = 0$.*

Proof Consider three replicas of (18) with parameters (γ_1, γ_2), (γ_2, γ_3), (γ_3, γ_1), respectively. Adding them together we have

$$\Xi^{(0,3)} = Y(\gamma_1, \gamma_2) + Y(\gamma_2, \gamma_3) + Y(\gamma_3, \gamma_1), \tag{21}$$

where the terms containing u, v are canceled. This yields Eq. (2).

Proposition 2 *Let $(u, v) : \mathbb{R} \times \mathbb{Z}^2 \to \mathbb{R}^2$ be a pair of functions such that (i) Eqs. (14a, 14b) have compatible solution χ for one value of λ; (ii) the system of Eqs. (15a, 15b) and (15d) has a solution W. Then W solves Eq. (3), i.e. $\Xi^{(1,2)} = 0$.*

Proof In light of (15d), the last term in (18) is equal to $-(\tilde{W} - \bar{W})_x$. Thus the proof is completed since $\Xi^{(1,2)} = -Y(\gamma_1, \gamma_2)$.

Proposition 3 *Let $(u, v) : \mathbb{R}^2 \times \mathbb{Z} \to \mathbb{R}^2$ be a pair of functions such that (i) Eqs. (7), (11a) and (14a) have compatible solution χ for one value of λ; (ii) the system of Eqs. (15a, 15d) has a solution W. Then W solves Eq. (4), i.e. $\Xi^{(2,1)} = 0$.*

Proof The compatibility condition $\partial_y \partial_x \chi = \partial_x \partial_y \chi$ gives rise to the NLS equations (9a), rewritten as

$$\Xi_1^{(2,0)} \equiv u_y - u_{xx} + 2u^2 v = 0, \tag{22a}$$

$$\Xi_2^{(2,0)} \equiv v_y + v_{xx} - 2uv^2 = 0. \tag{22b}$$

In fact,

$$\begin{aligned}(\partial_y \partial_x - \partial_x \partial_y)\chi &= (U_{1,y} - U_{2,x} + [U_1, U_2])\chi \\ &= \begin{pmatrix} 0 & \Xi_1^{(2,0)} \\ \Xi_2^{(2,0)} & 0 \end{pmatrix} \begin{pmatrix} \chi^{(1)} \\ \chi^{(2)} \end{pmatrix} = \begin{pmatrix} \chi^{(2)} & 0 \\ 0 & \chi^{(1)} \end{pmatrix} \begin{pmatrix} \Xi_1^{(2,0)} \\ \Xi_2^{(2,0)} \end{pmatrix}. \end{aligned} \tag{23}$$

Further, the compatibility condition $\partial_x T_1 \chi = T_1 \partial_x \chi$ yields the semi-discrete NLS equations

$$\Xi_1^{(1,1)} \equiv u_x - \tilde{u} - \gamma_1 u + u^2 \tilde{v} = 0, \tag{24a}$$

$$\Xi_2^{(1,1)} \equiv \tilde{v}_x + v + \gamma_1 \tilde{v} - u\tilde{v}^2 = 0, \tag{24b}$$

since the cross action leads to

$$
\begin{aligned}
(\partial_x T_1 - T_1\partial_x)\chi &= (D_x^{(\gamma_1)} - \tilde{U}_1 D^{(\gamma_1)} + D^{(\gamma_1)} U_1)\chi \\
&= \begin{pmatrix} \kappa_{11} & \varXi_1^{(1,1)} \\ \varXi_2^{(1,1)} & 0 \end{pmatrix} \begin{pmatrix} \chi^{(1)} \\ \chi^{(2)} \end{pmatrix} = \begin{pmatrix} \tilde{\chi}^{(2)} & u\chi^{(1)} \\ 0 & \chi^{(1)} \end{pmatrix} \begin{pmatrix} \varXi_1^{(1,1)} \\ \varXi_2^{(1,1)} \end{pmatrix},
\end{aligned} \tag{25}
$$

where $\kappa_{11} = (u\tilde{v})_x - \tilde{u}\tilde{v} + uv = \tilde{v}\varXi_1^{(1,1)} + u\varXi_2^{(1,1)}$, and the relation $\tilde{\chi}^{(2)} = \tilde{v}\chi^{(1)} + \chi^{(2)}$ has been used. By calculation we have

$$
\begin{aligned}
&\tilde{v}\varXi_1^{(2,0)} + u\tilde{\varXi}_2^{(2,0)} = (u\tilde{v})_y - (u_x\tilde{v} - u\tilde{v}_x)_x - 2u\tilde{v}(\tilde{u}\tilde{v} - uv), \\
&(\tilde{v}\varXi_1^{(1,1)} - u\varXi_2^{(1,1)})_x = (u_x\tilde{v} - u\tilde{v}_x)_x - (\tilde{u}\tilde{v} + uv + 2\gamma_1 u\tilde{v} - 2u^2\tilde{v}^2)_x.
\end{aligned}
$$

Adding them together we arrive at

$$
\varXi^{(2,1)} = -2(\tilde{v}\varXi_1^{(2,0)} + u\tilde{\varXi}_2^{(2,0)}) - 2(\tilde{v}\varXi_1^{(1,1)} - u\varXi_2^{(1,1)})_x, \tag{26}
$$

where the term $(u_x\tilde{v} - u\tilde{v}_x)_x$ is canceled and the variable W is introduced by Eqs. (15a, 15d). Thus $\varXi^{(2,1)} = 0$.

Let us back to the Eqs. (2), (3) and (4). We have seen that (2) is nothing but the lpKP equation (1). Besides, Eqs. (3) and (4) have close relations with the (N-2) and (N-3) models that were discovered by Date, Jimbo and Miwa Date et al. (1982), which are

$$
\varXi^{\mathrm{N2}} \equiv (\tilde{V} - \bar{V})_x - (e^{\bar{\tilde{V}}} - e^{\tilde{V}} - e^{\bar{V}} + e^{V}) = 0, \tag{27}
$$

$$
\varXi^{\mathrm{N3}} \equiv \Delta(V_y + \frac{2}{h}V_x - 2VV_x) - (\Delta + 2)V_{xx} = 0, \tag{28}
$$

where $\Delta f = \tilde{f} - f$ for arbitrary function f. In fact, for (3), introducing

$$
V = \ln[(\tilde{W} - \bar{W})/2 + \gamma_1 - \gamma_2], \tag{29}
$$

and then using (3), one finds

$$
V_x = \frac{(\tilde{W} - \bar{W})_x/2}{(\tilde{W} - \bar{W})/2 + \gamma_1 - \gamma_2} = \frac{1}{2}(\bar{\tilde{W}} - \tilde{W} - \bar{W} + W).
$$

It then follows that $(\tilde{V} - \bar{V})_x$ is equal to the second part in Eq. (27). Hence $\varXi^{\mathrm{N2}} = 0$. Thus, for any solution W of the Eq. (3), V defined by (29) provides a special solution for (27). For the Eq. (4), if W is a solution, then

$$
V = (\tilde{W} - W)/2 \tag{30}
$$

solves the (N-3) Eq. (28). Actually, it is easy to find (with $\gamma_1 = 1/h$)

$$\frac{1}{2}\Xi^{(2,1)} = \left(V_y + \frac{2}{h}V_x - 2VV_x\right) - (V+W)_{xx},$$

which implies $\Xi^{\mathrm{N3}} = \Delta\Xi^{(2,1)}/2 = 0$. In this sense, Eq. (4) is the potential version of (N-3). Note that (N-3) model was also derived by Kanaga Vel and Tamizhmani, with the help of quasi-difference operators (Kanaga Vel and Tamizhmani 1997), known as the DΔKP equation there. Besides, some properties of the DΔKP hierarchy, including symmetries, Hamiltonian structures and continuum limit, were investigated in Fu et al. (2013).

At the end of this subsection, we consider continuum limits of Eqs. (2, 3, 4). Let $\gamma_k = -1/\varepsilon_k, \varepsilon_k = c_k\varepsilon, (k = 1, 2, 3)$, where $c_1,\ c_2,\ c_3$ are arbitrary distinct non-zero constants. For any smooth function $W(x, y, t)$, define

$$T_k W = W(x + c_k\varepsilon,\ y - c_k^2\varepsilon^2/2,\ t + c_k^3\varepsilon^3/3), \quad (k = 1, 2, 3). \tag{31}$$

Denote $T_1 W = \tilde{W},\ T_2 W = \bar{W},\ T_3 W = \hat{W}$ for short. By straightforward calculations we have the following.

Proposition 4 *Under the Ansatz (31), in the neighborhood of $\varepsilon \sim 0$, the following Taylor expansions hold for any smooth function $W(x, y, t)$,*

$$\Xi^{(2,1)} = \Xi^{(3,0)}\frac{2c_1^2}{3}\varepsilon^2 + O(\varepsilon^3), \tag{32a}$$

$$\Xi^{(1,2)} = \Xi^{(3,0)}\frac{c_1c_2(c_1-c_2)}{3}\varepsilon^3 + O(\varepsilon^4), \tag{32b}$$

$$\Xi^{(0,3)} = \Xi^{(3,0)}\frac{1}{3}[c_1c_2(c_2-c_1) + c_2c_3(c_3-c_2) + c_3c_1(c_1-c_3)]\varepsilon^3 + O(\varepsilon^4). \tag{32c}$$

Thus, all the continuum limits of the lattice pKP equations (2), (3) and (4) give rise to the same pKP equation (5). The Ansatz (31) is crucial, which is proposed based on comparing the velocities of the Abel-Jacobi variable $\vec{\phi}$ along the discrete S_{γ_k}-flow and the continuous H_j-flow (see Appendix A).

3 The Integrable Hamiltonian System (H_1)

In Cao et al. (1999), Cao and Zhang (2012) an integrable Hamiltonian system is constructed from the ZS-AKNS spectral problem,

$$\partial_x \begin{pmatrix} p_j \\ q_j \end{pmatrix} = \begin{pmatrix} -\partial H_1/\partial q_j \\ \partial H_1/\partial p_j \end{pmatrix} = \begin{pmatrix} \alpha_j/2 & -<p,p> \\ <q,q> & -\alpha_j/2 \end{pmatrix} \begin{pmatrix} p_j \\ q_j \end{pmatrix}, \tag{33a}$$

$$H_1(p,q) = -\frac{1}{2} < Ap, q > +\frac{1}{2} < p, p >< q, q > . \tag{33b}$$

where $A = \text{diag}(\alpha_1, \ldots, \alpha_N)$, $< \xi, \eta >= \sum_{j=1}^{N} \xi_j \eta_j$. It can be regarded as N replicas of Eq. (7) with eigenvalues $\alpha_1, \ldots, \alpha_N$, respectively, under the constraint

$$(u, v) = f_{U_1}(p,q) = (- < p, p >, < q, q >). \tag{34}$$

The integrability requires enough number of involutive integrals. In deriving them, we use the Lax equation

$$\partial_x L(\lambda) = [U_1(\lambda), L(\lambda)], \tag{35}$$

which has a solution, the Lax matrix (Cao et al. 1999; Cao and Zhang 2012)

$$L(\lambda; p, q) = \begin{pmatrix} 1/2 + Q_\lambda(p,q) & -Q_\lambda(p,p) \\ Q_\lambda(q,q) & -1/2 - Q_\lambda(p,q) \end{pmatrix}, \tag{36}$$

where $Q_\lambda(\xi, \eta) =< (\lambda I - A)^{-1}\xi, \eta >$. By Eq. (35), $F(\lambda) = \det L(\lambda)$ is independent of the argument x. Three sets of integrals are derived from the expansions

$$F(\lambda) = -\frac{1}{4} + \sum_{k=1}^{N} \frac{E_k}{\lambda - \alpha_k} = -\frac{1}{4} + \sum_{j=0}^{\infty} F_j \lambda^{-j-1}, \tag{37a}$$

$$H(\lambda) = \sqrt{-F(\lambda)} = \frac{1}{2} - 2\sum_{k=0}^{\infty} H_k \lambda^{-k-1}, \tag{37b}$$

with $F_0 = - < p, q >$, $H_0 = - < p, q > /2$, H_1 exactly the same as in Eq. (33b), and

$$E_k = -p_k q_k + \sum_{1 \le j \le N;\, j \ne k} \frac{(p_j q_k - p_k q_j)^2}{\alpha_k - \alpha_j}, \tag{38a}$$

$$F_k = - < A^k p, q > + \sum_{\substack{i+j=k-1;\\ i,\, j \ge 0}} (< A^i p, p >< A^j q, q > - < A^i p, q >< A^j p, q >), \tag{38b}$$

$$H_k = \frac{1}{2} F_k + 2 \sum_{\substack{i+j=k-1;\\ i,\, j \ge 0}} H_i H_j. \tag{38c}$$

The functions $\{E_k\}$ are called confocal polynomials, satisfying

$$\sum_{k=1}^{N} \alpha_k^j E_k = F_j, \quad \sum_{k=1}^{N} E_k = F_0 = - < p, q > . \tag{39}$$

Further, we have the Lax equation along the $F(\lambda)$-flow,

$$\frac{\mathrm{d}}{\mathrm{d}t_\lambda} L(\mu) = \{L(\mu), F(\lambda)\} = \frac{2}{\lambda - \mu}[L(\lambda), L(\mu)], \tag{40}$$

which can be verified directly. It implies $\{F(\mu), F(\lambda)\} = \partial_{t_\lambda} \mathrm{det} L(\mu) = 0$. Here $\{A, B\}$ is the usual Poisson bracket defined as

$$\{A, B\} = \sum_{k=1}^{N} \left(\frac{\partial A}{\partial q_k} \frac{\partial B}{\partial p_k} - \frac{\partial A}{\partial p_k} \frac{\partial B}{\partial q_k} \right).$$

As a corollary we have

Lemma 2 *The members in the set* $\{E_j, F_k, H_l\}$ *are involutive in pairs.*

By Cao et al. (1999), there is an inner relation between the integral H_k and (X_k), the AKNS equation (8). The involutivity $\{H_1, H_k\} = 0$ implies the commutativity of the Hamiltonian phase flows $g_{H_1}^x, g_{H_k}^{\tau_k}$. This yields a compatible solution for (H_1), (H_k), and hence a solution to equation (X_k), respectively, as

$$(p(x, \tau_k), q(x, \tau_k)) = g_{H_1}^x g_{H_k}^{\tau_k}(p_0, q_0), \tag{41a}$$

$$(u(x, \tau_k), v(x, \tau_k)) = f_{U_1}(p, q) = (- < p, p >, < q, q >). \tag{41b}$$

Let $\alpha(\lambda) = \Pi_{k=1}^N (\lambda - \alpha_k)$. By Cao et al. (1999), Cao and Zhang (2012), a curve $\mathcal{R} : \xi^2 = R(\lambda)$, with genus $g = N - 1$, is constructed by the factorization of $F(\lambda) = -\Lambda(\lambda)/[4\alpha(\lambda)]$, with $R(\lambda) = \Lambda(\lambda)\alpha(\lambda)$. For non-branching λ, there are two points $\mathfrak{p}(\lambda)$, $\tau\mathfrak{p}(\lambda)$ on $\mathcal{R}$, with $\tau : \mathcal{R} \to \mathcal{R}$ the map of changing sheets. Consider two objects on the curve, the canonical basis $a_1, \ldots, a_g, b_1, \ldots, b_g$ of homology group of contours, and the basis of holomorphic differentials, written in the vector form as $\vec{\omega}' = (\omega_1', \ldots, \omega_g')^T$, $\omega_j' = \lambda^{g-j}\mathrm{d}\lambda/(2\xi)$. It is normalized into $\vec{\omega} = C\vec{\omega}'$, where $C = (a_{jk})_{g \times g}^{-1}$, with a_{jk} the integral of ω_j' along a_k. Near the infinities, the local expansions have simple relation as

$$\vec{\omega} = \begin{cases} +(\vec{\Omega}_1 + \vec{\Omega}_2 z + \vec{\Omega}_3 z^2 + \cdots)\mathrm{d}z, & \text{near } \infty_+, \\ -(\vec{\Omega}_1 + \vec{\Omega}_2 z + \vec{\Omega}_3 z^2 + \cdots)\mathrm{d}z, & \text{near } \infty_-. \end{cases} \tag{42}$$

Periodic vectors $\vec{\delta}_k$ and $\vec{B}_k$ are defined as integrals of $\vec{\omega}$ along a_k and b_k, respectively. They span a lattice $\mathscr{T}$, which defines the Jacobian variety $J(\mathcal{R}) = \mathbb{C}^g/\mathscr{T}$. The Abel map $\mathscr{A}(\mathfrak{p})$ is given as the integral of $\vec{\omega}$ from the fixed point $\mathfrak{p}_0$ to $\mathfrak{p}$. The matrix B, with $\vec{B}_k$ as columns, is used to construct the theta function $\theta(\vec{z}, B)$.

The elliptic variables μ_j, ν_j are given by the roots of the off-diagonal entries of the Lax matrix,

$$L^{12}(\lambda) = - < p, p > \frac{\mathfrak{m}(\lambda)}{\alpha(\lambda)}, \quad \mathfrak{m}(\lambda) = \Pi_{j=1}^{g}(\lambda - \mu_j), \tag{43a}$$

$$L^{21}(\lambda) = < q, q > \frac{\mathfrak{n}(\lambda)}{\alpha(\lambda)}, \quad \mathfrak{n}(\lambda) = \Pi_{j=1}^{g}(\lambda - \nu_j). \tag{43b}$$

They define the quasi-Abel-Jacobi and Abel-Jacobi variables, respectively, as

$$\vec{\psi}' = \sum_{k=1}^{g} \int_{\mathfrak{p}_0}^{\mathfrak{p}(\mu_k)} \vec{\omega}', \quad \vec{\psi} = C\vec{\psi}' = \mathscr{A}\left(\sum_{k=1}^{g} \mathfrak{p}(\mu_k)\right), \tag{44a}$$

$$\vec{\phi}' = \sum_{k=1}^{g} \int_{\mathfrak{p}_0}^{\mathfrak{p}(\nu_k)} \vec{\omega}', \quad \vec{\phi} = C\vec{\phi}' = \mathscr{A}\left(\sum_{k=1}^{g} \mathfrak{p}(\nu_k)\right). \tag{44b}$$

The evolution of these two variables along the $F(\lambda)$-flow is obtained by Eq. (40). Actually, in the component equation for $L^{21}(\mu)$, by letting $\mu \to \nu_k$, we calculate

$$\frac{1}{2\sqrt{R(\nu_k)}} \frac{\mathrm{d}\nu_k}{\mathrm{d}t_\lambda} = \frac{-\mathfrak{n}(\lambda)}{\alpha(\lambda)(\lambda - \nu_k)\mathfrak{n}'(\nu_k)}, \tag{45a}$$

$$\{\phi'_s, F(\lambda)\} = \frac{\mathrm{d}\phi'_s}{\mathrm{d}t_\lambda} = \sum_{k=1}^{g} \frac{\nu_k^{g-s}}{2\sqrt{R(\nu_k)}} \frac{\mathrm{d}\nu_k}{\mathrm{d}t_\lambda} = -\frac{\lambda^{g-s}}{\alpha(\lambda)}, \tag{45b}$$

where $\vec{\phi}' = (\phi'_1, \dots, \phi'_g)$. By the partial fraction expansion (37a), we get

$$\{\phi'_s, E_k\} = -\alpha_k^{g-s}/\alpha'(\alpha_k), \quad (k = 1, \dots, N), \tag{46a}$$

$$\{\phi'_s, E_1 + \dots + E_N\} = \{\phi'_s, F_0\} = 0. \tag{46b}$$

Proposition 5 *Each Hamiltonian system* (H_k), $k = 1, 2, \dots$, *is integrable in Liouville sense, sharing the same integrals* $E_1, \dots, E_N$, *which are involutive in pairs and functionally independent in* $\mathbb{R}^{2N} - \{0\}$.

Proof According to Lemma 2, it only needs to prove the functional independence of the confocal polynomials. Suppose $\sum_{k=1}^{N} c_k \mathrm{d}E_k = 0$. Then $\sum_{k=1}^{N} c_k\{\phi'_s, E_k\} = 0$. By Eq. (46b), we have

$$\sum_{k=1}^{g}(c_k - c_N)\{\phi'_s, E_k\} = 0, \quad (1 \le s \le g).$$

The coefficient matrix is non-degenerate since by Eq. (46a) it is of Vandermonde type. Hence we have $c_k - c_N = 0$ and $c_N \sum_{k=1}^{N} \mathrm{d}E_k = 0$. This implies $c_N = 0$ since

$$\sum_{k=1}^{N} \mathrm{d}E_k = -\mathrm{d} < p, q >= -\sum_{j=1}^{N} (q_j \mathrm{d}p_j + p_j \mathrm{d}q_j) \neq 0.$$

Lemma 3 *The Abel-Jacobi variables straighten out the $H(\lambda)$-flow as*

$$\{\vec{\phi}, H(\lambda)d\lambda\} = 2\vec{\omega}. \tag{47}$$

Proof Since $F(\lambda) = -H^2(\lambda)$, Eq. (45b) is transformed into the following formula, which, by multiplied the matrix C, leads to Eq. (47),

$$\{\phi_s', H(\lambda)\mathrm{d}\lambda\} = \frac{\lambda^{g-s}\mathrm{d}\lambda}{2H(\lambda)\alpha(\lambda)} = \frac{\lambda^{g-s}\mathrm{d}\lambda}{\sqrt{R(\lambda)}} = 2\omega_s'.$$

Proposition 6 *The Abel-Jacobi variables straighten out the H_k-flow, as $\{\vec{\phi}, H_0\} = 0$ and*

$$\frac{d\vec{\phi}}{d\tau_k} = \{\vec{\phi}, H_k\} = \vec{\Omega}_k, \quad (k = 1, 2, \ldots), \tag{48a}$$

$$\vec{\phi}(\tau_k) \equiv \vec{\phi}(0) + \tau_k \vec{\Omega}_k, \quad (\text{mod } \mathscr{T}). \tag{48b}$$

Proof By the Eqs. (37b) and (42), we have an expansion of Eq. (47) near ∞_+. Equation (48a) is then obtained as its coefficient.

For another Abel-Jacobi variable, by the Eq. (3.25) in Cao and Zhang (2012), we have

$$\vec{\psi} + \vec{\eta}_+ \equiv \vec{\phi} + \vec{\eta}_-, \quad (\text{mod } \mathscr{T}), \tag{49a}$$

$$\vec{\eta}_\pm = \int_{\infty_\pm}^{\mathfrak{p}_0} \vec{\omega}, \quad \vec{\Omega}_D = \vec{\eta}_+ - \vec{\eta}_- = \int_{\infty_+}^{\infty_-} \vec{\omega}, \tag{49b}$$

$$\vec{\psi}(\tau_k) \equiv \vec{\phi}(\tau_k) - \vec{\Omega}_D \equiv \vec{\psi}(0) + \tau_k \vec{\Omega}_k, \quad (\text{mod } \mathscr{T}). \tag{49c}$$

4 The Integrable Map S_γ

In Cao and Zhang (2012), an integrable symplectic map $S_\gamma : \mathbb{R}^{2N} \to \mathbb{R}^{2N}, (p, q) \mapsto (\tilde{p}, \tilde{q})$, is constructed with the help of N replicas of discrete ZS-AKNS equation (12),

$$\begin{pmatrix} \tilde{p}_j \\ \tilde{q}_j \end{pmatrix} = (\alpha_j - \gamma)^{-1/2} D^{(\gamma)}(\alpha_j; a, b) \begin{pmatrix} p_j \\ q_j \end{pmatrix}, \quad (1 \leq j \leq N), \tag{50}$$

under the discrete constraint $(a, b) = f_\gamma(p, q)$,

$$a = - < p, p >, \quad b = \frac{1}{Q_\gamma(p, p)}\Big(-\frac{1}{2} - Q_\gamma(p, q) \pm \frac{\sqrt{R(\gamma)}}{2\alpha(\gamma)}\Big). \tag{51}$$

It can be derived from the continuous constraint (34) through the relation $a = u, b = \tilde{v}$ in (13). In fact,

$$\begin{aligned}\tilde{v} - b &=< \tilde{q}, \tilde{q} > -b =< (A - \gamma I)^{-1}(bp + q), bp + q > -b \\ &= b^2 L^{12}(\gamma) - 2bL^{11}(\gamma) - L^{21}(\gamma) \equiv P^{(\gamma)}(b).\end{aligned}$$

Thus $P^{(\gamma)}(b) = 0$, whose roots lead to Eq. (51). The factor $(\alpha_j - \gamma)^{-\frac{1}{2}}$ in Eq. (50) is introduced so that the coefficient determinant equals to unity, which is necessary for making the resulting map S_γ symplectic.

As in the continuous case, the Liouville integrability of the map S_γ requires enough number of involutive integrals. Similarly, the discrete Lax equation, given as follows, plays a central role,

$$L(\lambda; \tilde{p}, \tilde{q})D^{(\gamma)}(\lambda; a, b) = D^{(\gamma)}(\lambda; a, b)L(\lambda; p, q). \tag{52}$$

By Cao and Zhang (2012), under the constraint (51), it has the same Lax matrix, given by Eq. (36), as its solution. Immediately we have $F(\lambda; \tilde{p}, \tilde{q}) = F(\lambda; p, q)$ by taking the determinant of (52). Thus $F(\lambda)$, together with $H(\lambda)$, E_j, F_k, H_l, are all invariant under the action of the map S_γ.

Proposition 7 *Cao and Zhang (2012) The map S_γ is symplectic and integrable, possessing $F(\lambda)$, $\{F_j\}$, $\{H_l\}$ and the confocal polynomials $E_1, \ldots, E_N$, as its integrals.*

Construct a discrete flow

$$(p(m), q(m)) = S_\gamma^m(p_0, q_0) \tag{53}$$

by iteration. It generates the finite genus potential functions for Eq. (12),

$$(a_m, b_m) = (u_m, v_{m+1}) = (- < p, p >, < \tilde{q}, \tilde{q} >). \tag{54}$$

Define $L_m(\lambda) = L(\lambda; p(m), q(m))$, $D_m^{(\gamma)}(\lambda) = D^{(\gamma)}(\lambda; u_m, v_{m+1})$. Rewrite Eq. (52) as

$$L_{m+1}(\lambda)D_m^{(\gamma)}(\lambda) = D_m^{(\gamma)}(\lambda)L_m(\lambda). \tag{55}$$

Consider the discrete ZS-AKNS problem (12) with finite genus potential functions as

$$h(m + 1, \lambda) = D_m^{(\gamma)}(\lambda)h(m, \lambda). \tag{56}$$

The solution space $\mathcal{E}_\lambda$ is invariant under the action of $L_m(\lambda)$ due to the commutativity relation (55). The linear operator $L_m(\lambda)$ has eigenvalues $\pm H(\lambda)$, with associated eigenvectors $h_\pm$ in $\mathcal{E}_\lambda$, satisfying

$$L_m(\lambda)h_\pm(m,\lambda) = \pm H(\lambda)h_\pm(m,\lambda), \tag{57a}$$

$$h_\pm(m+1,\lambda) = D_m^{(\gamma)}(\lambda)h_\pm(m,\lambda). \tag{57b}$$

Roughly speaking, the situation can be regarded as an algebro-difference analogue of the Burchnall-Chaundy's theory on commuting differential operators (Burchnall and Chaundy 1923, 1928). Actually, let $\mathcal{L}(\lambda) = 2\alpha(\lambda)L(\lambda)$. Then $\det \mathcal{L}(\lambda) = -R(\lambda)$ is a polynomial rather than a rational function. The commutativity relation (55) is rewritten as $\mathcal{L}_{m+1}D_m^{(\gamma)} = D_m^{(\gamma)}\mathcal{L}_m$. The algebraic spectral problem (57a) is revised as $\mathcal{L}_m h_\pm = \xi h_\pm$, with $\xi = \pm\sqrt{R(\lambda)}$. The algebraic problem and the difference problem share common eigenvectors $h_\pm$, with eigenvalues satisfying the algebraic relation, $\xi^2 = R(\lambda)$, exactly the same as the affine equation of the algebraic curve $\mathcal{R}$.

Let $M(m,\lambda)$ be fundamental solution matrix of Eq. (56). Under the normalization condition $h_\pm^{(2)}(0,\lambda) = 1$, the eigenvectors are determined uniquely as

$$h_\pm(m,\lambda) = \begin{pmatrix} h_\pm^{(1)}(m,\lambda) \\ h_\pm^{(2)}(m,\lambda) \end{pmatrix} = M(m,\lambda)\begin{pmatrix} c_\lambda^\pm \\ 1 \end{pmatrix}, \tag{58a}$$

$$c_\lambda^\pm = \frac{L_0^{11}(\lambda) \pm H(\lambda)}{L_0^{21}(\lambda)} = \frac{-L_0^{12}(\lambda)}{L_0^{11}(\lambda) \mp H(\lambda)}. \tag{58b}$$

Two meromorphic functions, the Baker functions, $\mathfrak{h}^{(\kappa)}(m,\mathfrak{p})$, $\mathfrak{p} \in \mathcal{R}$, $\kappa = 1, 2$, are defined as

$$\mathfrak{h}^{(\kappa)}(m,\mathfrak{p}(\lambda)) = h_+^{(\kappa)}(m,\lambda), \quad \mathfrak{h}^{(\kappa)}(m,\tau\mathfrak{p}(\lambda)) = h_-^{(\kappa)}(m,\lambda). \tag{59}$$

The commutativity relation (55) implies formulas of Dubrovin-Novikov's type (Cao and Zhang 2012). They are applied to calculate the divisors of the Baker functions. This leads to the straightening out of the flow S_γ^m on the Jacobian variety as Cao et al. (1999)

$$\vec{\psi}(m) \equiv \vec{\phi}(0) + m\vec{\Omega}_\gamma - \vec{\Omega}_D, \quad (\mathrm{mod}\ \mathscr{T}), \tag{60a}$$

$$\vec{\phi}(m) \equiv \vec{\phi}(0) + m\vec{\Omega}_\gamma, \qquad (\mathrm{mod}\ \mathscr{T}), \tag{60b}$$

$$\vec{\Omega}_\gamma = \int_{\mathfrak{p}(\gamma)}^{\infty_+} \vec{\omega}, \quad \vec{\Omega}_D = \int_{\infty_+}^{\infty_-} \vec{\omega}, \tag{60c}$$

where the Abel-Jacobi variables are given by (44) as

$$\vec{\psi}(m) = \mathscr{A}\Big(\sum_{j=1}^{g} \mathfrak{p}(\mu_j(m))\Big), \quad \vec{\phi}(m) = \mathscr{A}\Big(\sum_{j=1}^{g} \mathfrak{p}(\nu_j(m))\Big).$$

For any two distinct points $\mathfrak{q}, \mathfrak{r} \in \mathcal{R}$, there exists a dipole $\omega[\mathfrak{q},\mathfrak{r}]$, an Abel differential of the third kind, with residues 1 and -1 at the poles $\mathfrak{q}, \mathfrak{r}$, respectively, satisfying

(Toda 1981)

$$\int_{a_j} \omega[\mathfrak{q}, \mathfrak{r}] = 0, \quad \int_{b_j} \omega[\mathfrak{q}, \mathfrak{r}] = \int_{\mathfrak{r}}^{\mathfrak{q}} \omega_j, \quad (j = 1, \ldots, g). \tag{61}$$

With the help of these dipoles, the Baker functions can be reconstructed as Cao and Zhang (2012)

$$\mathfrak{h}^{(1)}(m, \mathfrak{p}) = d_m^{(1)} \frac{\theta[-\mathscr{A}(\mathfrak{p}) + \vec{\psi}(m) + \vec{K}]}{\theta[-\mathscr{A}(\mathfrak{p}) + \vec{\phi}(0) + \vec{K}]} e^{\int_{\mathfrak{p}_0}^{\mathfrak{p}} m\omega[\mathfrak{p}(\gamma), \infty_+] + \omega[\infty_-, \infty_+]}, \tag{62a}$$

$$\mathfrak{h}^{(2)}(m, \mathfrak{p}) = d_m^{(2)} \frac{\theta[-\mathscr{A}(\mathfrak{p}) + \vec{\phi}(m) + \vec{K}]}{\theta[-\mathscr{A}(\mathfrak{p}) + \vec{\phi}(0) + \vec{K}]} e^{\int_{\mathfrak{p}_0}^{\mathfrak{p}} m\omega[\mathfrak{p}(\gamma), \infty_+]}, \tag{62b}$$

where $d_m^{(1)}$, $d_m^{(2)}$ and $\vec{K}$ are constants, independent of $\mathfrak{p} \in \mathcal{R}$.

With these results in hand, we start to derive an explicit formula for the function $u\tilde{v}$. To this end we consider the local expression of the dipole near ∞_+, ($z = \lambda^{-1}$),

$$\omega[\mathfrak{p}(\gamma), \infty_+] = [-z^{-1} + \varphi(z)]\mathrm{d}z, \tag{63}$$

with $\varphi(z)$ holomorphic near $z \sim 0$. A simple calculation yields

$$\partial_z \log(z \exp \int_{\mathfrak{p}_0}^{\mathfrak{p}} \omega[\mathfrak{p}(\gamma), \infty_+]) = \varphi(z). \tag{64}$$

Recalling Eq. (42), we have

$$-\mathscr{A}(\mathfrak{p}) = \vec{\eta}_+ - \vec{\Omega}_1 z + O(z^2), \quad \vec{\eta}_+ = \int_{\infty_+}^{\mathfrak{p}_0} \vec{\omega}. \tag{65}$$

Then, from (62a) we get

$$\frac{z\tilde{h}_+^{(1)}}{h_+^{(1)}} = \frac{d_{m+1}}{d_m} \frac{\theta[-\vec{\Omega}_1 z + O(z^2) + \vec{\eta}_+ + \vec{\psi}(m+1) + \vec{K}]}{\theta[-\vec{\Omega}_1 z + O(z^2) + \vec{\eta}_+ + \vec{\psi}(m) + \vec{K}]} \cdot z e^{\int_{\mathfrak{p}_0}^{\mathfrak{p}} \omega[\mathfrak{p}(\gamma), \infty_+]}, \tag{66a}$$

$$\partial_z \log \frac{z\tilde{h}_+^{(1)}}{h_+^{(1)}} = \partial_z \log \frac{\theta[-\vec{\Omega}_1 z + O(z^2) + \vec{\eta}_+ + \vec{\psi}(m+1) + \vec{K}]}{\theta[-\vec{\Omega}_1 z + O(z^2) + \vec{\eta}_+ + \vec{\psi}(m) + \vec{K}]} + \varphi(z). \tag{66b}$$

On the other hand, since $h_\pm = (h_\pm^{(1)}, h_\pm^{(2)})^T$ satisfies Eq. (57b), we have

$$\frac{z\tilde{h}_+^{(1)}}{h_+^{(1)}} = 1 + (u\tilde{v} - \gamma)z + \frac{uh_+^{(2)}}{h_+^{(1)}} z = 1 + (u\tilde{v} - \gamma)z + O(z^2), \tag{67}$$

where the following estimation is used,

$$\frac{uh_+^{(2)}}{h_+^{(1)}} = \frac{L^{11}(\lambda) - H(\lambda)}{-L^{12}(\lambda)} =< q, q > \lambda^{-1}[1 + O(\lambda^{-1})] = O(z).$$

Now, taking derivative of the Eq. (67) with respect to z and comparing it with (66b) at $z = 0$, with the help of the relation $\vec{\psi} + \vec{\eta}_+ \equiv \vec{\phi} + \vec{\eta}_-$ in Eq. (49a), we obtain the following.

Proposition 8 *Let $(a, b) = (u, \tilde{v})$ be finite genus potential functions of Eq. (12), defined by Eq. (54). Then we have*

$$u\tilde{v} = -\partial_z|_{z=0}\log\frac{\theta[\vec{\Omega}_1 z + \vec{\phi}(m+1) + \vec{\eta}_- + \vec{K}]}{\theta[\vec{\Omega}_1 z + \vec{\phi}(m) + \vec{\eta}_- + \vec{K}]} + [\gamma + \varphi(0)]. \tag{68}$$

5 Finite Genus Solutions to the lpKP

Let $\gamma = \gamma_1, \gamma_2, \gamma_3$ be distinct and non-zero. We can apply the same theory we developed in Sect. 4 to the three corresponding cases, respectively. The resulting integrable maps $S_{\gamma_1}, S_{\gamma_2}, S_{\gamma_3}$ commute in pairs since they share the same integrals $E_1, \ldots, E_N$ (see Appendix in Cao and Zhang (2012)). By iteration we have discrete flows $S_{\gamma_1}^{m_1}, S_{\gamma_2}^{m_2}, S_{\gamma_3}^{m_3}$, and hence well-defined functions from any starting point

$$(p(m_1, m_2, m_3), q(m_1, m_2, m_3)) = S_{\gamma_1}^{m_1} S_{\gamma_2}^{m_2} S_{\gamma_3}^{m_3}(p_0, q_0), \tag{69a}$$

$$(u(m_1, m_2, m_3), v(m_1, m_2, m_3)) = (-< p, p >, < q, q >)|_{(m_1, m_2, m_3)}. \tag{69b}$$

Define $a = u$, and let b take $\tilde{v} = T_1 v,\ \bar{v} = T_2 v,\ \hat{v} = T_3 v$, respectively. By the commutativity of the flows, one can present the functions given by Eq. (69a) in three ways, respectively as

$$(p(m_k), q(m_k)) = S_{\gamma_k}^{m_k}(p_0^{(k)}, q_0^{(k)}), \quad (k = 1, 2, 3). \tag{70}$$

Thus, from Eq. (50) in the three special cases, the j-th component satisfies three equations simultaneously with $\lambda = \alpha_j$,

$$T_k\begin{pmatrix} p_j \\ q_j \end{pmatrix} = (\alpha_j - \gamma_k)^{-1/2} D^{(\gamma_k)}(\alpha_j; u, T_k v)\begin{pmatrix} p_j \\ q_j \end{pmatrix}, \quad (k = 1, 2, 3). \tag{71}$$

Introducing

$$\chi = (\alpha_j - \gamma_1)^{m_1/2}(\alpha_j - \gamma_2)^{m_2/2}(\alpha_j - \gamma_3)^{m_3/2}\begin{pmatrix} p_j \\ q_j \end{pmatrix}, \tag{72}$$

we then have

$$T_k\chi = D^{(\gamma_k)}(\alpha_j; u, T_k v)\chi, \quad (k = 1, 2, 3). \tag{73}$$

In other words, the overdetermined system of equations (14a–14c) has a compatible solution χ for the parameter $\lambda = \alpha_j$. Now, for the lpKP equation (2), recalling Proposition 1, we arrive at the following.

Proposition 9 *The lpKP equation (2), $\Xi^{(0,3)} = 0$, has a special solution*

$$\begin{aligned} W(m_1, m_2, m_3) =& 2\partial_z|_{z=0}\log\frac{\theta[z\vec{\Omega}_1 + \vec{\phi}(m_1, m_2, m_3) + \vec{\eta}_- + \vec{K}]}{\theta[z\vec{\Omega}_1 + \vec{\phi}(0, 0, 0) + \vec{\eta}_- + \vec{K}]} \\ & - 2\sum_{s=1}^{3} m_s[\gamma_s + \varphi_s(0)] + W(0, 0, 0), \end{aligned} \tag{74}$$

where

$$\vec{\phi}(m_1, m_2, m_3) = \sum_{s=1}^{3} m_s\vec{\Omega}_{\gamma_s} + \vec{\phi}(0, 0, 0), \tag{75}$$

and $\varphi_s(z)$ is defined by Eq. (63) in the case of $\gamma = \gamma_s$.

Proof We have $(k = 1, 2, 3)$

$$T_k W - W = 2\partial_z|_{z=0}\log\frac{\theta[z\vec{\Omega}_1 + T_k\vec{\phi}(m_1, m_2, m_3) + \vec{\eta}_- + \vec{K}]}{\theta[z\vec{\Omega}_1 + \vec{\phi}(m_1, m_2, m_3) + \vec{\eta}_- + \vec{K}]} - 2[\gamma_k + \varphi_k(0)].$$

It is equal to $-2u(T_k v)$ by Eq. (68). Thus W solves (15a–15c) simultaneously. According to Proposition 1, W solves Eq. (2).

6 Solutions of Other Equations

In solving pKP equation $\Xi^{(j,k)} = 0$ that contains at least one continuous argument x, we will derive an explicit analytic expression for uv, which is similar to (68) and also meets the auxiliary Eq. (15d). This can be done on the Liouville integrable platform as well, like in the discrete case. We list the main steps as follows.

Consider $(p(x), q(x)) = g^x_{H_1}(p_0, q_0)$. Hence $(u(x), v(x)) = (-<p, p>, <q, q>)$ provide the finite genus potential functions. For the ZS-AKNS equation (7) with these potential functions, the solution space $\mathcal{E}_\lambda$ is invariant under the action of $L(\lambda)$ due to the commutativity relation (35). The linear operator $L(\lambda)$ has eigenvalues $\pm H(\lambda)$, with associated eigenvectors $h_\pm$ in $\mathcal{E}_\lambda$, satisfying

$$L(\lambda)h_\pm(x, \lambda) = \pm H(\lambda)h_\pm(x, \lambda), \tag{76a}$$

$$\partial_x h_\pm(x, \lambda) = U_1(\lambda; u(x), v(x))h_\pm(x, \lambda). \tag{76b}$$

Let $M(x, \lambda)$ be basic solution matrix of Eq. (76b). The eigenvectors are uniquely determined under the normalized condition $h_{\pm}^{(2)}(0, \lambda) = 1$ and can be expressed as

$$h_{\pm}(x, \lambda) = \begin{pmatrix} h_{\pm}^{(1)}(x, \lambda) \\ h_{\pm}^{(2)}(x, \lambda) \end{pmatrix} = M(x, \lambda) \begin{pmatrix} c_{\lambda}^{\pm} \\ 1 \end{pmatrix}, \tag{77a}$$

$$c_{\lambda}^{\pm} = \frac{L^{11}(0, \lambda) \pm H(\lambda)}{L^{21}(0, \lambda)}. \tag{77b}$$

Two meromorphic functions $\mathfrak{h}^{(\kappa)}(x, \mathfrak{p})$, $\kappa = 1, 2$, are defined in $\mathcal{R} - \{\infty_+, \infty_-\}$ by

$$\mathfrak{h}^{(\kappa)}(x, \mathfrak{p}(\lambda)) = h_{+}^{(\kappa)}(x, \lambda), \quad \mathfrak{h}^{(\kappa)}(x, \tau\mathfrak{p}(\lambda)) = h_{-}^{(\kappa)}(x, \lambda).$$

A formula of Dubrovin-Novikov's type is derived from the commutativity relation (35). It is used to calculate the divisor of $\mathfrak{h}^{(2)}(x, \mathfrak{p})$, which is equal to $\sum_{j=1}^{g}[\mathfrak{p}(\nu_j(x)) - \mathfrak{p}(\nu_j(0))]$. By Eqs. (44b) and (48b), we have

$$\vec{\phi}(x) = \mathscr{A}\Big(\sum_{j=1}^{g} \mathfrak{p}(\nu_j(x))\Big) \equiv x\vec{\Omega}_1 + \vec{\phi}(0), \quad (\text{mod } \mathscr{T}). \tag{78}$$

On the two-sheeted Riemann surface $\mathcal{R}$, an Abel differential, $\omega^{(1)}[\infty_-, \infty_+]$, of the third kind is constructed, having only poles at ∞_-, ∞_+ with

$$\omega^{(1)}[\infty_-, \infty_+] = \begin{cases} [-z^{-2} - a^{(1)}(z)]\mathrm{d}z, & \text{near } \infty_+, \\ [+z^{-2} + a^{(1)}(z)]\mathrm{d}z, & \text{near } \infty_-, \end{cases} \tag{79}$$

where $a^{(1)}(z)$ is holomorphic near $z \sim 0$. Without loss of generality, it can be arranged to satisfy the condition

$$\int_{a_j} \omega^{(1)} = 0, \quad \int_{b_j} \omega^{(1)} = -4\pi \mathrm{i} \Omega_1^j, \quad (1 \le j \le g), \tag{80}$$

where $\vec{\Omega}_1 = (\Omega_1^1, \ldots, \Omega_1^g)^T$. Actually, by adding a linear combination of holomorphic differentials $\omega_1, \ldots, \omega_g$ to $\omega^{(1)}$, we can make the former formula in Eq. (80) valid. The latter is a corollary of the former, which can be verified by using the canonical representation of the Riemann surface $\mathcal{R}$ (Farkas and Kra 1992; Toda 1981). The form of local expressions (79) is invariant with adjusted $a^{(1)}(z)$. We adopt the same symbol, for short. Through a usual analysis we reconstruct (Arnold and Novikov 1990; Toda 1981)

$$\mathfrak{h}^{(2)}(x, \mathfrak{p}) = c^{(2)}(x) \frac{\theta[-\mathscr{A}(\mathfrak{p}) + \vec{\phi}(x) + \vec{K}]}{\theta[-\mathscr{A}(\mathfrak{p}) + \vec{\phi}(0) + \vec{K}]} \cdot \exp\Big(\frac{x}{2} \int_{\mathfrak{p}_0}^{\mathfrak{p}} \omega^{(1)}[\infty_-, \infty_+]\Big), \tag{81}$$

where $c^{(2)}$ is independent of $\mathfrak{p} \in \mathcal{R}$. Equations (80) are used to cancel the extra factors caused by the uncertain linear combination of the contours $a_1, \ldots, a_g, b_1, \ldots, b_g$ in the integration route from the point $\mathfrak{p}_0$ to $\mathfrak{p}$, both in $\mathscr{A}(\mathfrak{p})$ and in the integral of $\omega^{(1)}$.

By Eq. (42), near ∞_- we have ($z = \lambda^{-1} \sim 0$)

$$-\mathscr{A}(\mathfrak{p}) = \vec{\eta}_- + \vec{\Omega}_1 z + O(z^2), \quad \vec{\eta}_- = \int_{\infty_-}^{\mathfrak{p}_0} \vec{\omega}.$$

Exerting action $\partial_z\partial_x\log$ on Eq. (81), we obtain $\partial_z\partial_x\log h_-^{(2)}$, which is equal to

$$\partial_z\partial_x\log\theta[\vec{\Omega}_1 z + O(z^2) + \vec{\phi}(x) + \vec{\eta}_- + \vec{K}] + \frac{1}{2}[z^{-2} + a^{(1)}(z)]. \tag{82}$$

On the other hand, by Eq. (76a) we estimate

$$v\frac{h_-^{(1)}}{h_-^{(2)}} = v\frac{L^{11}(\lambda) - H(\lambda)}{L^{21}(\lambda)} = -uv\lambda^{-1} + O(\lambda^{-2}), \tag{83}$$

where the following estimations are employed,

$$\begin{aligned} L^{11}(\lambda) &= \frac{1}{2} + \langle p, q \rangle z + \langle Ap, q \rangle z^2 + O(z^3), \\ L^{21}(\lambda) &= \langle q, q \rangle z + O(z^2), \\ H(\lambda) &= \frac{1}{2} - 2H_0 z - 2H_1 z^2 + O(z^3). \end{aligned}$$

From Eq. (76b) and the estimation (83), we have

$$\partial_x\log h_-^{(2)} = -\frac{\lambda}{2} + v\frac{h_-^{(1)}}{h_-^{(2)}} = -\frac{1}{2}z^{-1} - uvz + O(z^2), \tag{84a}$$

$$\partial_z\partial_x\log h_-^{(2)} = z^{-2}/2 - uv + O(z). \tag{84b}$$

Then, equating Eq. (82) with (84b) to cancel the singular term $z^{-2}/2$, we obtain the following.

Proposition 10 *Let (u, v) be finite genus potential functions for Eq. (7). Then*

$$-2uv = 2\partial_z|_{z=0}\partial_x\log\theta[\vec{\Omega}_1 z + \vec{\phi}(x) + \vec{\eta}_- + \vec{K}] + a^{(1)}(0). \tag{85}$$

Next, we can recover W for the pKP equations with continuous arguments. In order to solve $\Xi^{(1,2)} = 0$, we consider the integrable maps $S_{\gamma_1}, S_{\gamma_2}$ and $g_{H_1}^x$, which commute in pairs since they share the same integrals $\{E_j\}$ (cf. Cao and Zhang (2012)). Well-defined functions are constructed as

$$(p(x,m_1,m_2),q(x,m_1,m_2)) = g_{H_1}^x S_{\gamma_1}^{m_1} S_{\gamma_2}^{m_2}(p_0,q_0), \tag{86a}$$

$$(u(x,m_1,m_2),v(x,m_1,m_2)) = (-<p,p>,<q,q>)|_{(x,m_1,m_2)}. \tag{86b}$$

By the commutativity of the flows, the functions in Eq. (86a) can be presented in three ways, respectively, as

$$(p(x),q(x)) = g_{H_1}^x(p_0',q_0'), \tag{87a}$$

$$(p(m_k),q(m_k)) = S_{\gamma_k}^{m_k}(p_0^{(k)},q_0^{(k)}), \quad (k=1,2). \tag{87b}$$

Thus the j-th component satisfies three equations simultaneously with $\lambda = \alpha_j$,

$$\partial_x\begin{pmatrix} p_j \\ q_j \end{pmatrix} = U_1(\alpha_j;u,v)\begin{pmatrix} p_j \\ q_j, \end{pmatrix}, \tag{88a}$$

$$T_k\begin{pmatrix} p_j \\ q_j \end{pmatrix} = (\alpha_j-\gamma_k)^{-1/2}D^{(\gamma_k)}(\alpha_j;u,T_kv)\begin{pmatrix} p_j \\ q_j \end{pmatrix}, \quad (k=1,2). \tag{88b}$$

Introducing $\chi = (\alpha_j-\gamma_1)^{m_1/2}(\alpha_j-\gamma_2)^{m_2/2}(p_j\,q_j)^T$, we have

$$\partial_x\chi = U_1(\alpha_j;u,v)\chi, \tag{89a}$$

$$T_k\chi = D^{(\gamma_k)}(\alpha_j;u,T_kv)\chi, \quad (k=1,2). \tag{89b}$$

Thus Eqs. (7) and (14a, 14b) have a compatible solution χ for the parameter $\lambda = \alpha_j$.

Proposition 11 *The semi-discrete pKP equation (3), i.e. $\Xi^{(1,2)} = 0$, has a solution*

$$\begin{aligned} W(x,m_1,m_2) =& 2\partial_z|_{z=0}\log\frac{\theta[z\vec{\Omega}_1+\vec{\phi}(x,m_1,m_2)+\vec{\eta}_-+\vec{K}]}{\theta[z\vec{\Omega}_1+\vec{\phi}(0,0,0)+\vec{\eta}_-+\vec{K}]} \\ & -2\sum_{s=1}^{2}m_s[\gamma_s+\varphi_s(0)]+a^{(1)}(0)x+W(0,0,0), \end{aligned} \tag{90}$$

where $\vec{\phi}(x,m_1,m_2) = x\vec{\Omega}_1+\sum_{s=1}^2 m_s\vec{\Omega}_{\gamma_s}+\vec{\phi}(0,0,0)$, $\varphi_s(z)$ defined by Eq. (63) with $\gamma=\gamma_s$, and $a^{(1)}(z)$ given by Eq. (79).

Proof From (90) we have $(k=1,2)$

$$\partial_x W = 2\partial_z|_{z=0}\partial_x\log\theta[z\vec{\Omega}_1+\vec{\phi}(x,m_1,m_2)+\vec{\eta}_-+\vec{K}]+a^{(1)}(0),$$

$$T_kW-W = 2\partial_z|_{z=0}\log\frac{\theta[z\vec{\Omega}_1+T_k\vec{\phi}(x,m_1,m_2)+\vec{\eta}_-+\vec{K}]}{\theta[z\vec{\Omega}_1+\vec{\phi}(x,m_1,m_2)+\vec{\eta}_-+\vec{K}]}-2[\gamma_k+\varphi_k(0)],$$

which are equal to $-2uv$ and $-2u(T_kv)$ according to Eqs. (85) and (68), respectively. Recalling Proposition 2, W solves Eq. (3).

By similar analysis we have the following.

Proposition 12 *The semi-discrete pKP equation (4), i.e.* $\Xi^{(2,1)} = 0$*, has a solution*

$$\begin{aligned} W(x, y, m_1) =& 2\partial_z|_{z=0}\log\frac{\theta[z\vec{\Omega}_1 + \vec{\phi}(x, y, m_1) + \vec{\eta}_- + \vec{K}]}{\theta[z\vec{\Omega}_1 + \vec{\phi}(0, 0, 0) + \vec{\eta}_- + \vec{K}]} \\ & - 2m_1[\gamma_1 + \varphi_1(0)] + a^{(1)}(0)x + W(0, 0, 0), \end{aligned} \tag{91}$$

where $\vec{\phi}(x, y, m_1) = x\vec{\Omega}_1 + y\vec{\Omega}_2 + m_1\vec{\Omega}_{\gamma_1} + \vec{\phi}(0, 0, 0)$.

Proposition 13 *The pKP equation (5), i.e.* $\Xi^{(3,0)} = 0$*, is solved by*

$$\begin{aligned} W(x, y, t) =& 2\partial_z|_{z=0}\log\frac{\theta[z\vec{\Omega}_1 + \vec{\phi}(x, y, t) + \vec{\eta}_- + \vec{K}]}{\theta[z\vec{\Omega}_1 + \vec{\phi}(0, 0, 0) + \vec{\eta}_- + \vec{K}]} \\ & + a^{(1)}(0)x + W(0, 0, 0), \end{aligned} \tag{92}$$

where $\vec{\phi}(x, y, t) = x\vec{\Omega}_1 + y\vec{\Omega}_2 + t\vec{\Omega}_3 + \vec{\phi}(0, 0, 0)$.

7 Concluding Remarks

In this paper we have shown that the lpKP equation, semi-discrete pKP equations and continuous pKP equation can be derived as compatibilities of Lax triads that originate from the ZS-AKNS spectral problems. The approach to constructing finite genus solutions for 2D lattice equations (Cao and Xu 2012; Cao and Zhang 2012) was extended to 3D cases. As a result, we obtained finite genus solutions for the discrete, semi-discrete and continuous pKP equations. Note that these solutions are different from the elliptic solitons that are genus-one solutions obtained by Nijhoff, et al. in Nijhoff and Atkinson (2010), Yoo-Kong and Nijhoff (2013).

In deriving those pKP equations, we employed the auxiliary relations (15). We note that usually W in (15) can not be exactly solved out for all arbitrarily given (u, v); therefore it is hard to say when Lax triads provide strict integrability for 3D equations in some cases (cf. Levi et al. (1994)). However, as for the case of finite genus solutions, since the finite-dimensional integrable flows $g_{H_j}^{T_j}$ and $S_{\gamma_k}^{m_k}$ share same Liouville integrals, same Lax matrix and same algebraic curve, it enables us to treat (15) on the same Liouville platform and obtain explicit expressions for W from (15) by algebro-geometric integration.

The lpKP equation is one of the five octahedron-type equations with 4D consistency (Adler et al. 2012). We believe our approach can be extended to other octahedron-type integrable equations. This will be a part of our future work.

Acknowledgements The authors are grateful to the referee for the invaluable comments. This work is supported by the National Natural Science Foundation of China (grant nos 10971200, 11501521, 11426206, 11631007 and 11875040 for the three authors, respectively).

Appendices

A An Heuristic Deduction of Ansatz (31)

The Abel-Jacobi variable $\vec{\phi}$ in the Jacobian variety $J(\mathcal{R})$ provides a favorable window to observe the evolution of the discrete symplectic flow $S_{\gamma_k}^{m_k}$ as well as the Hamiltonian flow $g_{H_j}^{T_j}$. The discrete velocity $\vec{\Omega}_{\gamma_k}$ of $\vec{\phi}$ is given by Eq. (60c), while the continuous velocity $\vec{\Omega}_j$ is calculated by Eq. (42) and (48a). They are bridged by the normalized basis $\vec{\omega}$ of holomorphic differentials. Let the parameter $\gamma = \gamma_k$ tend to be infinity in the way as $\gamma_k = -1/\varepsilon_k$, with $\varepsilon_k = c_k\varepsilon$, $\varepsilon \to 0$. By the local expression of $\vec{\omega}$ near ∞_+ given by Eq. (42), we have

$$\vec{\Omega}_{\gamma_k} = \int_{\mathfrak{p}(\gamma_k)}^{\infty_+} \vec{\omega} = \varepsilon_k\vec{\Omega}_1 - \frac{\varepsilon_k^2}{2}\vec{\Omega}_2 + \frac{\varepsilon_k^3}{3}\vec{\Omega}_3 + O(\varepsilon^4).$$

Substituting this into Eq. (75), we obtain

$$\vec{\phi} - \vec{\phi}_0 = \sum_{k=1}^{3} m_k\big(\varepsilon_k\vec{\Omega}_1 - \frac{\varepsilon_k^2}{2}\vec{\Omega}_2 + \frac{\varepsilon_k^3}{3}\vec{\Omega}_3\big) + O(\varepsilon^4).$$

On the other hand, by Eq. (92), the 3D continuous evolution of $\vec{\phi}$ reads

$$\vec{\phi} - \vec{\phi}_0 = (x - x_0)\vec{\Omega}_1 + (y - y_0)\vec{\Omega}_2 + (t - t_0)\vec{\Omega}_3.$$

Thus, up to $O(\varepsilon^4)$, we have

$$x - x_0 = \sum_{s=1}^{3} m_k\varepsilon_k, \quad y - y_0 = -\sum_{s=1}^{3} m_k\frac{\varepsilon_k^2}{2}, \quad t - t_0 - \sum_{s=1}^{3} m_k\frac{\varepsilon_k^3}{3},$$
$$T_k x = x + \varepsilon_k, \quad T_k y = y - \frac{\varepsilon_k^2}{2}, \quad T_k t = t + \frac{\varepsilon_k^3}{3}.$$

By substituting them into $T_k W = W(T_k x, T_k y, T_k t)$, we obtain Ansatz (31).

B Continuum Limit of the lNLS

The lNLS equation (17), i.e. $\Xi^{(0,2)} = 0$, is first obtained by Konopelchenko (1982). It is solved in Cao and Zhang (2012). At first glance, its relation with the NLS equation (22), $\Xi^{(2,0)} = 0$, is not clear. It turns out that there is a transformation of Nijhoff's type,

$$u = (-\gamma_1)^{m_1}(-\gamma_2)^{m_2}u', \quad v = (-\gamma_1)^{-m_1}(-\gamma_2)^{-m_2}v',$$

which reduces the lNLS equation into an equation of (u', v'),

$$(\Xi')_1^{(0,2)} \equiv (\gamma_1\gamma_2)^{-1}(\gamma_1\tilde{u}' - \gamma_2\bar{u}')u'\bar{\tilde{v}}' + \gamma_1(\tilde{u}' - u') - \gamma_2(\bar{u}' - u') = 0,$$
$$(\Xi')_2^{(0,2)} \equiv (\gamma_1\gamma_2)^{-1}(\gamma_2\tilde{v}' - \gamma_1\bar{v}')u'\bar{\tilde{v}}' + \gamma_1(\bar{\tilde{v}}' - \bar{v}') - \gamma_2(\bar{\tilde{v}}' - \tilde{v}') = 0.$$

Let $-\gamma_k^{-1} = \varepsilon_k = c_k\varepsilon$, ($k = 1, 2$), where c_1, c_2 are distinct non-zero constants. For any smooth functions $u'(x, y)$, $v'(x, y)$, define

$$\tilde{u}' = u'(x + \varepsilon_1, y - \varepsilon_1^2/2), \quad \bar{u}' = u'(x + \varepsilon_2, y - \varepsilon_2^2/2),$$
$$\bar{\tilde{u}}' = u'(x + \varepsilon_1 + \varepsilon_2, y - \varepsilon_1^2/2 - \varepsilon_2^2/2),$$

and similar expressions for $\tilde{v}'$, $\bar{v}'$, $\bar{\tilde{v}}'$. Then, as $\varepsilon \sim 0$, we have the following Taylor expansion which confirms that the continuum limit of the lNLS is NLS up to a Nijhoff's type transformation,

$$(\Xi')^{(0,2)} = \begin{pmatrix} u'_y - u'_{xx} + 2(u')^2v' \\ v'_y + v'_{xx} - 2u'(v')^2 \end{pmatrix} \frac{c_1 - c_2}{2}\varepsilon + O(\varepsilon^2).$$

Similarly, the semi-discrete NLS equation (24), $\Xi^{(1,1)} = 0$, is transformed as

$$(\Xi')^{(1,1)} \equiv \begin{pmatrix} u'_x + \gamma_1(\tilde{u}' - u') - \gamma_1^{-1}(u')^2\tilde{v}' \\ \tilde{v}'_x + \gamma_1(\tilde{v}' - v') + \gamma_1^{-1}u'(\tilde{v}')^2 \end{pmatrix} = 0,$$

by the transformation of Nijhoff's type, $u = (-\gamma_1)^{m_1}u'$, $v = (-\gamma_1)^{-m_1}v'$. Let

$$\tilde{u}' = u'(x + \varepsilon_1, y - \varepsilon_1^2/2), \quad \tilde{v}' = v'(x + \varepsilon_1, y - \varepsilon_1^2/2).$$

Then, as $\varepsilon \sim 0$, we have the following Taylor expansion, which confirms that the continuum limit of the time-discrete NLS equation (24) is the NLS equation up to a Nijhoff's type transformation,

$$(\Xi')^{(1,1)} = \begin{pmatrix} u'_y - u'_{xx} + 2(u')^2v' \\ v'_y + v'_{xx} - 2u'(v')^2 \end{pmatrix} \frac{c_1}{2}\varepsilon + O(\varepsilon^2).$$

References

Ablowitz, M.J., Ladik, J.: Nonlinear differential-difference equations. J. Math. Phys. **16**, 598–603 (1975). https://doi.org/10.1063/1.522558

Ablowitz, M.J., Ladik, J.: Nonlinear differential-difference equations and Fourier analysis. J. Math. Phys. **17**, 1011–1018 (1976). https://doi.org/10.1063/1.523009

Adler, V.E., Bobenko, A.I., Suris, Yu.B.: Classification of integrable discrete equations of octahedron type. Int. Math. Res. Not. **2012**, 1822–1889 (2012). https://doi.org/10.1093/imrn/rnr083

Adler, V.E., Yamilov, R.I.: Explicit auto-transformations of integrable chains. J. Phys. A: Math. Gen. **27**, 477–492 (1994). https://doi.org/10.1088/0305-4470/27/2/030

Arnold, V.I., Novikov, S.P. (eds.): Dynamical Systems IV: Symplectic Geometry and Its Applications. Springer, Berlin (1990)

Burchnall, J.L., Chaundy, T.W.: Commutative ordinary differential operators. Proc. Lond. Math. Soc. **s2-21**, 420–440 (1923). https://doi.org/10.1112/plms/s2-21.1.420

Burchnall, J.L., Chaundy, T.W.: Commutative ordinary differential operators. Proc. R. Soc. Lond. A **118**, 557–583 (1928). https://doi.org/10.1098/rspa.1928.0069

Cao, C.W., Wu, Y.T., Geng, X.G.: Relation between the Kadomtsev-Petviashvili equation and the confocal involutive system. J. Math. Phys. **40**, 3948–3970 (1999). https://doi.org/10.1063/1.532936

Cao, C.W., Xu, X.X.: A finite genus solution of the H1 model. J. Phys. A: Math. Theor. **45**, No.055213, 13pp (2012). https://doi.org/10.1088/1751-8113/45/5/055213

Cao, C.W., Zhang. G.Y.: Integrable symplectic maps associated with the ZS-AKNS spectral problem. J. Phys. A: Math. Theor. **45**, No.265201, 15pp (2012). https://doi.org/10.1088/1751-8113/45/26/265201

Chen, K., Deng, X., Zhang, D.J.: Symmetry constraint of the differential-difference KP hierarchy and a second discretization of the ZS-AKNS system. J. Nonl. Math. Phys. **24**(sup 1), 18–35 (2017). https://doi.org/10.1080/14029251.2017.1418051

Date, E., Jimbo, M., Miwa, T.: Method for generating discrete soliton equations II. J. Phys. Soc. Jpn. **51**, 4125–4131 (1982). https://doi.org/10.1143/JPSJ.51.4125

Farkas, H.M., Kra, I.: Riemann Surfaces. Springer, New York (1992)

Fu, W., Huang, L., Tamizhmani, K.M., Zhang, D.J.: Integrability properties of the differential-difference Kadomtsev-Petviashvili hierarchy and continuum limits. Nonlinearity **26**, 3197–3229 (2013). https://doi.org/10.1088/0951-7715/26/12/3197

Grammaticos, B., Kosmann-Schwarzbach, Y., Tamizmani, T. (eds.): Discrete Integrable Systems. Springer, Berlin (2004)

Hietarinta. J., Joshi, N., Nijhoff, F.W.: Discrete Systems and Integrability. Cambridge University Press, Cambridge (2016). https://doi.org/10.1017/CBO9781107337411

Kanaga Vel, S., Tamizhmani, K.M.: Lax pairs, symmetries and conservation laws of a differential-difference equation-Sato's approach. Chaos Solitons Fractals **8**, 917–931 (1997). https://doi.org/10.1016/S0960-0779(96)00142-7

Konopelchenko, B.: Elementary Bäcklund transformations, nonlinear superposition principle and solutions of the integrable equations. Phys. Lett. A **87**, 445–448 (1982). https://doi.org/10.1016/0375-9601(82)90754-X

Konopelchenko, B., Sidorenko, J., Strampp, W.: (1+1)-dimensional integrable systems as symmetry constraints of (2+1)-dimensional systems. Phys. Lett. A **157**, 17–21 (1991). https://doi.org/10.1016/0375-9601(91)90402-T

Levi, D., Ragnisco, O., Shabat, A.B.: Construction of higher local (2+1) dimensional exponential lattice equations. Can. J. Phys. **72**, 439–441 (1994). https://doi.org/10.1139/p94-059

Matveev, V.B.: 30 years of finite-gap integration theory. Phil. Trans. R. Soc. A **366**, 837–875 (2008). https://doi.org/10.1098/rsta.2007.2055

Merola, I., Ragnisco, O., Tu, G.Z.: A novel hierarchy of integrable lattices. Inverse Probl. **10**, 1315–1334 (1994). https://doi.org/10.1088/0266-5611/10/6/009

Nijhoff, F.W., Atkinson, J.: Elliptic N-soliton solutions of ABS lattice equations. Int. Math. Res. Not. **2010**, 3837–3895 (2010). https://doi.org/10.1093/imrn/rnq010

Nijhoff, F.W., Capel, H.W., Wiersma, G.L., Quispel, G.R.W.: Bäcklund transformations and three-dimensional lattice equations. Phys. Lett. A **105**, 267–272 (1984). https://doi.org/10.1016/0375-9601(84)90994-0

Suris, Yu.B.: The Problem of Integrable Discretization: Hamiltonian Approach. Birkhäuser Verlag, Basel (2003)

Toda, M.: Theory of Nonlinear Lattices. Springer, Berlin (1981)

Yoo-Kong, S., Nijhoff, F.W.: Elliptic (N, N)-soliton solutions of the lattice Kadomtsev-Petviashvili equation. J. Math. Phys. **54**, No.043511, 20pp (2013). https://doi.org/10.1063/1.4799274

GPSR Compliance
The European Union's (EU) General Product Safety Regulation (GPSR) is a set of rules that requires consumer products to be safe and our obligations to ensure this.

If you have any concerns about our products, you can contact us on

ProductSafety@springernature.com

In case Publisher is established outside the EU, the EU authorized representative is:

Springer Nature Customer Service Center GmbH
Europaplatz 3
69115 Heidelberg, Germany

www.ingramcontent.com/pod-product-compliance
Ingram Content Group UK Ltd.
Pitfield, Milton Keynes, MK11 3LW, UK
UKHW021907280526
471597UK00003B/40
9783030569990